Conceptual basis, formalisations and parameterization of the STICS crop model

Nadine Brisson
Marie Launay
Bruno Mary
Nicolas Beaudoin

Éditions Quæ
c/o Inra, RD 10, 78026 Versailles Cedex

Collection *Update Sciences & Technologies*

L'élevage en mouvement. Flexibilité et adaptation des exploitations d'herbivores
Benoît Dedieu, Eduardo Chia, Bernadette Leclerc, Charles-Henri Moulin, Muriel Tichit
2008, 296 p.

Landscape: from Knowledge to Action
Martine Berlan-Darqué, Yves Luginbühl, Daniel Terrasson
2008, 308 p.

Multifractal Analysis in Hydrology. Application to Time Series
Pietro Bernardara, Michel Lang, Éric Sauquet, Daniel Schertzer, Ioulia Tchiriguyskaia
2008, 58 p.

Analyse multifractale en hydrologie. Application aux séries temporelles
Pietro Bernardara, Michel Lang, Éric Sauquet, Daniel Schertzer, Ioulia Tchiriguyskaia
2008, 58 p.

Paysages : de la connaissance à l'action
Martine Berlan-Darqué, Yves Luginbühl, Daniel Terrasson
2007, 316 p.

Estimation de la crue centennale pour les plans de prévention des risques d'inondations
Michel Lang, Jacques Lavabre
2007, 232 p.

Territoires et enjeux du développement régional
Amédée Mollard, Emmanuelle Sauboua, Maud Hirczak
2007, 240 p.

This book was supported by the Inra department "Environment and agronomy" as well as the Agroclim research unit.

© Éditions Quæ, 2008   ISBN: 978-2-7592-0169-3   ISSN: 1773-7923

To Yves Crozat

His understanding of agronomy, of the balance between experimentation and modelling, his scientific curiosity, his open-mindedness, his capacity for listening, his organizational skills and his extreme kindness…

we will miss greatly

Contributors to STICS formalisations

R. Antonioletti[1], N. Beaudoin[1], P. Bertuzzi[1], T. Boulard[1], N. Brisson[1], S. Buis[1], P. Burger[1], F. Bussière[1], Y.M. Cabidoche[1], P. Cellier[1], Y. Crozat[2], P. Debaeke[1], F. Devienne-Barret[1], C. Durr[1], M Duru[1], B. Gabrielle[1], I. Garcia de Cortazar Atauri[1], C. Gary[1], F. Gastal[1], P. Gate[3], J.P. Gaudillère[1], S. Génermont[1], M. Guérif[1], C. Hénault[1], G. Helloux[2], B. Itier[1], M.H. Jeuffroy[1], E. Justes[1], M. Launay[1], S. Lebonvallet[1], G. Lemaire[1], F. Maraux[4], B. Mary[1], T. Morvan[1], B. Nicolardot[1], B. Nicoullaud[1], H. Ozier-Lafontaine[1], L. Pagès[1], B. Rebière[4], S. Recous[1], G. Richard[1], R. Roche[1], J. Roger-Estrade[1], F. Ruget[1], C. Salon[1], B. Seguin[1], J. Sierra[1], H. Sinoquet[1], J. Tournebize[4], R. Tournebize[1], C. Valancogne[1], A.S. Voisin[1], D. Zimmer[4]

(1) : Inra
(2) : ESA-Angers
(3) : Arvalis Institut du végétal
(4) : Cemagref

Dominique Ripoche is responsible for the STICS Fortran code, its development and maintenance

English revised by Alan Scaife (alan@ascaife.plus.com)

Notations

The illustrations and equations are numbered by chapter. The variable names used in the equations are listed in annex 1. and the parameters are identifiable by indices: "T" for technical parameters, "S" for soil parameters, "C" for climate parameters, "P" for genotype-independent plant parameters, "V" for genotype-dependent plant parameters and "G" for general parameters. The index "I" is used for the initial status of the key variables. When used in the text those variables are written in capital letters. In the equations the variables (VAR) are referred to the current day, I, by writing them as VAR (I) while this reference is omitted in the text to lighten it. In a summation over time, the time daily variable is named J.

Table of contents

Preface

What is a crop model? 'Snake oil' (Passioura 1996), i.e. an impossible (and moderately honest) challenge to fit the current scientific knowledge into a single framework? A mechanistic view of plant growth and development which represent causality between component processes and yield (Yin et al. 2004)? Robust empirical relations between plant behaviour and the main environmental variables (Passioura 1996)? A tool for analysing plant behaviour and its genetic variability which bypasses, but may help to increase the knowledge about underlying mechanisms (Tardieu 2003, Hammer 2006)? All these definitions are partly true, all are potentially misleading.

Considering the achievements of crop models is perhaps the best way to understand what they are. STICS and other crop models have profoundly changed the vision that the agronomic community had of the soil – plant – atmosphere system and of its interactions with cultivation techniques. It has also changed the way agronomists design experiments and test hypotheses. Important and legitimate questions such as "which is the best sowing density for a crop?", "is an early cultivar better than a late one?", "what is the best fertilisation strategy?" have been the subject of hundreds of experiments in the 60's and 70's. Nobody would now imagine answering them without a model because "try it and see" experiments may well be the worst method for answering them, due to experimental errors and to the variability of behaviour of each genotype in different environments. Although our current knowledge is often poor for detailed processes, the behaviour of soil-plant-atmosphere systems is surprisingly predictable in relation to what could be expected from the synthesis of all mechanisms involved in it (Tardieu 2003). STICS, like other crop models, can therefore help to answer the above questions for a wide range of conditions which could never be tested experimentally. The role of experiments has changed, and is now to check whether experimental results, obtained in a limited number of environmental conditions, are consistent with those of the model in a wide range of situations to verify the credibility of the model in the studied range of environments (Lyon et al. 2003, Corre Hellou et al. 2007). Lack of agreement between the model and the experiments may suggest ways for improving some aspects of the model.

Is this science or engineering (Passioura, 1996)? This lengthy debate has been largely fruitless. The same model can be used for good or unexciting science, for good or inappropriate engineering. The important point is that the user is able to be critical with the model, so that his/her judgement or decisions after using STICS will be the result of some personal input and understanding of the model. This is the objective, hopefully fulfilled, of this book.

Making it clear, that STICS is a tool for reasoning and not a magic wand for prediction, is one of the main aims of this book. The model is by no means an exact representation of all the processes involved in a virtual experiment. It is therefore essential that the user has access to its workings, i.e. its architecture, equations and parameters, and that the robustness of equations is discussed and compared with that of other models. The reader can find every single process used in the STICS model, with its equations and parameters, and with figures which explain the meaning of equations and their consequences on model outputs. This gives several possibilities to the user. Most skilled users can go into the detail of some processes, check the consistency of hypotheses with their own ideas, and interpret results according to this information ("I get this output with that hypothesis, would I get a different output with this other hypothesis? "). Less skilled users will use the book for understanding the reasoning which accompanies the equations of a particular module. For instance the observations of Figure 5.2 and 5.3 clearly suggest that the objective is not to compare the root systems of rape seed, corn and wheat, which vary widely between fields, but to investigate what happens if the characteristics of the root system change with the species or with the soil ("examples are given for 3 species. What would be the behaviour of my favourite species in my soil?").

STICS is based on simple processes, essentially the same as in other crop models, but with some appreciable differences in method. This book clearly presents the basis for computing the progression of phenological stages from temperature, the light interception by leaves following Monteith's equation, the transpiration following Penman Monteith's equation, and the water and nutrient uptakes following Gardner's pioneering work. To my knowledge, these fundamentals do not differ essentially from those of other models (Yin and Van Laar 2005, Keating et al. 2003) except that the equations used in STICS have been chosen in a more "physics-oriented" way than those of other models. In STICS, as in any other model, things become less straightforward for simulations of growth and of distribution of assimilates and responses to environmental stresses. The STICS group was successful in representing complex networks of interactions without generating scores of equations and parameters which can never be checked. Are the methods used in STICS better than those of other models? Another book could be written to compare the respective value of the algorithms used in different models. For most users, it is enough to know that methods and algorithms are coarse but useful representations of reality and that they can vary substantially between models, so it may be useful for some purposes to compare the output of STICS with those of other models.

An important side effect of the work of the STICS group has been to provide a common "meeting place" for scientists of several agronomic disciplines (plant science, soil science and cropping systems), for social scientists and for people working in extension services. This book should help to provide a bridge between scientific communities. It is a necessary tool for scientists who use the STICS model, for agronomists who are curious about the different topics which can be covered with crop models, and for modellers of different disciplines who wish to copy the methods of the STICS group. Will geneticists and molecular physiologists join the community of plant modellers? This is a major challenge for the years to come. Progress has been made (Hammer et al. 2006, Struik et al. 2007, Chenu et al. 2008), but these two groups seem reluctant to employ modelling methods (see e.g. Benfey and Mitchell-Olds 2008).

In conclusion, we have to be grateful to the authors, especially Nadine Brisson, for carrying out the huge and difficult task of explaining the detail of all that is involved in the STICS model.

François Tardieu

François Tardieu is a crop scientist and an ecophysiologist who works to fill the gap between agronomy and genetics. He was involved in projects in which crop modelling had an essential role. This, together with his role in scientific management in Inra (France) gives him a wide overview of the uses and concerns of crop modelling.

1 Introduction

1.1 Purpose

The aims of STICS (Simulateur mulTIdisciplinaire pour les Cultures Standard) are similar to those of a large number of existing models (Whisler *et al.*, 1986), while paying attention to cropping system diversity. It is a crop model with a daily time-step and input variables relating to climate, soil and the crop system. Its output variables relate to yield in terms of quantity and quality and to the environment in terms of drainage and nitrate leaching. The simulated object is the crop situation for which a physical medium and a crop management schedule can be determined. The main simulated processes are crop growth and development as well as the water and nitrogen balances. A full description of crop models with their fundamental concepts is available in Brisson *et al.* (2005).

STICS has been developed since 1996 at INRA (French National Institute for Agronomic Research) in collaboration with other research (CIRAD[1], CEMAGREF[2], École des Mines de Paris, ESA[3], LSCE[4]) or professional (ARVALIS[5], CETIOM[6], CTIFL[7], ITV[8], ITB[9], Agrotransferts[10], etc.) and teaching institutes. For more than 10 years STICS has been used and regularly improved thanks to a close link between development and application, involving scientists and technicians from various disciplines.

---

[1] Centre de coopération internationale en recherche agronomique pour le développement.
[2] Centre du machinisme agricole, du génie rural et des eaux et forêts.
[3] École supérieure d'agriculture d'Angers.
[4] Laboratoire des sciences du climat et de l'environnement.
[5] Arvalis, institut du végétal.
[6] Centre technique interprofessionnel des Oléagineux métropolitains.
[7] Centre technique interprofessionnel des fruits et légumes.
[8] Institut technique de la vigne.
[9] Institut technique de la betterave.
[10] Agrotransferts for the regions Poitou-Charentes and Picardie.

When STICS began to be developed, many well-known models were available (CERES: Ritchie and Otter, 1984; ARCWHEAT: Weir *et al.*, 1984; EPIC: Williams *et al.*, 1989; SUCROS: van Keulen and Seligman, 1987, etc.) that were developed from the pioneer works by de Wit (1978) or Duncan (1971 cited in Baker, 1980). However new models appear regularly in the literature (Amir and Sinclair, 1991a,b; Brisson *et al.*, 1992a; Hunt and Pararajasingham, 1995; Kanneganti and Fick, 1991; Maas, 1993; McMaster *et al.*, 1991; Teittinen *et al.*, 1994). As Sinclair and Seligman (1996) explained, this is because no one universal model can exist in the field of agricultural science and it is necessary to adapt system definitions, simulated processes and model formalisations to specific environments or to new problems (technical, genetic, environmental, etc.). These same authors emphasize the heuristic potential of modelling, a determining element in the development of STICS.

From a conceptual point of view, STICS is made up of a number of original parts compared with other crop models (e.g. simulation of crop temperature, simulation of many techniques) but most of the remaining parts are based on conventional formalisations or have been taken from existing models. Its strong points are the following:

– its "crop" generality: adaptability to various crops (wheat, maize, soybean, sorghum, flax, grassland, tomato, beetroot, sunflower, vineyard, pea, rapeseed, banana, sugarcane, carrot, lettuce, etc.)

– its robustness: ability to simulate various soil-climate conditions without too much error in the outputs (Brisson *et al.*, 2002a) and easy availability of its soil and technical inputs. Yet, this robusness can jeopardise accuracy on a local scale.

– its "conceptual" modularity: the possibility of adding new modules or complementing the system description (e.g.: ammonia volatilisation, symbiotic nitrogen fixation, plant mulch, stony soils, many organic residues, etc.). The purpose of such modularity is to facilitate subsequent development.

Around 50 scientists of various disciplines participated in the STICS formalisations, most of them from INRA (Institut National de la Recherche Agronomique). Thus the model can be regarded as a synthesis of the French agronomic knowledge on the field and crop cycle scales, which motivated this book. It presents the formalisations of the STICS model (version 6.2), which can be considered as references used in the framework of crop sciences, helping professionals and students in the partitioning and understanding of the complex agronomic system. The book arrangement relies on the way the model designs the crop-soil system functioning, each chapter being devoted to one important function such as growth initiation, yield onset, water uptake, transformation of organic matter etc. One chapter is devoted to the cropping system and long term simulations and the final chapter is about the involvement of the user in terms of option choices and parameterization.

1.2 Overall description of the system with its components

1.2.1 The system

STICS simulates the behaviour of the soil-crop system, in one dimension, over one crop cycle or several successive cycles. The upper boundary of the system is the atmosphere, characterised by standard weather variables (radiation, minimum and maximum temperatures, rainfall, reference evapotranspiration and possibly wind and humidity) and the lower boundary corresponds to the soil/sub-soil interface.

Crops are generally perceived in terms of their above-ground biomass and nitrogen content, leaf area index, and the number and biomass (and nitrogen content) of harvested organs. Vegetative organs (leaves, stems, branches or tillers, roots) are functionally separated in terms of radiation, water and nutrient sensors or reservoir role. Soil is described as a sequence of horizontal layers, each of which is characterised in terms of its water content and mineral and organic nitrogen contents. Soil and crop interact via the roots, and these roots are defined in terms of root density distribution in the soil profile.

STICS can also simulate intercropping, i.e. two crops (annual or perennial) growing simultaneously as a mixture, each crop developing and growing with its own rhythm resulting from the resource partitioning. In this case the soil-plant-atmosphere system is divided into three sub-systems at the canopy level. There is the dominant canopy and the understorey canopy that is divided into two parts: a shaded part and a sunlit part, each of them being defined by a light microclimate that drives the different behaviour of the sub-systems.

1.2.2 Simulated processes

Crop growth is driven by the plant carbon accumulation (de Wit, 1978): solar radiation intercepted by the foliage and then transformed into aboveground biomass that is directed to the harvested organs during the final phase of the crop cycle. The crop nitrogen content depends on the carbon accumulation and on the nitrogen availability in the soil. According to the plant type, crop development is driven either by a thermal index (degree-days), a photothermal index or a photothermal index taking into account vernalisation. The development module is used to make the leaf area index and the roots evolve and define the harvested organ filling phase. Water stress and nitrogen stress, if any, reduce leaf growth and biomass accumulation. This reduction is based on stress indices that are calculated in water and nitrogen balance modules. Other stresses such as waterlogging and thermal stresses (frost or high temperatures) are also taken into account.

Particular emphasis is placed on the effect of crop management on the dynamics of the soil-crop-microclimate system, knowing that crop peculiarities influence both ecophysiology and crop management (e.g. accounting for the various forms of forage cutting, fertiliser composition, plastic or crop residue mulching, etc.).

1.2.3 Modules and options

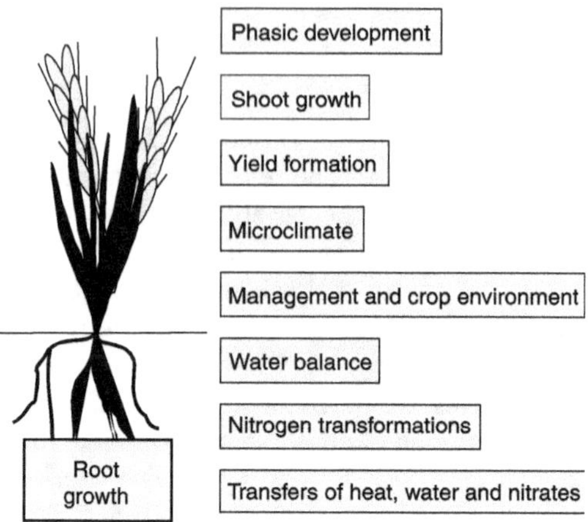

Figure 1.1. The main modules of the STICS crop model.

The STICS model is organised into modules (*Figure 1.1*), with each module composed of sub-modules dealing with specific mechanisms. A first set of three modules deals with the ecophysiology of above-ground plant parts (phenology, shoot growth, yield formation). A second set of four modules deals with how the soil responds in interaction with underground plant parts (root growth, water balance, nitrogen balance, soil transfers). The crop management module deals with the interactions between the applied techniques and the soil-crop system. The microclimate module simulates the combined effects of climate and water balance on the temperature and air humidity within the canopy.

Within each module, there are options that can be used to extend the scope with which STICS can be applied to various crop systems. These options relate to ecophysiology and to crop management, for example:
– competition for assimilate between vegetative organs and reserve organs (hereafter referred to as trophic competition);
– considering the geometry of the canopy when simulating radiation interception;
– the description of the root density profile;
– using a resistive approach to estimate the evaporative demand by plants;
– the mowing of forage crops;
– plant or plastic mulching under vegetation.

Certain options depend on data availability. For example, the use of a resistive model is based on availability of additional climatic driving variables: wind and air humidity.

2 Development

2.1 The simulated events

2.1.1 Phenological stages

The phenological stages (Table 2.1) are used as steps for simulating vegetative dynamics (leaf area index and roots) and harvested organ filling (grain, fruit, tuber). The two phenological scales are independent of each other: for example, the onset of filling of the harvested organs (IDRP) can occur before or after the "maximal leaf area index" stage (ILAX).

Table 2.1. List of the phenological stages of STICS. Some stages are required as a function of the options chosen : * for sown crops, ** for determinate crops, *** for indeterminate crops.

Vegetative stages (leaf area index)	Harvested organs stages
IPLT : sowing or planting (annuals)	
IGER* : germination	
IDEBDORM and IFINDORM : beginning and break of dormancy (woody plants)	
ILEV : emergence or budding	
ILET : end of the plantlet frost sensitive stage	ILAT** : beginning of the critical phase for grain number onset
IAMF : maximum acceleration of leaf growth, end of juvenile phase	IFLO : flowering (start of fruit sensitivity to frost)
	IDRP : onset of filling of harvested organs
ILAX : maximum leaf area index, end of leaf growth	INOU*** : end of setting (indeterminate option)
	IDEBDES ; onset of water dynamics in fruits
	IMAT : physiological maturity
IREC : harvest	

As in most crop models, the development stages simulated by STICS can differ from the stages defined in classical agronomic scales. The development stages in STICS are growth stages rather than organogenetic stages (Brisson and Delécolle, 1991). Stages correspond in fact to changes in the trophic or morphological strategy of the crop that influence the evolution of leaf area index or grain filling (Figure 2.1). Using generic terms to name the various stages allows different species to be simulated, exhibiting either determinate growth (vegetative and reproductive growth occur successively) or indeterminate growth (vegetative and reproductive growth occur simultaneously, at least partly). The IAMF stage equates to the beginning of stem elongation and is generally not far from the end of leaf initiation: it is the "1cm ear" stage for wheat and graminaceous forage crops, just slightly later than the double-ridge stage for most varieties, whereas it is the floral induction for corn. For indeterminate crops like tomato and vines, it is more difficult to find an equivalent in organogenesis and this stage is instead regarded as a number of leaves (3 or 4). The stage ILAX must be regarded as a growth stage since it is the end of leaf onset, that can occur before or after the IDRP stage. The beginning of grain filling (IDRP) is always preceded by a key stage for the onset of the number of harvested organs (grains or fruits) that can be ILAT for determinate crops and INOU for indeterminate crops. At physiological maturity (IMAT) the harvested organs stop growing in dry matter terms and the IMAT-IREC period depends on the required quality for the final product (see § 4.3).

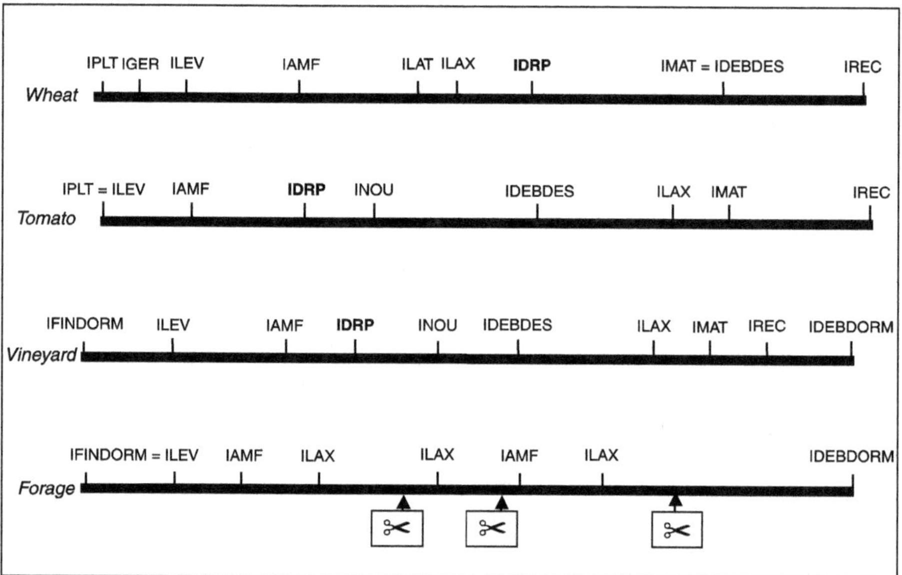

Figure 2.1. Illustration of the stages of interest for crops of various types such as wheat (annual determinate), tomato (annual indeterminate), vine (perennial indeterminate) and forage crop (perennial determinate interrupted by cuts symbolised by ✂). The flowering stage IFLO is mostly confounded with the IDRP stage (in bold).

2.1.2 Leaf development

The developmental component of foliage onset is included in the logistic relationship given in § 3.1.1 as the x-axis and uses the above-mentioned vegetative stages. Thus the notion of phyllotherm is not used to build up the LAI mainly because nor leaves nor stems are individualized Nevertheless it is used to calculate an early stage of frost sensitivity quantified in leaf number (see § 3.1.5).

The model uses the notion of lifespan to manage foliage senescence (§ 3.1.2). Thus the fraction of foliage formed on a given day disappears as green functional surface after a certain period of time which depends on temperature and environmental stresses.

2.2 Emergence and initiation of crop development and growth

This chapter concerns i) the emergence of sown annual crops, ii) the onset of crop development after planting for transplanted annual crops and iii) the onset of crop development after winter rest for perennial crops (bud growth of trees and the beginning of herbaceous growth).

2.2.1 Emergence of sown crops

In the first generation of crop models, the sowing-emergence phase was approached in a general way and related only to air temperature, as in the models CERES, ARCWHEAT, and SUCROS. Later on, the effect of the soil water status on the duration of emergence was also taken into account (Kanneganti and Fick, 1991). Recent work on germination and the beginning of shoot[1] growth (Durr *et al.*, 2001; Itabari *et al.*, 1993; Hucl, 1993; Weaich *et al.*, 1996) now distinguishes two phases in emergence, e.g. in the model SHOOTGRO of McMaster *et al.* (1991), and its derivatives (MODWTH3 of Rickman *et al.*, 1996). Such an approach allows the simulated duration of emergence to vary with three main factors - temperature, water status of the soil, and sowing depth. The effect of the soil water status has been shown to be particularly important (Bouaziz and Bruckler, 1989, Alm *et al.*, 1993, Bradford, 2002). These papers link the simulation of emergence to the good simulation of soil water status in the surface soil layers, especially when sowing is shallow. Generally the soil structure (size, amount and distribution of soil aggregates) is not accounted for in crop models, while models specifically dedicated to crop establishment do so (Durr *et al.*, 2001). In addition, the effects of waterlogging, through its physiological impact of anoxia on the embryo or through rooting effects, are not directly introduced.

In STICS, the emergence phase is broken down into three subphases: seed imbibition, followed by germination and lastly, shoot elongation. The soil physical conditions influence not only the duration of emergence but also the number of emerged plants, in particular in dry conditions or when there is a surface crust.

---

[1] Shoot: in this chapter "shoot" must be understood as the part of the seedling stem growing from the grain beneath the soil.

2.2.1.a Moistening

Seed moistening can be regarded as a passive process starting at a species-dependent -water potential prevailing in the seed bed (POTGERMI$_p$ in MPa). The relationship from Clapp and Hornberger (1978), parameterized by the characteristic soil water contents of field capacity and wilting point, was used to convert POTGERMI$_p$ into water content (see § 9.4.3). Once the seed is moistened, it has a limited number of days of autotrophy[2] (NBJGRAUTO) due to its reserves (eq. 2.1). This number has a species-dependent component (NBJGERLIM$_p$) but also a thermal one, since it is thought that at low temperature (i.e the average soil temperature in the seed bed, SB, from the beginning of moistening, IMB), respiration processes and the consumption of reserves are slower (the minimum at high temperature is PROPNBJGERLIM$_G$ × NBJGERLIM$_p$). When the temperature is lower than the germination base temperature, TGMIN$_p$, then the day number is maximal (NBJGERLIM$_p$). Above TDMAX$_p$, the seed uses up its reserves in the least time, parameterized by default to 20% of the maximum (PROPNBJGERLIM$_G$=0.2)

eq. 2.1

$$NBJGRAUTO(I) = -\frac{1 - PROPNBJGERMIN_G}{TDMAX_P - TGMIN_P} \times \left[\frac{\sum_{J=IMB}^{I} TSOL(SB,J)}{I - IMB + 1} - TGMIN_P \right] + 1$$

and $PROPNBJGERLMIN_G \times NBJGERLIM_P \leq NBJGRAUTO(I) \leq NBJGERLIM_P$

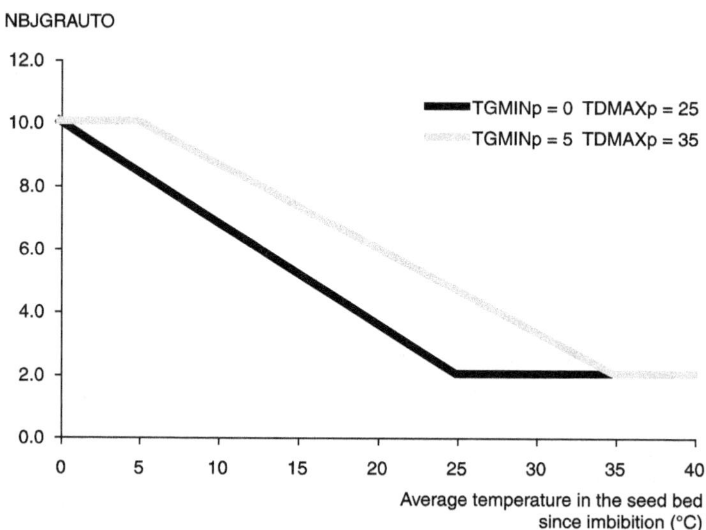

Figure 2.2. Evolution of the number of days of autotrophy as a function of temperature for two sets of cardinal temperatures.

---

[2] Autotrophy is here used to express the maximal delay between the grain imbibition and the outing of the rootlet as the first visible signal of growth.

2.2.1.b Germination

Germination is achieved when the growing degree-days from planting in the seed bed (SOMGER) reaches a given threshold (STPLTGER$_p$), with a condition as to the dryness of the soil (eq. 2.2 and eq. 2.3).

$$\text{eq. 2.2}$$

$$= \text{IGER if SOMGER}(I) = \sum_{J=IPLT}^{I} \left[(\text{TSOL}(SB, J) - \text{TGMIN}_P) \cdot \text{HUMIRAC}(SB, J) \right] = \text{STPLTGER}_P$$

$$\text{and} \quad SB = \text{PROFSEM} \pm 1\text{cm}$$

TSOL is the soil temperature and TGMIN$_p$ is the base temperature for germination. Soil moisture in the seedbed (SB=depth of sowing $\pm$ 1 cm) influences germination through the HUMIRAC variable (eq. 2.3).

$$\text{eq. 2.3}$$

$$\textit{if } HUMSOL(SB, I) > HN_S$$

$$\textit{then} \quad HUMIRAC(SB, I) = SENSRSEC_P + (1 - SENSRSEC_P) \frac{HUMSOL(SB, I) - HN_S}{HX_S - HN_S}$$

$$\textit{if } HUMSOL(SB, I) \leq HN_S$$

$$\textit{then} \quad HUMIRAC(SB, I) = \frac{SENSRSEC_P}{HN_S} HUMSOL(SB, I)$$

Where HUMSOL, HN$_S$ and HX$_S$ are the actual water content, the wilting point and the field capacity in the seed bed, respectively, and SENSRSEC$_p$ is a plant parameter which can be given a value between 0 and 1. If SENSRSEC$_p$=1 the effect of soil dryness on all the functions of root growth is only effective for water contents below the wilting point (Figure 2.3).

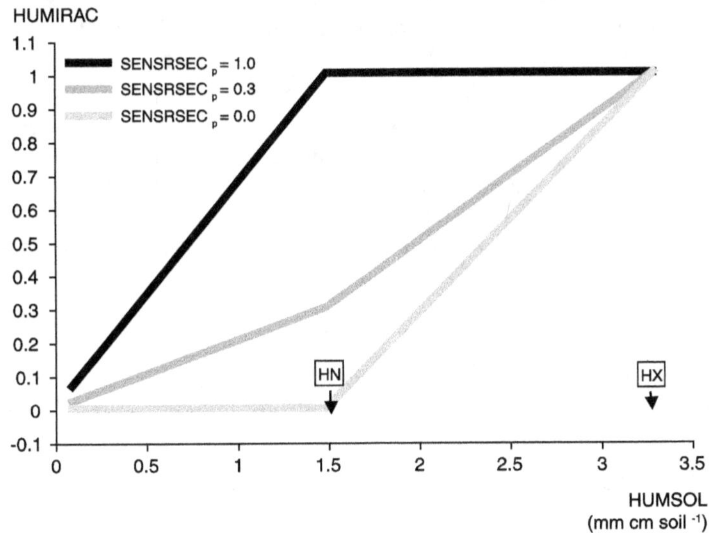

Figure 2.3. Evolution of the variable HUMIRAC as a function of the parameter SENSRSEC$_p$ and the values of seedbed water contents at field capacity (HX$_S$) and at wilting point (HN$_S$).

If the seedbed dries out, it may delay germination significantly. This does not impair grain viability as long as the grain has not already imbibed water. If however the soil water content has been high enough to allow grain moistening, grain viability is reduced. To account for this effect, we relied on Bradford's (1990, 2002) work showing that too long a time for germination after moistening reduces the germination rate if the number of days of moistening (NBJHUMEC) is higher than a plant- and temperature-dependent threshold duration (NBJGRAUTO: see eq. 2.1). It is assumed that germination occurs (IGER being the germination day) but at a reduced plant density (ratio between density of germinated plants, DENSITE, to sowing density, $DENSITE_T$) proportional to the thermal time deficit (eq. 2.4). An illustration of the chronology of germination in various soil conditions is given in Figure 2.4.

2.2.1.c Subsoil plantlet growth

Germination initiates the growth of the root and then of the shoot (see § 5). The growth rate of the shoot is assumed to be a logistic function (eq. 2.5) of soil degree-days that may slow down with unsuitable soil moisture (HUMIRAC). The parameterization of eq. 2.5 can be significantly different in actual soil conditions when compared to laboratory (finely sieved soil) conditions because the presence of clods or compacted earth slows down the shoot's vertical upward growth. Emergence occurs when elongation (ELONG) is greater than sowing depth ($PROFSEM_T$) as shown in Figure 2.5. HUMIRAC is calculated as described in eq. 2.3 by using the average soil moistures between the seedbed and the root front ZRAC (layer denoted HB). The variable CRUST stands for soil crusting conditions and will be explained in the following paragraph. In eq. 2.5 (see p. 26), $ELMAX_p$, $BELONG_p$ and $CELONG_p$ are species-dependent parameters.

As for germination, if the duration, between germination (IGER) and emergence (ILEV), is too long ($NLEVLIM1_p$ and $NLEVLIM2_p$ parameters in Figure 2.6, p. 26), there may be emergence deficiencies represented by the variable COEFLEV, i.e. the ratio of the emerged to the germinated density.

The effect of frost on young plantlets can be simulated and causes an additional reduction in population density. The plantlet stage (ILET) is assumed to end at a defined number of leaves ($NBFGELLEV_p$), calculated from the plastochrone ($PHYLLOTHERME_p$). The frost damage function for emergence (FGELLEV) is calculated in the same way as other frost functions (see § 3.4.4) with thresholds of specific sensitivity for the plantlet stage ($TGELLEV10_p$ and $TGELLEV90_p$) and reduces the plant density in a multiplicative way (eq. 2.7).

eq. 2.7: $DENSITE(I) = DENSITE(ILEV) \cdot FGELLEV(I)$ *where* $ILEV < I < ILET$

It may be necessary to modify the threshold values according to differential genetic tolerances and forms of frost occurrence (thermal amplitude, frost and thaw cycles).

2.2.1.d Influence of soil crusting on emergence

In the particular case of loamy soils, a crust may occur after sowing, creating a physical obstacle to emergence (Duval and Boiffin, 1990). In addition to the textural characteristics of the surface soil layer, the development of such a crust depends on soil

eq. 2.4:

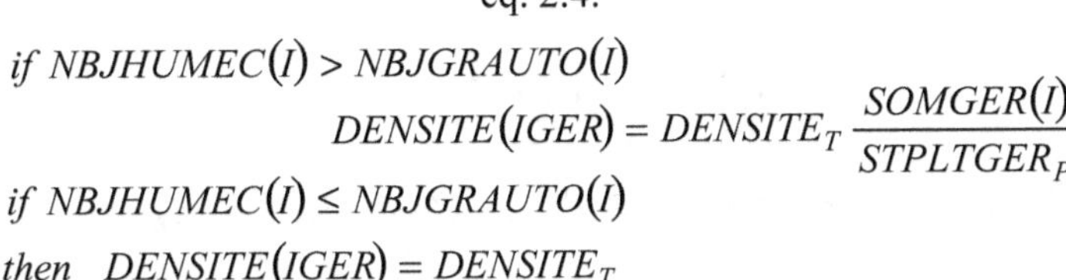

$$\textit{if } NBJHUMEC(I) > NBJGRAUTO(I)$$
$$DENSITE(IGER) = DENSITE_T \; \frac{SOMGER(I)}{STPLTGER_P}$$
$$\textit{if } NBJHUMEC(I) \leq NBJGRAUTO(I)$$
$$\textit{then } \; DENSITE(IGER) = DENSITE_T$$

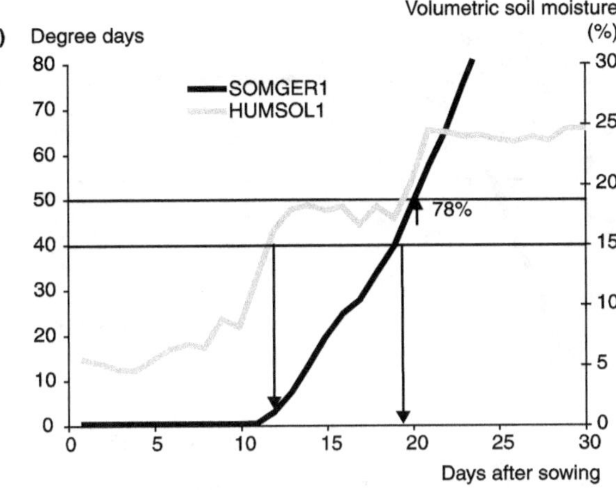

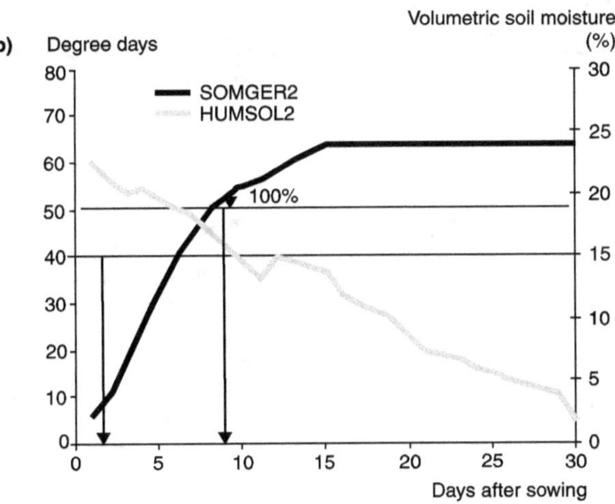

Figure 2.4. Chronology of germination represented for two different soil conditions: a) soil wetting and b) soil drying. The first arrow indicates the moistening date (soil above POTGERMI$_p$) and the second arrow the germination date. In the first case the required thermal time for germination (STPLTGER$_p$=50 degree days) is not reached by 6 days (NBJGERLIM$_p$) of moistening, which causes a decrease in density (78%).

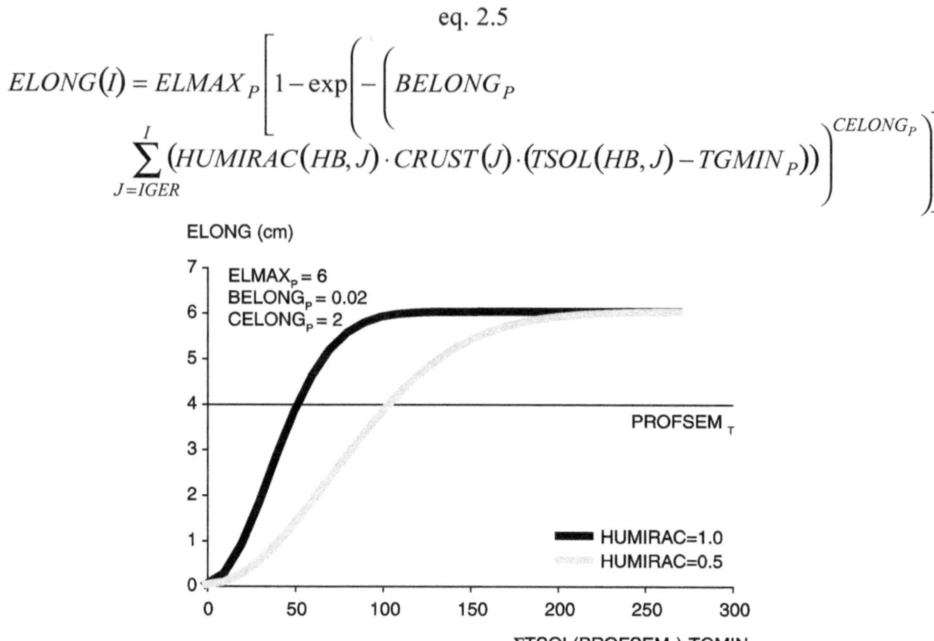

eq. 2.5

$$ELONG(I) = ELMAX_P\left[1 - \exp\left(-\left(BELONG_P \sum_{J=IGER}^{I}(HUMIRAC(HB,J)\cdot CRUST(J)\cdot(TSOL(HB,J) - TGMIN_P))\right)^{CELONG_P}\right)\right]$$

Figure 2.5. Elongation of the coleoptile (ELONG) as a function of soil temperature (TSOL) and water status (HUMIRAC) and occurrence of emergence when ELONG > PROFSEM$_T$.

eq. 2.6

$$\text{if} \quad ILEV - IGER < NLEVLIM1_P \quad \text{then} \quad DENSITE(ILEV) = DENSITE(IGER)$$
$$\text{if} \quad NLEVLIM1_P \leq ILEV - IGER \leq NLEVLIM2_P$$
$$\text{then} \quad DENSITE(ILEV) = DENSITE(IGER)\cdot COEFLEV(ILEV)$$
$$\text{if} \quad ILEV - IGER > NLEVLIM2_P \quad \text{then} \quad DENSITE(ILEV) = 0.0$$

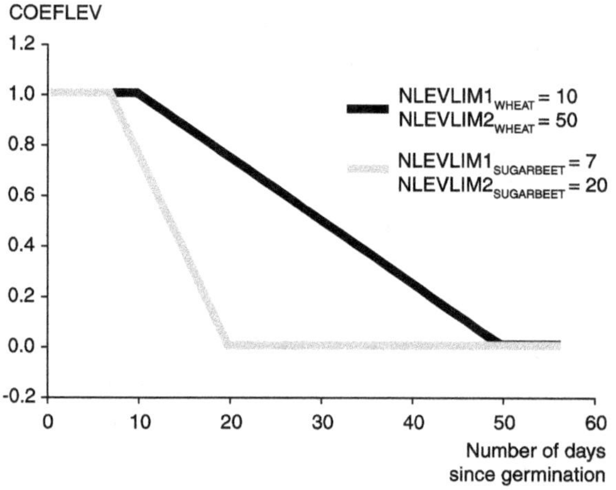

Figure 2.6. Simulation of emergence density proportion, COEFLEV(ILEV), according to the length of the germination-emergence period (ILEV-IGER).

fragmentation following seedbed preparation and on the weather at the time. Indeed, post-sowing rainfall may destroy soil fragments and then drought renders this layer almost impenetrable for the plantlets, since the resistance to emergence depends on the weather through its evaporative demand and on the force exerted by the plantlet.

The formalisation of these processes in STICS relies partly on the work of Durr *et al.* (2001). Surface crusting is assumed to occur only after sowing once a certain amount of rainfall (soil-dependent parameter $PLUIEBAT_S$) has occurred. The crust is assumed to be dry when the natural mulch depth (XMULCH: variable calculated from the soil evaporation formulations: see § 7.1) is greater than the threshold parameter $MULCHBAT_G$, in which case XMULCH is considered as the thickness of the crusted layer.

The subsequent delay in emergence can, just like the water deficit in the seedbed, reduce plant density. Yet not all the plantlets will be affected because of the heterogeneity of the crust and the differences in individual plantlet vigour. In STICS it is assumed that the ease of penetrating the crust is accounted for by a plant-dependent parameter ($VIGUEURBAT_P$). The delay in emergence is formalised by stopping the accumulation of thermal time in eq. 2.5) when the shoot reaches the base of the crust (CRUST=0.0 calculated in eq. 2.8).

$$\text{eq. 2.8}$$

$$if \quad \sum_{J=IPLT}^{I} PRECIP(J) \geq PLUIEBAT_S \quad then$$

$$if \quad XMULCH(I) \geq MULCHBAT_G \quad and \quad ELONG(I) \geq PROFSEM_T - XMULCH(I)$$
$$then \quad CRUST(I) = 0.0$$

$$if \quad XMULCH(I) < MULCHBAT_G \quad or \quad ELONG(I) < PROFSEM_T - XMULCH(I)$$
$$then \quad CRUST(I) = 1.0$$

The density reduction law is specific to the crusting phenomenon (COEFLEVB) but analogous to the other constraint law (COEFLEV depicted in Figure 2.6) with a minimum threshold corresponding to the $VIGUEURBAT_P$ parameter: if $VIGUEURBAT_P$ is greater than 0, which means that when the soil is crusted a proportion of plants succeed in emerging, the COEFEVB function is less effective than the water content and temperature-dependent COEFLEV function. The combination of both relationships is made dynamically by calculating the daily derivatives of both laws: if "CRUST=0", which means a crust obstacle occurs the current day, the density reduction is done according to the COEFLEVB law; otherwise it is the COEFLEVB law that prevails (Figure 2.7, see p. 28).

Thus as soon as significant rainfall occurs, growth of the shoot continues normally. Table 2.2 shows the sensitivity of the formalisations described above of the effect of soil crusting by varying the three required parameters.

2.2.2 Onset of crop development and growth after planting

For transplanted crops, a latency phase between planting and the onset of crop development can be simulated in the same way as the germination phase, based on accumulated growing degree. days. In this case, the simulated date of actual onset is the date corresponding to planting, to which is added the interval corresponding to the STPLTGER$_p$ parameter, calculated from soil temperatures at the depth of planting and taking into account the effect of soil dryness, as in eq. 2.2. The leaf area index of the plantlet (LAIPLANTULE$_p$) serves to initialise the dynamics of the leaf area index. If the "coverage rate" option is selected rather than the "LAI" option (see § 3.1.4), the LAIPLANTULE$_p$ parameter must be given in terms of percentage of soil cover; otherwise it is expressed in LAI units (i.e. m²m⁻²). It is also possible to specify the number of leaves per plant (NBFEUILPLANT$_p$) which enables initialisation of the calculation of the number of leaves. In a similar way biomass and rooting depth are initialized using the plant parameters MASECPLANTULE$_p$ and ZRACPLANTULE$_p$. The plantlet nitrogen content is calculated assuming no nitrogen storage, i.e. as responding to the critical nitrogen curve for a low biomass canopy (§ 8.6.1) involving ADIL$_p$ and the initial biomass (MASECPLANTULE$_p$) according to eq. 2.9.

$$\text{eq. 2.9:}\quad QNPLANTULE(IPLT) = 10\,ADIL_p \cdot MASECPLANTULE_p$$

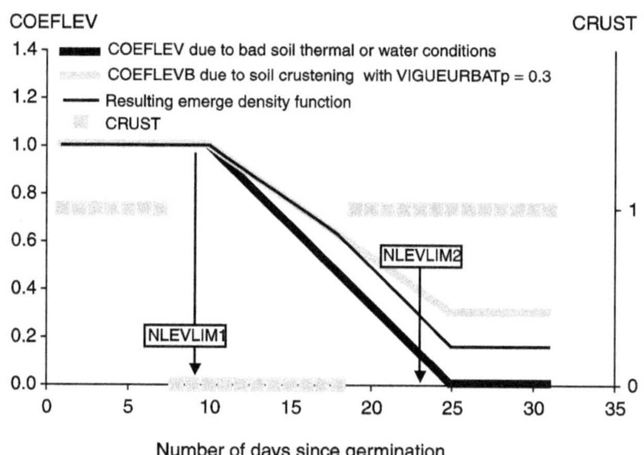

Figure 2.7. Combination of the two laws (COEFLEV depending on non-optimal water content and temperature conditions and COEFLEVB depending on the crust layer) affecting the emerged density as a function of the occurrence of the soil crust factor "CRUST=0.0", which means a crust obstacle occurs, and the plantlet vigour (VIGUEUBAT$_p$). The parameters NLEVLIM1$_p$ and NLEVLIM2$_p$ are defined in eq. 2.6.

Table 2.2. Sensitivity analysis of the soil crusting parameters on the emergence variables (example of a maize crop in the western France).

Sensitivity to crusting (SC)	No SC	High SC	Low SC	High SC
Plantlet vigour (PV)	–	High PV	Low PV	Low PV
PLUIEBAT$_S$ (mm)	50	3	9	3
MULCHBAT$_G$ (cm)	0.5	0.5	1.5	0.5
VIGUEURBAT$_p$	–	0.8	0.15	0.15
Sowing – emergence duration (days)	12	27	24	27
Emerged density relative to sown density (%)	77	64	31	19

2.2.3 Onset of crop development and growth in perennial plants

For perennial plants, the active onset of vegetative development generally occurs after a period of winter rest (if this is not the case vernalisation or chilling requirements are set to 0). The dormancy or vernalisation duration is calculated by meeting the chilling (or vernalisation) requirements.

If the simulation is initialized at the IDOR stage, the model then assumes that this is the onset of dormancy (IDEBDORM) and that the chilling requirements are not met. If the model is initialized at the ILEV stage, the model assumes that the chilling requirements are met (N.B.: this does not apply to annuals). Concerning the growth status, because the rest period is not complete for herbaceous crops, an initialisation in terms of both LAI and shoot biomass ($LAI0_I$ and $MASEC0_I$) is required while it is not the case for woody plants since the wood biomass (i.e. the accumulation of lignified biomass) is not taken into account by the model, assuming it is reduced by pruning. For both types of plant, it is necessary to give a value to three other initial variables: $RESPERENNE0_I$ (carbon reservoir assumed to be stored in the root system at the beginning of the rest period and remobilised for the spring growth), $QNPLANTE0_I$ (nitrogen reservoir) and $ZRAC0_I$ (rooting depth and densities if the "true density" option for describing root system is chosen: see § 5).

When the model is run for several years, the phasic and trophic status of the plant is saved from one year to the next (see §10.1 Crop successions).

2.3 Above-ground development

2.3.1 Time scale

The periods separating successive stages are specific to the species and variety. These periods are evaluated in development units, reproducing the phenological time of the plant.

Relying on the long-accepted concept of growing degree days (Bonhomme *et al.*, 1994; Durand, 1967), temperature is always used in crop models as the driving variable of the phenological time. Yet authors like Ong (1983) and Pararajasingham and Hunt (1991) showed that it is better not to use the temperature of the air but rather a temperature closer to the plant (soil or organ) to explain the phasic chronology. In particular, this can explain the acceleration of the cycle in case of drought (Seghieri *et al.*, 1995; Desclaux and Roumet, 1996; Casals, 1996). Indeed, soil drying at the surface as well as at depth causes temperature increases at the plant level (Cellier *et al.*, 1993; Friend, 1991), which affect the progress of the cycle. Consequently, as in the model by Jamieson *et al.* (1995), we adopted the idea of Idso *et al.* (1978) who suggested linking phenological time to the crop temperature rather than to the temperature of the air. The other factors affecting the rate of development are modeled as brakes or accelerators on that rate per unit thermal time (Brisson and Delécolle, 1991). These factors generally include the photoperiod and vernalisation (e.g. CERES as described by Ritchie and Otter, 1984 or ARCWHEAT by Weir *et al.*, 1984) and sometimes water deficit (e.g. CROPGRO by Jones *et al.*, 2003). Through the use of crop temperature, the effect of the water deficit on development is linked directly to the thermal units and not to a reducing factor. Of course, what is simulated by the use of crop temperature is an acceleration of the cycle, while some authors speak of delay in the case of early stress acting upon floral induction (Seghieri *et al.*, 1995; Blum, 1996). Nitrogen nutrition conditions can also have an effect on the progress of the cycle (Girard, 1997), as well as light conditions through plant density (cryptochrome).

In STICS, crop temperature (UDEVCULT) drives development. It may be slowed by sub-optimal photoperiod conditions (RFPI<1), by non-compliance with vernalisation requirements (RFVI <1) or by the effect of water or nitrogen stress (STRESSDEV$_P$ >0 and TURFAC<1 or INNLAI <1). Thus, each day, the phasic course of the crop (UPVT) is given by the 2.10 equation:

$$\text{eq. } 2.10$$

$$UPVT(I) = UDEVCULT(I) \cdot RFPI(I) \cdot RFVI(I)$$
$$\cdot \left[STRESSDEV_P \cdot \min\left(TURFAC(I), INNLAI(I)\right) + 1 - STRESSDEV_P \right]$$

As far as the emergence period is concerned a specific calculation is made using the conditions prevailing in the soil (see § 2.2) as for the root life duration (DEBSENRAC$_P$). Leaf lifespan is expressed in exponential type time (also called Q10 time) for reasons explained in § 3.1.2.

Most phasic courses between two successive stages are regarded as variety-specific (Table 2.3), as are the parameters indicating the sensitivity to the photoperiod and vernalisation requirements.

Table 2.3. Table summarizing the various parameters of developmental duration and the driving variables used to calculate those durations. TCULT is the crop temperature and TSOL is the soil temperature at the root front level.

Parameter of developmental duration	Positive thermal response			Cold requirement[1]	Action of photo-period[1]	Slowing water stress effect[1]	Slowing nitrogen stress effect[1]
	TCULT	$2^{TCULT/10}$	TSOL				
STPLTGER$_P$			X			X	
STDORDEBOUR$_P$	X				X	X	X
STLEVAMF$_V$	X			X	X	X	X
STAMFLAX$_V$	X			X	X	X	X
STLEVDRP$_V$	X			X	X	X	X
STDRPMAT$_V$	X						
STDRPNOU$_P$	X						
STDRPDES$_P$	X						
STFLODRP$_V$	X				X	X	X
DUREEFRUIT$_V$	X						
DURVIEF$_V$		X					
PHYLLOTHERME$_P$	X						
DEBSENRAC$_P$			X				

(1) If appropriate, the option is activated according to the plant sensitivity to the relevant factor.

2.3.2 Positive effect of temperature

In STICS temperature positively affects plant phasic development from the emergence stage for annuals (ILEV) or from dormancy break for woody plants (IFINDORM) until physiological maturity (IMAT). For herbaceous perennials there is always a positive effect of temperature despite a rest period during winter. Crop temperature is calculated from the crop energy balance (see § 6.6.2 on microclimate). As has been shown in the article by Brisson *et al.* (2002a), use of the crop temperature may modify the standard values used routinely with the air temperature. Consequently multiplicative plant-dependent coefficients (COEFLEVAMF$_p$, COEFAMFLAX$_p$, etc.) make it possible to modify "air temperature" standards so that the crop temperature can be used, which has the advantage of representing shortenings in the cycle induced by drought.

The effect of temperature (eq.2.11), achieved at a daily time step, is linearly increasing between the TDMIN$_p$ and TDMAX$_p$ thresholds, and linearly decreasing between the TDMAX$_p$ and TCXSTOP$_p$ thresholds, as illustrated in the Figure 2.8. Affecting the parameters TDMAX$_p$ and TCXSTOP$_p$ are not easy because they correspond to occasional thermal conditions. Nevertheless including this decrease in developmental and leaf growth (§ 3.1), in agreement with experiments in hot conditions, is worthwhile to use the model in future climate conditions.

eq.2.11

$$\textit{if } TCULT(I) \leq TDMIN_P \quad UDEVCULT(I) = 0$$

$$\textit{if } TDMIN_P < TCULT(I) < TDMAX_P \quad UDEVCULT(I) = TCULT(I) - TDMIN_P$$

$$\textit{if } TDMAX_P \leq TCULT(I) < TCXSTOP_P$$

$$UDEVCULT(I) = \frac{TDMAX_P - TDMIN_P}{TDMAX_P - TCXSTOP_P}(TCULT(I) - TCXSTOP_P)$$

$$\textit{if } TCULT(I) \geq TCXSTOP_P \quad UDEVCULT(I) = 0$$

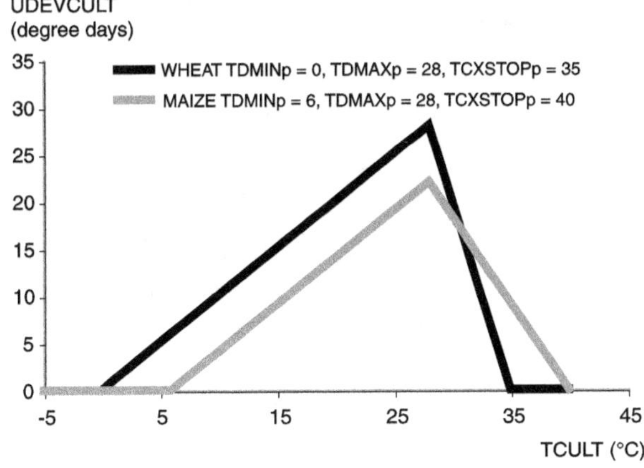

Figure 2.8. Development response to crop temperature.

The base temperature (TDMIN$_p$) is assumed constant throughout the crop cycle (from ILEV to IMAT). However, it was shown that this threshold could vary (Angus *et al.*, 1981) because the relationship between phasic development rates and temperature is not linear (Brisson *et al.*, 2005). For example, in the model ARCWHEAT (Weir *et al.* (1984) or in Hunt and Pararajasingham (1995), various temperature thresholds are used according to the stages. However, since there is a correlation between the duration and the temperature threshold, these parameters are difficult to calibrate.

2.3.3 Effect of photoperiod

For photoperiodic plants, the photoperiodic slowing effect, RFPI, applies between the threshold photoperiods PHOBASE$_p$ and PHOSAT$_p$. In the case of wheat, PHOBASE$_p$ is lower than PHOSAT$_p$: wheat is a long-day plant. In the case of soybean, PHOBASE$_p$ is higher than PHOSAT$_p$: soybean is a short-day plant (Figure 2.9 b). The current photo-

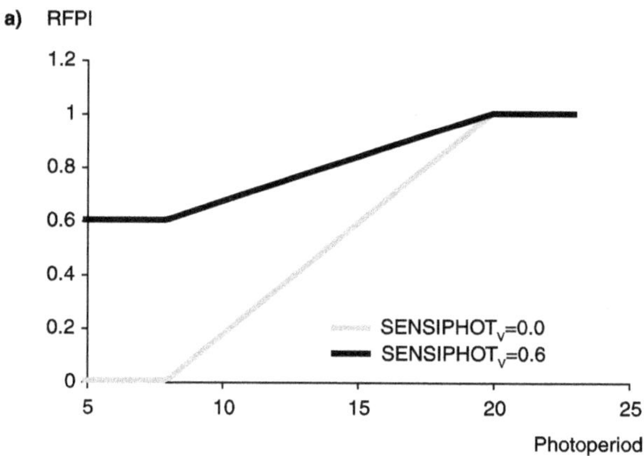

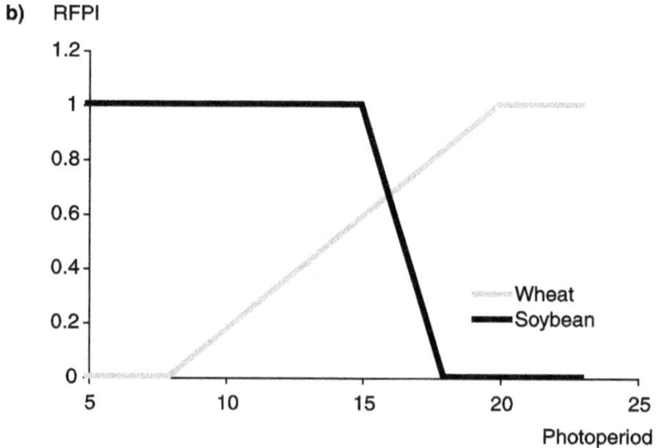

Figure 2.9. Photoperiodic limiting factor for phasic development (RFPI) when varying the sensitivity to photoperiod (a: with response type of wheat) or the photoperiodic response type (b: with sensiphot=0.0, for wheat [PHOBASE$_p$, PHOSAT$_p$]=[8,20] and for soybean [PHOBASE$_p$, PHOSAT$_p$]=[18,15]).

period (PHOI) is calculated on the basis of calendar days and latitude (Figure 2.10) using classic astronomical functions (Sellers, 1965). The photoperiod is calculated by assuming that light is perceptible until the sun is at 6° below the horizon, which corresponds to a duration 50 to 70 minutes longer than the strictly defined daylength.

The amplitude of sensitivity to the photoperiod is given by the SENSIPHOT$_V$ parameter: a value of 0 gives maximum sensitivity and a value of 1 cancels out this sensitivity (Figure 2.9 a). The effect of the photoperiod is exerted between the ILEV (herbaceous) or IFINDORM (ligneous) stages and IDRP. This formalisation allows the sensitivity to photoperiod of different varieties to be characterised.

eq. 2.12

$$RFPI(I) = (1 - SENSIPHOT_V)\frac{(PHOI(I) - PHOSAT_P)}{PHOSAT_P - PHOBASE_P} + 1$$

and $\quad$ SENSIPHOT$_V$ $\leq$ RFPI $(I) \leq 1$

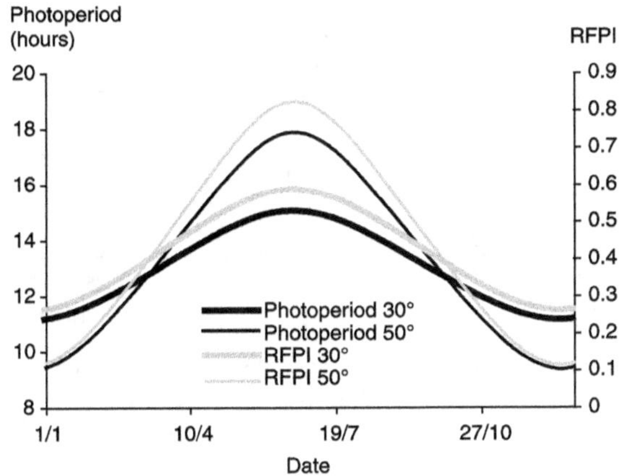

Figure 2.10. Annual variation of the photoperiod for two north latitudes and the consequence on the corresponding limiting factor for phasic development (RFPI) calculated for wheat crop (Figure 2.9 b).

2.3.4 Cold requirements

Winter crops and perennial crops in temperate climate zones have vernalisation or chilling requirements. The formalisations classically applied and used in STICS differ for herbaceous plants (vernalisation) and woody plants (dormancy). For herbaceous plants, the resting state is considered not to be total, and the "vernalisation" formalisation which applies to herbaceous plants allows a partial accumulation of development units during winter rest. For woody plants the "dormancy" formalisations are much severe, and development units are only active when all chilling requirements have been met. Consequently, non-compliance with vernalisation requirements slows (RFVI <1 for herbaceous plants) or stops (RFVI = 0 for ligneous plants) the development of crops. For woody plants the post-dormancy period is characterized by the phasic course between dormancy break (IFINDORM) and budding (ILEV), i.e. STDORDEBOUR$_p$.

2.3.4.a Vernalisation

Vernalisation requirements are defined by the genotype-dependent number of vernalising days (JVC$_V$) and the vernalising value of a given day (JVI) depends on crop temperature (Figure 2.11). Vernalising days are counted from germination (IGER) for annual crops because an active metabolism is required to initiate the process, or from the JULVERNAL$_p$ day for perennial crops. A minimum number of vernalising days, JVCMINI$_p$ is required (eq. 2.13). The progress in crop vernalisation, RFVI, gradually increases until it reaches the value of 1.

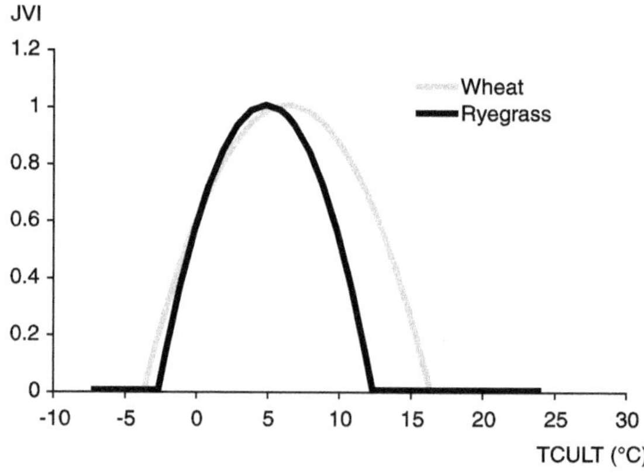

Figure 2.11. Vernalising value of a given day (JVI) as a function of the mean crop temperature (TCULT) for wheat [TFROID$_p$, AMPFROID$_p$]=[6.5,10] and for ryegrass [TFROID$_p$, AMPFROID$_p$]=[5.0, 7.5].

eq. 2.13

$$JVI(I) = \max\left(1 - \left[\frac{TFROID_P - TCULT(I)}{AMPFROID_P}\right]^2 ; 0.0\right)$$

$$RFVI(I) = \frac{\sum_{J=IGER\ or\ JULVERNAL_P}^{I} \left(JVI(J) - JVCMINI_P\right)}{JVC_V - JVCMINI_P}$$

TFROID$_p$ (optimum vernalisation temperature) and AMPFROID$_p$ (thermal semi-amplitude of the vernalising effect) are parameters which provide the range of vernalising activity of temperatures (Figure 2.11). AMPFROID$_p$ indicates the sensitivity of the species to vernalisation: if it is low, the range of vernalising temperatures is narrow and a long period will be necessary to meet the requirements; if it is high, the temperature range is broader and results in more rapid vernalisation. Figure 2.12 illustrates the sensitivity of the model to this parameter and its effects on leaf growth dynamics (details of calculation in § 3.1.1).

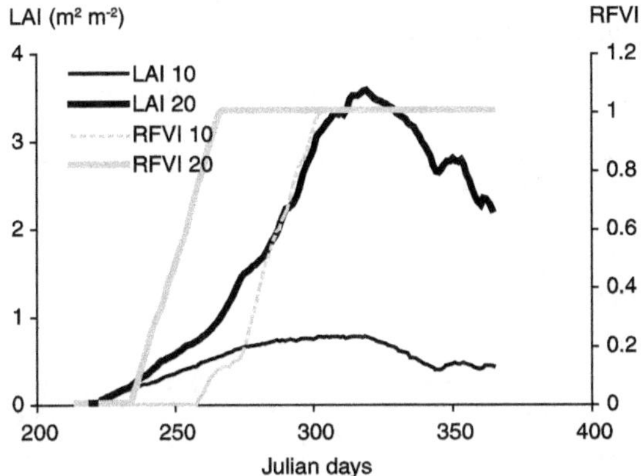

Figure 2.12. Sensitivity to the AMPFROID$_p$ parameter (assumptions of 10 and 20°C) on the calculation of the period of vernalisation (RFVI) and its consequences on leaf growth (LAI) for a ryegrass catch crop sown in late summer.

2.3.4.b Dormancy

This chapter deals with the perennial dormancy and not with the dormancy break of annual crop grains such as wheat, barley or pea that can lead to germination of the grain on the plants before harvest.

The aim is to calculate the day of break of dormancy (IFINDORM), which makes it possible to change the RFVI variable from 0 to 1, bearing in mind that it is always possible to impose this date and ignore the following dormancy calculations. Two options are available to calculate the break of dormancy: i) relying on well known formulae used for fruit trees for both vegetative or reproductive buds, ii) using minimal and maximal values of the air temperature (TMIN and TMAX).

In 1965 Bidabe proposed a formulation to calculate dormancy and post-dormancy durations for apple trees, based on the Q10 notion which corresponds to exponential type responses to temperature. In STICS, we just use what concerns the dormancy period, since the post-dormancy period (i.e. from IFINDORM to ILEV) is accounted for by the positive responses to temperature, according to a Q10 law (eq. 2.14). The daily responses are accumulated (CU) until the current day (I) from a starting date (IDEBDORM$_p$) generally taken to be during the autumn or the summer (Garcia de Cortazar, 2006). It has little effect on the calculation. The genetic-dependent parameter for the amount of chilling requirement is JVC$_V$.

eq. 2.14

$$CU(I) = \sum_{J=IDEBDORM}^{I} Q10_P^{\frac{-TMAX(J)}{10}} + Q10_P^{\frac{-TMIN(J)}{10}}$$

$$and \quad \begin{array}{l} if \quad CU(I) < JVC_V \quad RFVI(I) = 0.0 \\ if \quad CU(I) \geq JVC_V \quad RFVI(I) = 1.0 \end{array}$$

The other dormancy calculation included in STICS is the one by Richardson *et al.* (1974), developed for peach trees. It is based on accumulated chilling hourly units (CUH) effective to ensure break of dormancy. The relationship between CUH and hourly temperature is a "stepped" function, given in Table 2.4.

Table 2.4. Chilling hourly units (CUH) as a step function of temperature from Richardson (1974).

Hourly temperature (TH in °C)	CUH
TH ≤ 1.4	0
1.4 < TH ≤ 2.4	0.5
2.4 < TH ≤ 9.1	1
9.1 < TH ≤ 12.4	0.5
12.4 < TH ≤ 15.9	0
15.9 < TH ≤ 18.0	− 0.5
TH > 18.0	− 1

In order to be able to use bibliographic references to chilling requirements, we reconstitute hourly temperatures in accordance with Richardson's proposals (Richardson *et al.*, 1974): TMIN at 0 hour, TMAX at 12 hours and linear interpolation between the two. The active sum of CUH starts in the autumn as soon as the CUH are positive, defining IDEBDORM (eq. 2.15). The instability which may be generated by alternating positive and negative CUH has no effect on the final result for dormancy break.

eq. 2.15

$$CU(I) = \sum_{J=IDEBDORM}^{I} \sum_{H=1}^{24} CUH(J,H) \quad and \quad \begin{array}{ll} if & CU(I) < JVC_V \quad RFVI(I) = 0.0 \\ if & CU(I) \geq JVC_V \quad RFVI(I) = 1.0 \end{array}$$

A comparison between Bidabe and Richardson is presented in Figure 2.13 showing the very similar dynamics of the two methods. In terms of robustness of the parameterisation over sites and years, the Bidabe's formulation (Bidabe, 1965) seems to give better results (Liennard, 2002; Carcia de Cortazar, 2005).

2.3.5 Effect of stress

Early stresses can generate delays in the development of some crops. This effect counteracts the "acceleration" effect induced by using the crop temperature. It is active up to the IDRP stage, and can be modulated using a plant-dependent sensitivity parameter (STRESSDEV$_p$=0: crop insensitive to stress), as described in eq. 2.10. The lower of the two values of water stress (TURFAC) and nitrogen stress (INNLAI) is applied. For instance this effect causes a 5-8 day delay between a fertilised and an unfertilised situation in the Parisian basin for wheat (STRESSDEV$_p$=0.2). This effect is also accounted for in the calculation of leaf life span (§3: eq. 3.10).

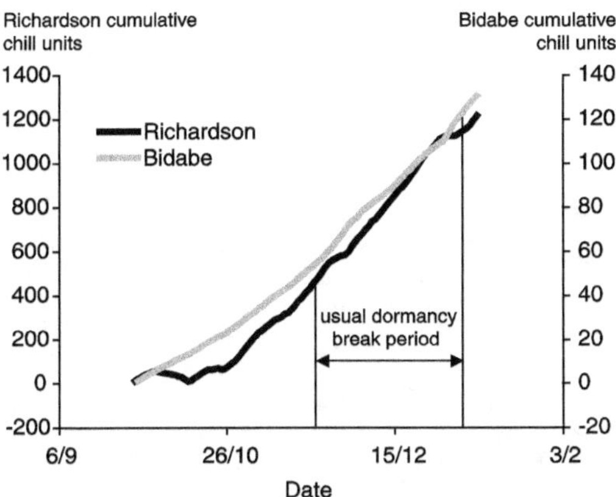

Figure 2.13. Comparison of the chilling responses by Bidabe ($Q10_p = 2$) and Richardson for a typical winter in the south of France.

3 Shoot growth

As in all crop models, the plant sub-system in STICS is characterized by its shoot biomass and leaf area index (LAI). Once calculated, the shoot biomass is partitioned into the various organs and feed-back occurs between this partitioning and shoot growth for "indeterminate" plants. In STICS, "indeterminate" denotes species for which there is significant trophic competition between vegetative organs and harvested organs. This definition is different from the botanical one and species like rapeseed or pea are considered in STICS as "determinate" since the assumption of independence between vegetative growth and reproductive growth is acceptable, though the two developmental scales (vegetative and reproductive) can overlap. On the other hand species like sugarbeet are regarded as "indeterminate" because the growing tuber greatly influences shoot growth. The harvested organs (grains, fruits or tuber) are the only ones characterized in terms of number and biomass (see § 4). This chapter deals with various interrelated processes, themselves depending on other chapters. The linkage between paragraphs of chapter 3 and with other chapters is shown in Figure 3.1.

3.1 Leaf dynamics

3.1.1 Leaf area growth

In most models, temperature is the main variable explaining potential leaf growth according to the crop's development stage (Weir *et al.*, 1984, Williams *et al.*, 1984, Hansen *et al.*, 1990, Amir et Sinclair, 1991a). Yet in some models, the increase in the leaf surface area is derived from their increase in mass by means of the specific leaf area (Van Keulen et Seligman, 1987). However, the specific leaf area is not a constant. It depends on the ratio between structural and non-structural mass (Thornley, 1996) which varies according to leaf age, temperature (Gary *et al.*, 1993), and the stresses experienced. Consequently, this kind of formalism is generally not very robust (Tardieu *et al*, 1999).

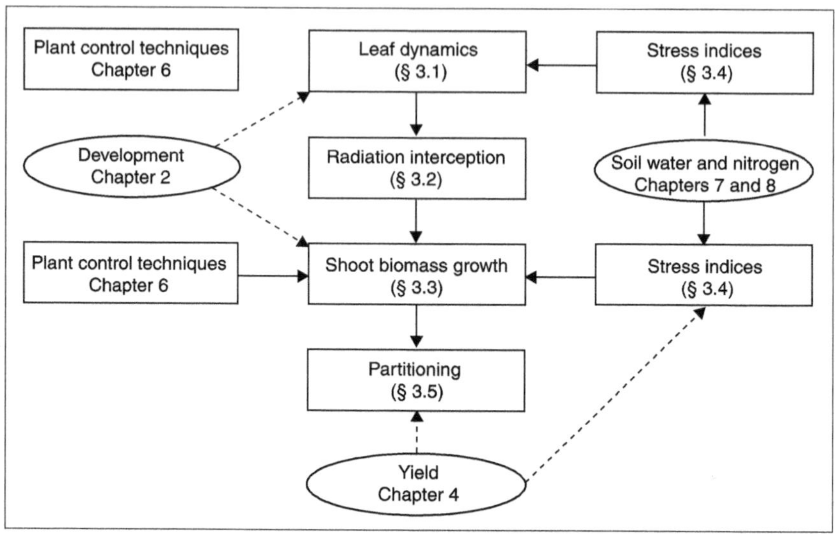

Figure 3.1. Main functional links between paragraph of chapter 3 and with other chapters.

Many models have a marked preference for "leaf to leaf" simulation (Ritchie et Otter, 1984, Amir et Sinclair, 1991a), using classic notions such as the phyllotherm and duration of leaf life (Muchow *et al.*, 1990). However Milroy and Goyne (1995) quoted several studies that showed that simulating LAI directly on a canopy scale gives as good results as a "leaf to leaf" model. Baret (1986), Milroy and Goyne (1995), and Chapman *et al.* (1993) worked on a canopy scale and they suggested splitting the evolution of LAI into two curves. The first one represents the growth (always a logistic curve) and the other curve is the senescence (logistic or exponential). In the first version of STICS (Brisson *et al.*, 1998a), the net leaf growth was directly simulated, without splitting the evolution of the LAI into gross growth and senescence, leading to a crude representation of LAI, with a plateau that does not exist in reality. However, when thinking in terms of efficiency of radiation interception, it appears that there is a plateau (Allen and Richardson, 1968, Cowan, 1968, Varlet Grancher et Bonhomme, 1979, Otegui *et al.*, 1995).

The assumption of a direct link between the evolution of LAI and crop development has been proposed by several authors (Nelder, 1961, Dale *et al.*, 1980, Dwyer and Stewart, 1986, Teittinen *et al.*, 1994; Hammer *et al.*, 1994) and in the model of Jamieson *et al.* (1995) four stages of evolution can be found for LAI.

In STICS, leaf area growth is driven by phasic development, temperature and stresses. An empirical plant density-dependent function represents inter-plant competition. For indeterminate plants, trophic competition is taken into account through a trophic stress index, while for determinate plants a maximal expansion rate threshold is calculated to avoid unrealistic leaf expansion.

3.1.1.a Valid calculations for all types of crop

The calculation of leaf growth rate (DELTAI$_1$ in m^{2}m^{-2} d^{-1}) is broken down in eq. 3.1. A first calculation of the LAI growth rate (DELTAI$_{dev}$ in m^2 plant^{-1} degree-day^{-1})

describes a logistic curve, related to the ILEV, IAMF and ILAX phenological stages. This value is then multiplied by the effective crop temperature (DELTAI$_T$ in degree-days), the plant density combined with a density factor, supposed to stand for inter-plant competition, that is characteristic for the variety (DELTAI$_{dens}$ in plant m^{-2}), and the water and nitrogen stress indices (DELTAI$_{stress}$).

eq. 3.1

$$DELTAI_1(I) = DELTAI_{dev}(I) \cdot DELTAI_T(I) \cdot DELTAI_{dens} \cdot DELTAI_{stress}(I)$$

The phasic development function (eq. 3.2) is comparable to that of the model PUTU (Singels and Jagger, 1991), i.e. a logistic function with DLAIMAXBRUT$_p$ as asymptote and PENTLAIMAX$_p$ as the slope at the inflexion point. It is driven by a normalized leaf development unit (ULAI) equal to 1 at ILEV and 3 at ILAX. At the end of the juvenile stage (IAMF), it is equal to VLAIMAX$_p$ when the inflexion of the dynamics (point of maximal rate) also occurs. Between the stages ILEV, IAMF and ILAX, the model performs linear interpolation based on development units (UPVT) which include all the environmental effects on phasic development (see § 2.3). As the ILAX stage approaches, it is possible to introduce a gradual decline in growth rate using the UDLAIMAX$_p$ parameter – the ULAI value beyond which there is a decline in the leaf growth rate. If UDLAIMAX$_p$=3 it has no effect and the leaf stops growing at once at ILAX (Figure 2.3).

eq. 3.2

$$if \ ULAI(I) < UDLAIMAX_P$$

$$DELTAI_{dev}(I) = \frac{DLAIMAXBRUT_P}{1 + \exp[PENTLAIMAX_P(VLAIMAX_P - ULAI(I))]}$$

$$if \ ULAI(I) \geq UDLAIMAX_P$$

$$DELTAI_{dev}(I) = DELTAI_{dev} \ MAX \left(1 - \frac{ULAI(I) - UDLAIMAX_P}{3 - UDLAIMAX_P}\right)^2$$

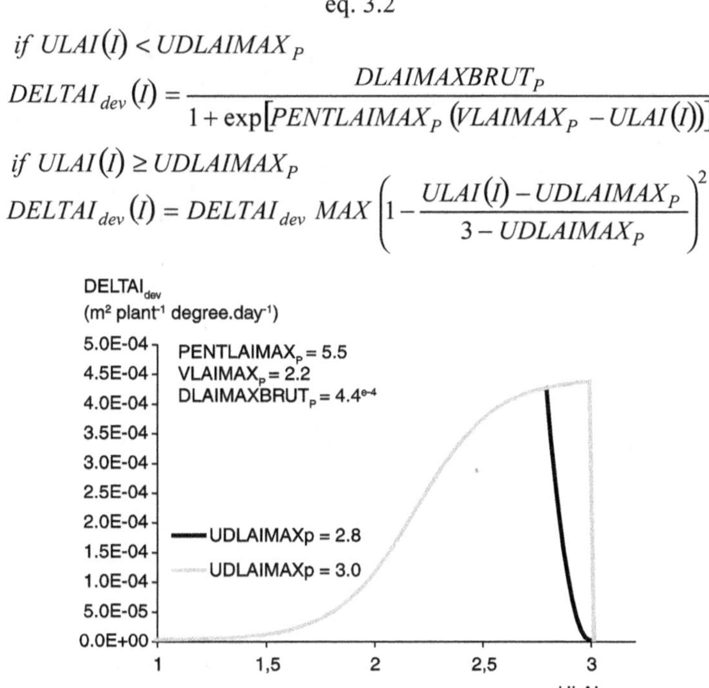

Figure 3.2. Leaf growth rate as a function of phasic development with the parameterization corresponding to wheat crop as given in Singels and Jagger (1991) with two hypotheses for leaf growth slowing at ILAX through the parameter UDLAIMAX$_p$ and consequences for the LAI curve shape.

The thermal function relies on crop temperature and cardinal temperatures ($TCMIN_p$ and $TCMAX_p$) which differ from the ones used for the phasic development. The extreme threshold $TCXSTOP_p$ is the same as for development.

eq. 3.3

$$if\ TCULT(I) \le TCMIN_P \qquad DELTAI_T(I) = 0.0$$

$$if\ TCMIN_P < TCULT(I) < TCMAX_P \quad DELTAI_T(I) = TCULT(I) - TCMIN_P$$

$$if\ TCMAX_P \le TCULT(I) < TCXSTOP_P$$

$$DELTAI_T(I) = \frac{TCMAX_P - TCMIN_P}{TCMAX_P - TCXSTOP_P}\,(TCULT(I) - TCXSTOP_P)$$

$$if\ TCULT(I) \ge TCXSTOP_P \qquad DELTAI_T(I) = 0$$

The density function ($DELTAI_{dens}$), is active solely after a given LAI threshold occurs ($LAICOMP_p$ parameter) if the plant density (DENSITE in plant m^{-2} calculated as explained in § 2.2 and possibly decreased by early frost: § 3.4.4) is greater than the $BDENS_p$ threshold, below which plant leaf area is assumed independent of density. Beyond this density value, leaf area per plant decreases exponentially. The $ADENS_V$ parameter represents the ability of a plant to withstand increasing densities. It depends on the species and may depend on the variety (Figure 3.3). For branching or tillering plants, $ADENS_V$ represents the plant's branching or tillering ability (e. g. wheat or pea). For single-stem plants, $ADENS_V$ represents competition between plant leaves within a given stand (e.g. maize or sunflower).

eq. 3.4

$$if\ LAI(I) \ge LAICOMP_P\ and\ DENSITE(I) \ge BDENS_P$$

$$DELTAI_{dens}(I) = DENSITE(I)\left(\frac{DENSITE(I)}{BDENS_P}\right)^{ADENS_V}$$

$$if\ LAI(I) < LAICOMP_P\ or\ DENSITE(I) < BDENS_P$$

$$DELTAI_{dens}(I) = DENSITE(I)$$

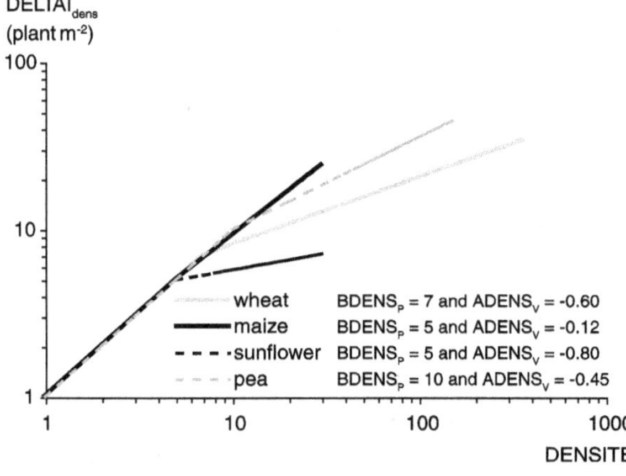

Figure 3.3. Density function ($DELTAI_{dens}$) for various species (wheat, maize, pea and sunflower).

Water and nitrogen affect leaf growth as limiting factors, i.e. stress indices whose values vary between 0 and 1 (see § 3.4 for their calculation). Water (TURFAC) and nitrogen deficits (INNLAI) are assumed to interact, justifying the use of the more severe of the two stresses. Meanwhile at the whole plant level the water-logging stress index is assumed to act independently.

$$\text{eq. 3.5: } DELTAI_{stress}(I) = \min(TURFAC(I),\ INNLAI(I)) \cdot EXOLAI(I)$$

3.1.1.b Features of determinate crops

Failure to account for trophic aspects in the calculation of leaf growth may cause problems when the radiation intercepted by the crop is insufficient to ensure leaf expansion (e.g. for crops under a tree canopy or crops growing in winter). Consequently, from the IAMF stage, we have introduced a trophic effect to calculate the definitive LAI growth rate (DELTAI$_2$) in the form of a maximum threshold for leaf expansion (DELTAIMAXI in $m^2m^{-2}d^{-1}$) using the notion of the maximum leaf expansion allowed per unit of biomass accumulated in the plant (SBVMAX in $cm^2\ g^{-1}$) and the daily biomass accumulation (DLTAMS in $t.ha^{-1}day^{-1}$ possibly complemented by remobilized reserve REMOBILJ). SBVMAX is calculated using the SLAMAX$_P$ and TIGEFEUILLE$_P$ parameters (eq. 3.6).

$$\text{eq. 3.6}$$

$$SBVMAX = \frac{SLAMAX_P}{1 + TIGEFEUILLE_P}$$

$$and \quad DELTAIMAXI(I) = \left[DLTAMS(I-1) + REMOBILJ(I-1)\right] \cdot SBVMAX \cdot 10^{-2}$$

$$if\ DELTAI_1(I) < DELTAIMAXI(I)\ or\ I < IAMF \qquad DELTAI_2(I) = DELTAI_1(I)$$

$$if\ DELTAI_1(I) \geq DELTAIMAXI(I)\ and\ I \geq IAMF \quad DELTAI_2(I) = DELTAIMAXI(I)$$

Eq. 3.6 is illustrated in Figure 3.4 for a wheat crop receiving reduced radiation (20% of incoming) as can happen under a tree canopy compared to a crop in the open air.

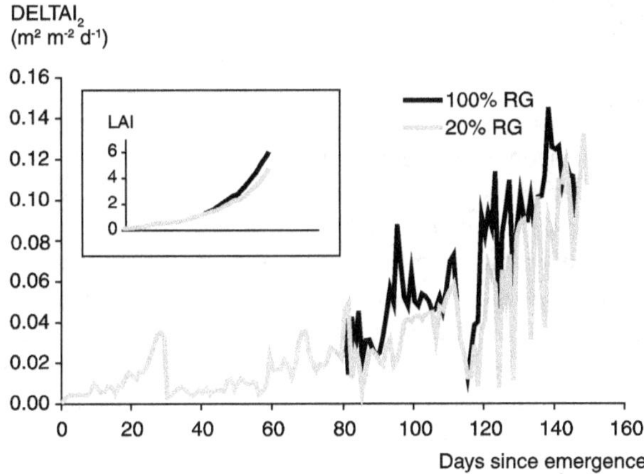

Figure 3.4. Dynamics of LAI growth rate (DELTAI$_2$) of a durum wheat crop in southern France with 100 and 20% of the incoming radiation (RG) without any stresses and the consequences on LAI values during its growing phase.

3.1.1.c Features of indeterminate crops

It has been possible to test the robustness of the above formalisation on a variety of crops, including crops where there is an overlap between the vegetative phase and the reproductive phase (soybean and flax for example). However, when trophic competition between leaves and fruits is a driving force for the production and management of the crop (for example tomato, sugarbeet), this formalisation is unsuitable. We therefore calculate the $DELTAI_2$ variable (eq. 3.7) so as to take trophic monitoring into consideration in the case of crops described as 'indeterminate', by introducing a trophic stress index (SPLAI explained in § 3.4.3).

$$\text{eq. 3.7:} \quad DELTAI_2(I) = DELTAI_1(I) \times SPLAI(I)$$

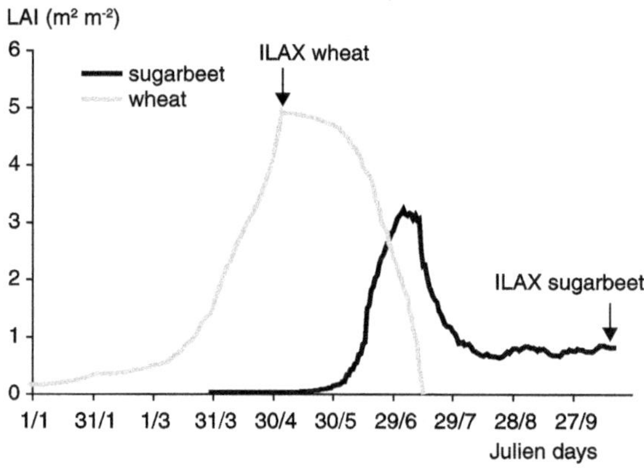

Figure 3.5. Comparison of determinate (wheat) and indeterminate (sugarbeet) LAI dynamics. The ILAX stage indicates the end of leaf onset.

As a consequence, the LAI can decrease markedly during the growth phase if the crop is experiencing severe stresses during the harvested organs filling phase: this case is illustrated in Figure 3.5 for sugarbeet.

3.1.2 Senescence

In STICS shoot senescence only concerns leaves: dry matter and LAI. For cut crops, it also affects residual biomass after cutting. While in the first versions of the model senescence was implicit (Brisson *et al.*, 1998a), it is now explicit, with a clear distinction between natural senescence due to the natural ageing of leaves, and senescence accelerated by stresses (water, nitrogen, frost). The concept of leaf lifespan, used for example by Maas (1993), is applied to the green leaf area and biomass produced. The leaf area and part of the leaf biomass (see § 3.5) produced on a given day is therefore lost through senescence once the lifetime has elapsed (Duru *et al.*, 1995). This part corresponds to the RATIOSEN$_p$ parameter, taking into account the part which was remobilised during its senescence.

3.1.2.a Calculation of lifespan

The natural lifespan of leaves (DURAGE) is defined by two values: the lifespan of early leaves, or $DURVIEI_p$ (expressed as a proportion of $DURVIEF_V$) and the lifespan of the last leaves emitted, or $DURVIEF_V$ (assumed genotype-dependent). Until the IAMF stage, the natural lifespan, calculated for the day when the leaves are emitted (I0) is $DURVIEI_p$; from IAMF to ILAX, the natural lifespan increases between $DURVIEI_p$ and $DURVIEF_V$ as a function of the leaf development variable ULAI (eq. 3.8).

Because of water or nitrogen stress, the current lifespan may be shortened if the stress undergone is more intense than previous stresses (during the period from I_0 to I in eq. 3.9). Specific stress indices for senescence are introduced (SENFAC and INNSENES see § 3.4.4.a). Frost (FSTRESSGEL that can be either FGELJUV or FGELVEG: see § 3.4) may also reduce or even cancel lifespan. In the event of over-fertilisation with nitrogen (INN >1), the foliage lifespan is increased from the IAMF stage up to a maximum given by the $DURVIESUPMAX_p$ parameter (eq. 3.9).

eq. 3.8

$$DURAGE(I_0) = DURVIEI_P + \frac{ULAI(I_0) - VLAIMAX_P}{3 - VLAIMAX_P}(DURVIEF_V - DURVIEI_P)$$

eq. 3.9

$$SENSTRESS(J) = \min\left(SENFAC(J), INNSENES(J), FSTRESSGEL(J)\right)$$

$$DURVIE(I) = DURVIESUP + DURAGE(I_0) \cdot \min_{J=I_0}^{I}\left(SENSTRESS(J)\right)$$

$$if\ \ INN(I) > 1\ \ DURVIESUP(I) = DURVIEF_V \cdot \min\left[DURVIESUPMAX_P, (INN(I) - 1)\right]$$

$$if\ \ INN(I) \leq 1\ \ DURVIESUP(I) = 0.0$$

The lifespan of leaves is not expressed in degree.days (like phasic development), because this has the disadvantage of stopping any progression as soon as the temperature is lower than the base temperature ($TDMIN_p$). To remedy this problem, the senescence course between I_0 and I (SOMSEN) is expressed by cumulative Q10 units (with Q10=2), i.e. an exponential type function (eq. 3.10).

eq. 3.10

$$SOMSEN(I) = \sum_{J=I_0}^{I} 2^{\frac{UDEVCULT(J) \cdot \left[STRESSDEV_P \cdot min(TURFAC(J), INNLAI(J)) + 1 - STRESSDEV_P\right]}{10}}$$

The senescence course is affected by the same cardinal temperatures as phasic development and can be slown down by stresses (§ 2.3). The lifespan parameter of the leaf ($DURVIEF_V$) expressed in Q10 units represents about 20% of the same lifespan expressed in degree.days (Figure 3.6).

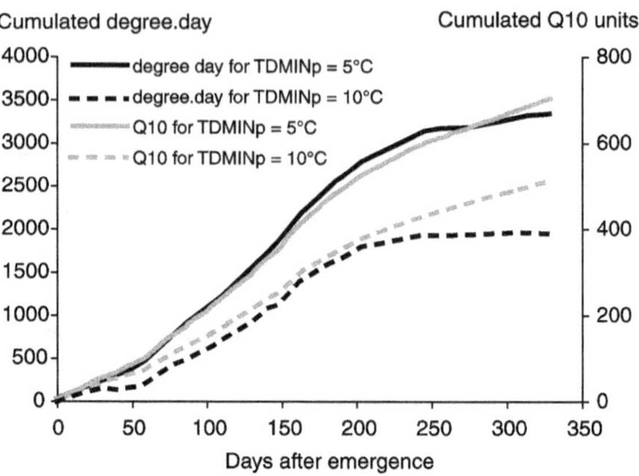

Figure 3.6. Comparison between phasic development courses expressed in degree.days and in Q10 units as defined in eq. 3.10 for 2 values of TDMIN$_\mathrm{p}$.

3.1.2.b Calculation of senescence

Material produced on day I_0 disappears by senescence after a period corresponding to DURVIE(I_0). Depending on the evolution of temperature and of lifespan as a function of phenology and stresses, senescence can vary from one day to another and affect several days of production (J=I_0, I_0+1…) or, on the contrary, none (DURVIE(I_0)>SOMSEN(I)) as explained in eq. 3.11. This principle is applied to the biomass (DLTAMSEN) and leaf area index (DLTAISEN). In general, the leaf biomass produced does not completely disappear (remobilisation): the RATIOSEN$_\mathrm{P}$ (<1) parameter enables the definition of the senescent proportion with reference to production. It is the PFEUILVERTE ratio (see § 3.5.2.a on partitioning) which defines the proportion of leaves in the biomass produced.

eq. 3.11

$$As\ long\ as\ SOMSEN(I) \geq \sum_{J=I_0}^{I} DURVIE(J)$$

$$then\quad DLTAISEN(I) = \sum_{J=I_0}^{I} DELTAI(J)$$

$$and\quad DLTAMSEN(I) = \sum_{J=I_0}^{I} DLTAMS(J) \cdot RATIOSEN_P \cdot PFEUILVERTE(J)$$

The cumulative senescent foliage is LAISEN. If the crop is a forage crop benefiting from residual dry matter from the previous cycle (MSRESIDUEL$_\mathrm{T}$ parameter), the senescence of residual dry matter (DELTAMSRESEN) starts as from cutting. It occurs at a rate (eq. 3.12) estimated from the relative daily lifespan course and weighted by the remobilisation (RATIOSEN$_\mathrm{p}$).

$$\text{eq. 3.12}$$

$$DLTAMSRESEN(I) = RATIOSEN_P \cdot MSRESIDUEL_T \cdot \frac{2^{\frac{UDEVCULT(I)}{10}}}{DURVIEI_P}$$

3.1.3 Photosynthetic function of storage organs

During the maturation of storage organs, the chlorophyll function of the organs or their envelopes may induce a significant accumulation of biomass. Such processes have been demonstrated for wheat ears (Abbad *et al.*, 2004, Araus *et al.*, 1993, Casals, 1996) and also exist in rapeseed siliquae, pea pods or grapes during their green period. To take account of this effect, we have introduced a parameter, SEA_P (cm^2 g^{-1}) which converts the biomass of these membranes (MAENFRUIT defined in § 3.5) into their equivalent leaf surface area (EAI in eq. 3.13).

$$\text{eq. 3.13:} \quad EAI = MAENFRUIT\frac{SEA_P}{100}$$

It is assumed that the chlorophyll function of storage organs lasts from the IDRP till the IDEBDES (beginning of dehydration) stages.

3.1.4 Use of ground cover instead of the leaf area index

Given the complexity and the numerous parameters required for LAI calculation, de Tourdonnet (1999) proposed a simple alternative by the direct calculation of the ground cover, which can be used as a state variable in calculations for radiation interception and water requirements. This can be particularly useful for plants with a complex foliage structure such as lettuce, or for a first modelling approach. It is programmed in STICS as an alternative option to all the previous calculations. This formalisation is particularly interesting when leaves have complex spatial arrangement or when the individual plant foliage is abundant.

To calculate the ground cover (TAUXCOUV), a temporal scale similar to that of LAI is used, and called ULAI; this varies from 0 to 2, depending on the phenological time. At the IAMF stage, ULAI is equal to $INFRECOUV_P$. TAUXCOUV is calculated by a logistic curve (eq. 3.14), using $TRECOUVMAX_P$ as the asymptote, which represents the proportion of the soil covered by an isolated plant, $INFRECOUV_P$ as the abscissa of the inflexion point and $PENTRECOUV_P$ as the slope at the inflexion point. The competitive effect linked to population growth is simulated in the same way (eq. 3.4) as for the leaf area index and uses the same parameters, $ADENS_V$, $BDENS_P$ and $LAICOMP_P$ (expressed as ground cover).

$$\text{eq. 3.14}$$

$$TAUXCOUV(I) = \frac{TRECOUVMAX_P \cdot DELTAI_{dens}(I)}{1 + \exp(PENTRECOUV_P\,(INFRECOUV_P - ULAI(I)))} + LAIPLANTULE_P$$

$$if\ TAUXCOUV(I) > 1 \quad TAUXCOUV(I) = 1$$

LAIPLANTULE$_p$ is the ground cover of plants at planting if the crop is transplanted rather than sown. The graph in Figure 3.7 shows the simulated evolution of ground cover for a lettuce crop with two planting densities.

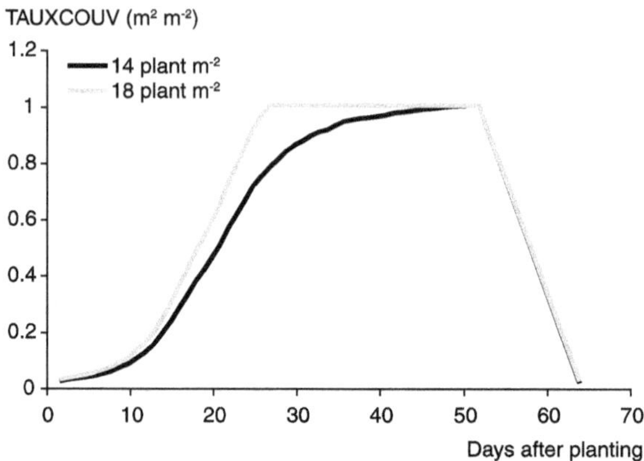

Figure 3.7. Ground cover dynamics for a lettuce crop comparing two plant densities. The parameters are as follows: TRECOUVMAX$_p$=0.072, INFRECOUV$_p$=0.85, PENTRECOUV$_p$=4.5, ADENS$_V$=−0.4, BDENS$_p$=5, LAICOMP$_p$=0.14.

Water and nitrogen shortage and waterlogging stresses are applied to the rate of growth of ground cover, calculated as the derivation of eq. 3.14, and the method of combining stresses is the same as for the leaf area index: DELTAI$_{stress}$(I) described in eq. 3.5.

3.1.5 Number of leaves

Calculation of the number of leaves (NBFEUILLE) is mainly indicative. Its only active role is to define the duration of the plantlet phase when calculating frost risks (see § 3.4). Indeed the plantlet stage is calculated as a leaf-number stage (2 or 3). NBFEUILLE is calculated up to the ILAX stage from the phyllotherm (PHYLLOTHERME$_p$) (the thermal period separating the emission of two successive leaves) expressed in crop degree.days as for the phasic development.

3.1.6 Green leaf specific area

Although STICS does not use the specific leaf area as a driving variable to directly calculate leaf area from the carbon balance, it is useful for certain tests and can at least be valuable as an output.

$$\text{eq. 3.15: } SLA(I) = \min\left(TURSLA(I) \cdot SLAMAX_P, SLAMIN_P\right)$$

TURSLA is the mean water stress TURFAC experienced since emergence, and SLAMAX$_p$ and SLAMIN$_p$ are two parameters which define the limits of variation in specific leaf area between a satisfactory water level and a state of extreme stress.

3.2 Radiation interception

Since most crop models are devoted to industrial crops the canopy is assumed to be a homogenous environment with leaves being randomly distributed over the area. A consequence of this random, homogeneous representation is that it allows the use of an optical analogy (Beer's law) to estimate the interception of photosynthetically active radiation. This law, having only one parameter (the extinction coefficient), has been thoroughly studied for many crops (Varlet-Grancher *et al.*, 1989): the more erect the plant, the smaller is the extinction coefficient.

This approach is very successful for homogenous crops, but poorly suited to canopies in rows or during the first stages of an annual crop because the homogeneity hypothesis cannot apply. Consequently, like CROPGRO (Boote and Pickering, 1994) the STICS model can simulate canopies in rows, with prediction of light interception dependent not only on LAI, but also on plant height and width, row spacing, plant spacing and direct and diffuse light absorption. Such capabilities are also required to simulate intercropping (see § 10.2).

Thus in STICS two options are available to calculate radiation interception: a simple Beer's law, recommended for homogenous crops, and a more complex calculation for radiation transfers within the canopy, recommended for crops in rows. If the leaf status variable is the ground cover and not the leaf area index, then only the Beer's law option is permitted.

3.2.1 Beer's law and calculation of height

The radiation intercepted by the crop (RAINT) is expressed according to a Beer's law function of LAI (eq. 3.16). $EXTIN_P$ is a daily extinction coefficient and $PARSURRG_C$ is a climatic parameter corresponding to the ratio of photosynthetically active radiation to the global radiation, TRG (around 0.48, Varlet Grancher *et al.*, 1982).

eq. 3.16

Using explicit LAI:

$$RAINT(I) = 0.95 \cdot PARSURRG_C \cdot TRG(I) \cdot \left[1 - \exp\left(- EXTIN_P \cdot \left(LAI(I) + EAI(I)\right)\right)\right]$$

Using ground cover (TAUXCOUV):

$$RAINT(I) = 0.95 \cdot PARSURRG_C \cdot TRG(I) \cdot TAUXCOUV(I)$$

For homogenous crops, crop height is deduced from the leaf area index or the ground cover (eq. 3.17). It serves particularly in the calculation module for water requirements via the resistive option. $KHAUT_G$ is assumed to be plant-independent (a general value of 0.7 is proposed) while the potential height of foliage growth is mostly plant-dependent and defined by the two limits $HAUTBASE_P$ and $HAUTMAX_P$.

eq. 3.17

Using explicit LAI:

$$HAUTEUR(I) = HAUTMAX_P \left[1 - \exp\left(- KHAUT_G \times \left(LAI(I) + LAISEN(I)\right)\right)\right] + HAUTBASE_P$$

Using ground over (TAUXCOUV):

$$HAUTEUR(I) = HAUTMAX_P \cdot TAUXCOUV(I) + HAUTBASE_P$$

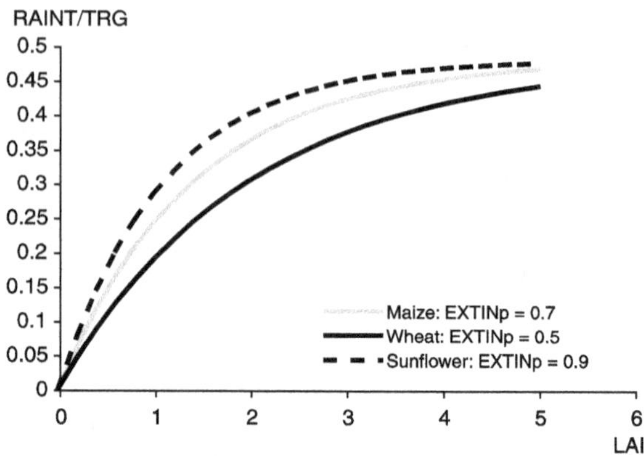

Figure 3.8. Proportion of global radiation (RAINT/TRG) intercepted for wheat, maize and sunflower (EXTIN$_p$ values of respectively 0.5, 0.7 and 0.9).

3.2.2 Radiation transfers and plant shape

A calculation of radiation transfer enables an estimate of the radiation intercepted by a crop in rows, taking account of its geometry in a simple fashion. The objective is to estimate, on a daily time step, the fraction of radiation intercepted by the crop and fraction part transmitted to the layer below, which can be either the soil or another crop (case of intercropping). To calculate those two components, the soil surface is split into a shaded part and a sunlit part and by convention the shaded part corresponds to the vertical projection of the crop foliage onto the soil surface. The available daily variables are the Leaf Area Index (LAI), calculated independently and the global radiation (TRG).

3.2.2.a Radiation transfers

The simplest method of calculating the radiation received at a given point X (located on the soil in the inter-row: Figure 3.9) is to calculate angles H1 and H2 corresponding to the critical angles below which point X receives the total radiation directly. At angles below H1 and above H2, point X receives an amount of radiation smaller than the total radiation value, due to absorption by the crop. Within those angle windows, Beer's law is used to estimate the fraction of transmitted radiation.

It is assumed that a canopy can be represented by a simple geometric shape (rectangle or triangle) and that it is isotropically infinite. We can therefore describe the daily radiation received at point X as the sum of the radiation not intercepted by the crop (RDROIT) (sun at an angle between H1 and H2) and the radiation transmitted (RTRANSMIS). The "infinite canopy" hypothesis allows us to assume that when the sun is at an angle below H1 and H2, all the radiation passes through the crop.

Each part of the radiation received at X includes a direct component and a diffuse component. Let us assume that, for the transmitted part, the same extinction coefficient (KTROU$_p$) applies to both components (which is generally accepted to be the case when the general Beer law is used with a daily time scale). In eq. 3.18 the parameter KTROU$_p$ corresponds to a gap fraction (Baret *et al.*, 1993).

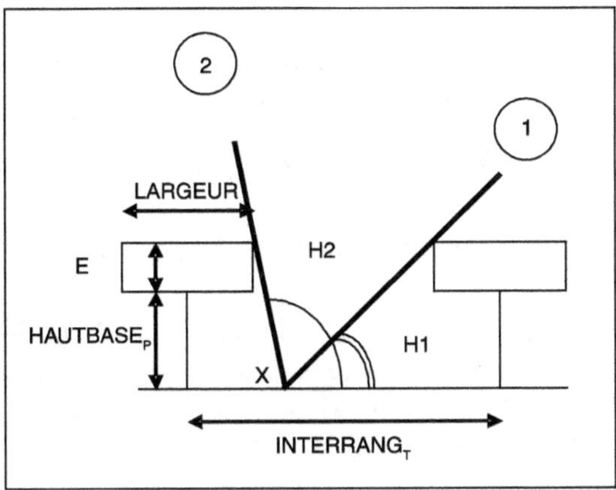

Figure 3.9. Simplified representation of plant canopy and the principles used for calculating daily radiation received by the inter-row ($INTERRANG_T$): $HAUTBASE_p$ is the base height of the canopy, E its thickness, LARGEUR its width, X is any point located in the inter-row and h1 and h2 are the two sun height angles corresponding to the daily positions 1 and 2 of the sun between which X is directly illuminated.

eq. 3.18

$$RTRANSMIS(I,X) = \left[1 - RDROIT(I,X)\right] \exp\left[-KTROU_P\left(LAI(I) + EAI(I)\right)\right]$$

In contrast, for RDROIT, direct and diffuse components should be separated because of the directional character of the direct component, which requires the calculation of separate proportions of radiation reaching the soil (KGDIFFUS and KGDIRECT are the proportions of diffuse radiation, RDIFFUS, and direct radiation, RDIRECT, respectively, reaching the soil)

eq. 3.19

$$RDROIT(I,X) = KGDIFFUS(I,X) \cdot RDIFFUS(I) + KGDIRECT(I,X) \cdot RDIRECT(I)$$

• *The case of direct components*

If θ1 and θ2 are the hourly angles (the actual angles that are zero at 12 h TSV) corresponding to H1 and H2, and assuming sinusoidal variation in the direct radiation during the day, we can write:

eq. 3.20

$$KGDIRECT(I,X) = 0.5\left(\cos\left(\pi/2 + \theta1(I,X)\right) + \cos\left(\pi/2 + \theta2(I,X)\right)\right)$$

In order to calculate the θ angles, it is necessary to solve the following set of equations:

eq. 3.21

$$\sin(H) = \sin(LAT_C)\sin(DEC) + \cos(LAT_C)\cos(DEC)\cos(\theta)$$

$$\cos(A) = \left[-\cos(LAT_C)\sin(DEC) + \sin(LAT_C)\cos(DEC)\cos(\theta)\right]/\cos(H)$$

$$\tan(H) = G\sin(A + ORIENTRANG_T)$$

where H is the height of the sun, A its azimuth, LAT_C is the latitude of the location and DEC the declination angle which depends on the day (Varlet-Grancher *et al.*, 1993), and $ORIENTRANG_T$ is the azimuth angle of the rows. G, the apparent tangent of H on Figure 3.9), depends on canopy geometry (LARGEUR, E and $HAUTBASE_p$ defined in Figure 3.9) and the position of the given point within the inter-row (X).

For example, assuming $X > $ LARGEUR/2 and the angle $H2$, $\quad G = \dfrac{HAUTBASE_P + E}{X - \text{LARGEUR}/2}$

The borderline between sun (SURFAS) and shade (SURFAO) is arbitrarily taken to be LARGEUR/2. The above set of equations cannot be solved by analytical methods, and must therefore be solved numerically (loop over θ with a basic variation of 3 degrees followed by linear interpolation).

- *The case of diffuse components*

We take 46 directions (azimuth, height) and the corresponding percentage of diffuse radiation (SOC standard). For each direction, the point X is checked to see if it is directly illuminated, depending on canopy geometry. The variable KGDIFFUS corresponds to the cumulative proportion of radiation received at point X for the 46 directions.

The diffuse to total radiation ratio (RDIF) is calculated according to Spitters *et al.* (1986) on the basis of the total to extraterrestrial radiation ratio (RSRSO), the extraterrestrial radiation being calculated from the classical astronomical formula (Varlet-Grancher *et al.*, 1993) represented in Figure 3.10.

eq. 3.22

$$if \quad RSRSO(I) < 0.07 \quad RDIF(I) = 1$$

$$if \quad RSRSO(I) \geq 0.07 \quad RDIF(I) = 1 - 2.3 \times \left(RSRSO(I) - 0.07\right)^2$$

$$if \quad RSRSO(I) > 0.35 \quad RDIF(I) = 1.33 - 1.46\,RSRSO(I)$$

$$if \quad RSRSO(I) > 0.75 \quad RDIF(I) = 0.23$$

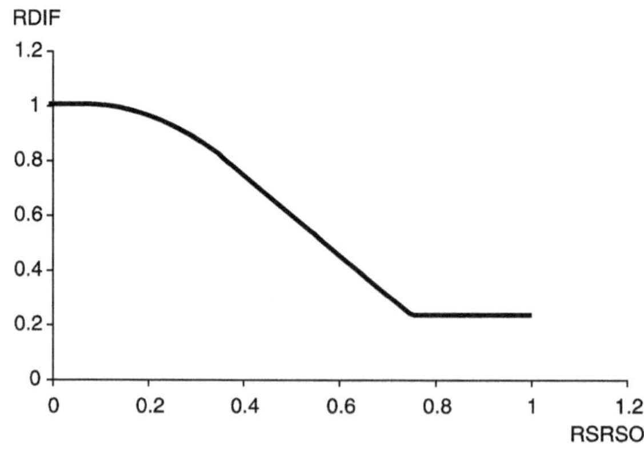

Figure 3.10. Relationship between the diffuse to total radiation ratio (RDIF) and the total to extraterrestrial radiation ratio (RSRSO).

The above equations are applied to 20 points spread equally along the inter-row, and the transmitted radiation values are then averaged for the shaded fraction (ROMBRE) and the sunlit fraction (RSOLEIL). The complementary part to the global radiation corresponds to the radiation intercepted by the crop (RAINT: eq. 3.23).

eq. 3.23

$$RAINT(I) = PARSURRG_C \cdot TRG(I)\left[1 - ROMBRE(I) \cdot SURFAO(I) \\ - RSOLEIL(I) \cdot SURFAS(I)\right]$$

3.2.2.b Crop geometry

LARGEUR and E are calculated using the following assumptions:

• The volume of the crown (or the group of crop leaves) has a simple shape. We assume that its cross-section is rectangular or triangular (parameter $FORME_p$ as a code).

• This volume can be evaluated on the basis of LAI, the inter-row value ($INTERRANG_T$), the leaf density (DFOL), and the RAP_p=E/LARGEUR ratio (thickness/width) of the shape. DFOL is a "within the shape" leaf density, which differs from the classical definition of leaf density as a ratio of leaf surface to 1 m^3 of air. DFOL can vary between two limits ($DFOLBAS_p$ and $DFOLHAUT_p$) depending on the foliage produced (FP accounting for LAI, EAI, LAISEN and leaves suppressed by specific techniques such as topping (LAIROGNECUM) or leaf removal (LAIEFFCUM): see § 6.1.3) and according to a slope, $ADFOL_p$. If we assume a constant foliage density, then $DFOLBAS_p$= $DFOLHAUT_p$.

eq. 3.24

$$FP(I) = LAI(I) + EAI(I) + LAIROGNECUM(I) + LAIEFFCUM(I) + LAISEN(I)$$
$$if\ ADFOL_P > 0 \quad DFOL(I) = ADFOL_P \cdot FP(I)$$
$$if\ ADFOL_P < 0 \quad DFOL(I) = ADFOL_P \cdot FP(I) + DFOLBAS_P + DFOLHAUT_P$$
$$if\ DFOL(I) \le DFOLBAS_P \quad DFOL(I) = DFOLBAS_P$$
$$if\ DFOL(I) \ge DFOLHAUT_P \quad DFOL(I) = DFOLHAUT_P$$

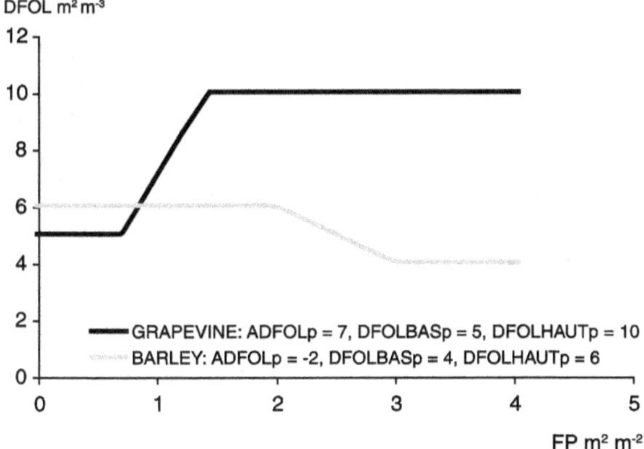

Figure 3.11. Leaf density (DFOL) evolution for grapevine and barley according to the cumulative foliage produced (FP), as two opposite examples.

This formalisation of leaf density makes it possible to represent both foliage getting denser while growing (e.g. grapevine) or conversely becoming less dense while growing (e.g. cereals).

eq. 3.25

$$LARGEUR(l) = \sqrt{\frac{LAI(l) \times INTERRANG_T}{DFOL(l) \times RAPFORME_P}} \quad \text{for the rectangle}$$

$$LARGEUR(l) = \sqrt{2\frac{LAI(l) \times INTERRANG_T}{DFOL(l) \times RAPFORME_P}} \quad \text{and for the triangle}$$

• Two types of triangle can be chosen: "right way up" or "upside down". The more appropriate shape for radiative transfer is "right way-up" triangles (Brisson *et al.*, 2004) suggesting that the low leaf density (in the classical sense: leaf area per m^3) measured in the upper parts allows more radiation to be transmitted than in the lower parts where the leaf density is higher. With our simple model based on a constant leaf density within the shape, this can be accounted for only by a triangle. Thus the shape required as a parameter in the model is far more linked to the leaf density profile than to the external shape of the plant foliage.

• A maximal limit, HAUTMAX$_P$, is imposed on the plant height value (HAUTBASE$_P$+E). Thereby, in the first stage, the shape of the plant evolves isotropically. Once the HAUTMAX$_P$ value is reached, the only way in which the shape can evolve is in terms of width. Height and width can also be limited by topping.

3.3 Shoot biomass growth

The linear relationship between accumulated biomass in the plant and radiation intercepted by foliage, demonstrated by Monteith (1972), defines the radiation use efficiency (or RUE) as the slope of this relationship. This parameter has become a concept widely employed in crop models (Bonhomme *et al.*, 1982; Ritchie and Otter, 1984; Jeuffroy and Recous, 1999), because it synthesizes (very economically in terms of the number of parameters involved) the processes of photosynthesis and respiration. Its calculation (ratio between above-ground biomass and absorbed radiation) implies that this parameter also takes account of a carbon allocation coefficient between above-ground and below-ground parts of the plant. Obviously, because of underlying physiological processes this ratio varies somewhat, due to stresses, temperature and phenology (Trapani *et al.*, 1992; Muchow *et al.*, 1990b; Sinclair *et al.*, 1993). To take account of these effects, Sinclair (1986) proposed that RUE should be considered as a physiological function, to which stress indices should be applied. In other models (Boote *et al.*, 1998; Weir *et al.*, 1984) the photosynthesis and respiration processes are calculated separately and a specific allocation to roots is assumed. In view of the increasing atmospheric CO_2 concentration, crop models now need to take this factor into account (Stockle *et al.*, 2003).

In STICS the calculation of the daily production of shoot biomass (DLTAMS) relies on the RUE concept (though the relationship between DLTAMS and RAINT is slightly parabolic) taking into account various factors known to influence the elementary photosynthesis and respiration processes, mostly as stresses defined in § 3.4 (FTEMP,

SWFAC, INNS and EXOBIOM). DLTAMS accumulated day by day gives the shoot biomass of the canopy, MASEC.

$$eq.\ 3.26$$

$$DLTAMS(I) = \left[EBMAX(I) \cdot RAINT(I) - COEFB_G \cdot RAINT(I)^2\right]$$
$$FTEMP(I) \cdot SWFAC(I-1) \cdot INNS(I-1) \cdot EXOBIOM(I-1) \cdot FCO2$$
$$+ DLTAREMOBIL(I-1)$$

Because of the consecutive nature of the calculations and modules, some variables are those of the previous day.

3.3.1 Influence of radiation and phasic development

The accumulation of shoot biomass depends on the intercepted radiation (RAINT) (Varlet Grancher *et al.*, 1981), and is almost linear but slightly asymptotic at high intercepted light values. It is simulated in STICS by a parabolic function involving a maximum radiation use efficiency specific to each species, EBMAX (eq. 3.26). The parameter $COEFB_G$ stands for the radiation saturating effect. This effect is the result, even buffered, of the saturation occurring within a short time step on the individual leaf scale and is easily observed when daily calculations are made with instantaneous formulae of canopy photosynthesis (Boote and Jones, 1987); such calculations lead to a value of 0.0815. The efficiency, EBMAX, may differ during the juvenile (ILEV-IAMF), vegetative (IAMF-IDRP) and reproductive (IDRP-IMAT) phases (corresponding respectively to the parameters $EFCROIJUV_p$, $EFCROIVEG_p$ and $EFCROIREPRO_p$). Classically, $EFCROIJUV_p = 1/2\ EFCROIVEG_p$ is used to take account of the preferential migration of assimilates towards the roots at the beginning of the cycle. The difference between $EFCROIVEG_p$ and $EFCROIREPRO_p$ arises from the biochemical composition of storage organs: e.g. for oil or protein crops $EFCROIREPRO_p$ is less than $EFCROIVEG_p$ because the respiratory cost to make oil and protein is higher than for glucose (Figure 3.12).

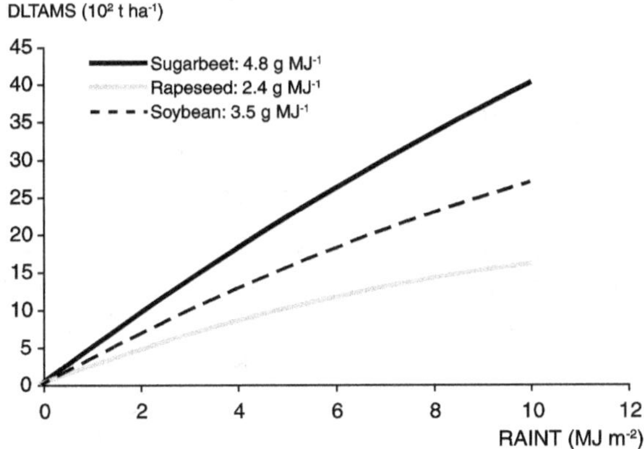

Figure 3.12. Potential daily accumulation of biomass (DLTAMS without any stress at $[CO_2]=350\text{ppm}$) as a function of intercepted radiation (RAINT) for three species during their filling stage: sugarbeet, soybean and rapeseed with values of $EFCROIREPRO_p$ of 4.8, 3.5 and 2.4 g MJ⁻¹ respectively.

3.3.2 Effect of atmospheric CO_2 concentration

The $CO2_C$ parameter stands for the atmospheric CO_2 concentration, which can be higher than the current value, assumed to be 350 ppm. The formalisation chosen in STICS was adapted from Stockle *et al.* (1992): the effect of $CO2_C$ on the radiation use efficiency is expressed by an exponential relationship, for which the parameter is calculated so that the curve passes through the point (600, ALPHACO2$_p$).

eq. 3.27

$$FCO2 = 2 - \exp\left[\log(2 - ALPHACO2_P) \cdot \frac{CO2_C - 350}{600 - 350}\right]$$

The parameter ALPHACO2$_p$ mainly varies with the plant's C3/C4 metabolism, being around 1.1 for C4 crops and 1.2 for C3 crops (from Ruget *et al.*, 1996, Stockle *et al.*, 1992, Peart *et al.*, 1989. The effect of CO2 on stomatal resistance will be covered in the paragraph on water requirements.

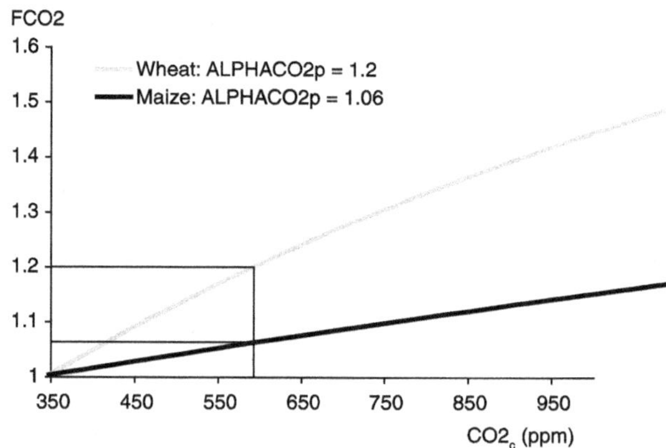

Figure 3.13. Calculation of the CO_2 effect (FCO2) for a species as a function of its C3/C4 metabolism: example of wheat (C3) and maize (C4).

3.3.3 Remobilisation of reserves

3.3.3.a Perennial reserve available from one cycle to the next

DLTAREMOBIL is obtained by the remobilisation of winter reserves in perennial plants. Each day the maximal proportion of the reserves that can be remobilised is REMOBRES$_p$, until perennial stocks (parameter RESPERENNE0$_I$ given as an initialisation at the beginning of the growth cycle) are exhausted. RESPERENNE0$_I$ only represents carbon reserves, and nitrogen reserves can only be added through initiation of the QNPLANTE0$_I$ parameter. The nitrogen remobilisation rate of the QNPLANTE0$_I$ stock is assumed to equal the nitrogen demand (see § 8.6) until it is exhausted. These reserves are only called upon if the newly formed assimilates (DLTAMS) fail to satisfy the sink strength (FPV and FPFT explained in eq. 3.31 and § 4.2), which leads to a first calculation of the source/sinks variable (SOURCEPUITS$_1$: eq. 3.28).

eq. 3.28

$$SOURCEPUITS_1(I) = \frac{DLTAMS(I)}{FPV(I) + FPFT(I)}$$

eq. 3.29

$$\textit{if } SOURCEPUITS_1(I) < 1 \quad \textit{as long as} \quad \sum RESPERENNE(I) \leq RESPERENNE0_I$$

$$\textit{if } FPV(I) + FPFT(I) - DLTAMS(I) \leq REMOBRES_P \cdot RESPERENNE(I) \quad \textit{then}$$
$$DLTAREMOBIL(I) = FPV(I) + FPFT(I) - DLTAMS(I)$$
$$\textit{if } FPV(I) + FPFT(I) - DLTAMS(I) > REMOBRES_P \cdot RESPERENNE(I) \quad \textit{then}$$
$$DLTAREMOBIL(I) = REMOBRES_P \cdot RESPERENNE(I)$$

These remobilisations contribute to increasing the source/sink ratio the following day because they are counted in the variable DLTAMS (eq. 3.26).

3.3.3.b Reserve built up and used during the cycle

Reserves built up during the vegetative cycle (variable RESPERENNE see § 3.5.4) and reused later on simply contribute to the estimation of the source/sink ratio for indeterminate crops. The maximum quantity which can be remobilised per day (REMOBILJ) is calculated similarly to DLTAREMOBIL (eq. 3.29). If the plant is both perennial and indeterminate, the reserves originating from the previous cycle are first used (DLTAREMOBIL) and when exhausted the current cycle's reserves (REMOBILJ) can be used.

3.3.4 Calculation of the source/sink ratio

A second value of the source/sink ratio (SOURCEPUITS) is calculated to account for possible carbon remobilisation. It is this variable which drives trophic stresses, useful for simulating indeterminate crop competition between vegetative and reproductive sinks.

eq. 3.30

$$SOURCEPUITS(I) = \frac{DLTAMS(I) + REMOBILJ(I)}{FPV(I) + FPFT(I)}$$

The sink strength of vegetative organs, FPV, is defined as the ratio between daily foliage growth ($DELTAI_2$) and the minimum ratio between leaf surface area and shoot biomass, calculated from $SLAMIN_P$ and $TIGEFEUILLE_P$ (eq. 3.31).

eq. 3.31

$$FPV(I) = \frac{DELTAI_2(I) \cdot 10^4}{\dfrac{SLAMIN_P}{1 + TIGEFEUILLE_P}}$$

The calculations of the variable FPFT (the fruit sink strength) will be explained in § 4.2 on yield components but it is important to indicate here that FPFT and FPV are not exactly of the same nature since FPFT relates to a potential growth while FPV corresponds to the real growth. Such a difference causes by construction a priority to fruits and can generate a day to day instability of the variable SOURCEPUITS by the feedback of SOURCEPUITS on FPV via the stress index SPLAI (§ 3.4.3).

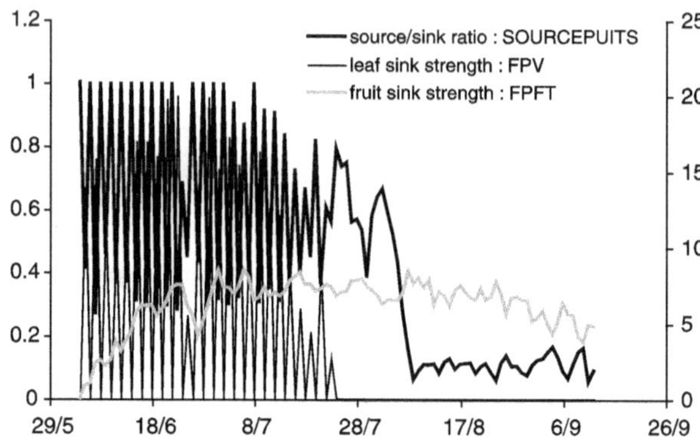

Figure 3.14. Dynamics of the variable SOURCEPUITS for grapevine depending upon the vegetative sink strength (FPV) and the fruit sink strength (FPFT).

3.3.5 Height-biomass conversion

For forage crops, it may be necessary to estimate an initial biomass value after each cutting on the basis of canopy height, and if this is the case the proportionality coefficient COEFMSHAUT$_p$ is used (e.g. it is 25 t.ha^{-1}m^{-1} for grass).

3.4 Stress indices

Stresses accounted for in most crop models are only of an abiotic nature. They are functions, varying between 0 and 1, that reduce plant processes depending on stress variables such as fraction of transpirable soil water, nitrogen nutrition index, fraction of root system in waterlogged conditions etc. These stress variables must therefore also be calculated.

The reduction functions are empirical relationships based on the limiting factor principle (an overview of the concept was given by Gary *et al.*, 1996). Nonetheless, they are based on what we know about the effects of these stresses on plant growth and development. For example, water stress acts via a hormonal or hydraulic signal on stomatal conductance, which causes a reduction in photosynthesis and hence in radiation use efficiency. The empirical function relates the reduction in radiation use efficiency directly to water stress. Similarly, water stress slows down cell division and expansion, phenomena which cause a reduction in the appearance and expansion of leaves and hence in the rate of increase of leaf area index. The empirical function then directly relates the reduction in leaf area index increase to water stress. Yet as demonstrated by Bradford and Hsiao

(1982) for water stress, the sensitivity of the various physiological functions can vary which requires calculating several stress indices for the same stress state variable.

The regulation involved in interactions between stresses is poorly understood on the whole plant scale, and is therefore modelled very simply by using either the product or the minimum of the reduction factors. Improved physiological approaches (Farqhuar *et al.*, 1980, for example) could lead to more realistic models for photosynthetic processes, but raise the problem of parameterization.

In STICS most of the relationships are simple bilinear functions, i.e. equal to a constant until a critical level of the state stress variable is reached and then linearly decreasing. For frost and waterlogging the relationships are more complex. The soil water content available to roots is the water deficit stress variable, the nitrogen nutrition index is the nitrogen stress variable, the source/sink ratio is the trophic stress variable, the minimal crop temperature is the frost stress variable and the proportion of roots flooded is the water logging stress variable. The sensitivity to the various stresses can be represented by appropriate parameterisations of the stress functions or by a sensitivity parameter (e.g. for waterlogging or for roots sensitivity to water deficiency).

3.4.1 Water deficiency

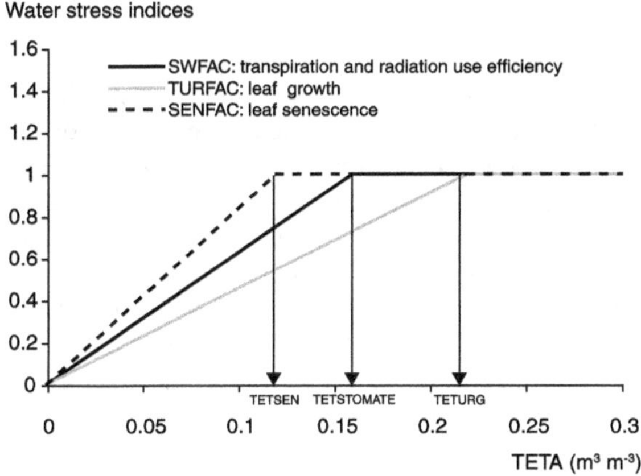

Figure 3.15. Water stress indices (TURFAC, SWFAC, SENFAC) as a function of the available water content in the root zone (TETA).

The stress variable (TETA) is the available water content (water content above the wilting point) in the root zone. The stress indices are SWFAC, TURFAC and SENFAC; they depend on TETA according to bilinear laws (Figure 3.15) and differ by specific thresholds (TETSTOMATE, TETURG and TETSEN). An example of the calculation is given for TURFAC and SWFAC in eq. 3.32:

$$\text{eq. 3.32}$$

$$\begin{aligned} &\text{if } TETA(I) < TETURG \quad TURFAC = \frac{TETA(I)}{TETURG} \\ &\text{if } TETA(I) \geq TETURG \quad TURFAC = 1 \end{aligned} \quad \text{and} \quad \begin{aligned} &\text{if } TETA(I) < TETSTOMATE \quad SWFAC = \frac{TETA(I)}{TETSTOMATE} \\ &\text{if } TETA(I) \geq TETSTOMATE \quad SWFAC = 1 \end{aligned}$$

The calculation of the TETSTOMATE and TETURG thresholds is explained in the paragraph on transpiration (see § 7.3). TETSEN is proportional to TETURG thanks to the $RAPSENTURG_p$ parameter ($TETSEN = RAPSENTURG_p \cdot TETURG$). The hierarchy between the three stress indices is generally that indicated in Figure 3.15, with $RAPSENTURG_p < 1$. The functions of these three stress indices are summarised in Table 3.1. We should also remember that the germination and epicotyl growth phases can also be affected by water shortage in response to soil moisture in the seed bed (HUMIRAC index).

Table 3.1. Impact of water stress on physiological functions through the various water stress indices.

Physiological function	Water stress index
Emergence (delay)	HUMIRAC
Root growth in depth (slowing)	HUMIRAC
Development (delay)	TURFAC
Leaf growth (slowing)	TURFAC
Leaf senescence (acceleration)	SENFAC
Radiation use efficiency (decrease)	SWFAC
Transpiration (decrease)	SWFAC

3.4.2 Nitrogen deficiency

The nitrogen status of a crop can be characterized using the concept of 'dilution curves' which relate the N concentration in plant shoots to the dry matter accumulated in them (Lemaire and Salette, 1984; Greenwood *et al.*, 1991). For a given species, a 'critical dilution curve' can be defined, which can be used to make a diagnosis of nitrogen nutrition (Justes *et al.*, 1994; Lemaire et Gastal., 1997): plants below this curve are or have been N deficient, whereas plants above the curve have an optimal growth, i.e. are not limited by nitrogen (Figure 3.16). The critical dilution curve is the basis for defining a nitrogen nutrition index (INN) which is the ratio of the actual nitrogen concentration (CNPLANTE, in % of dry matter) to the critical concentration (NC) corresponding to the same biomass (MASECABSO, in t ha^{-1}).

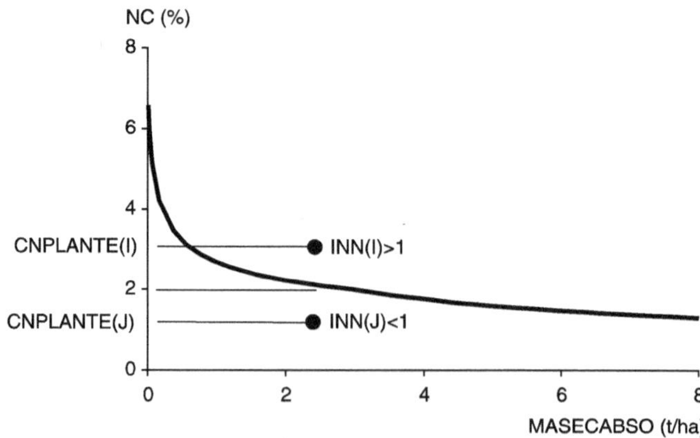

Figure 3.16. Critical dilution curve and INN calculation as the ratio between CNPLANTE and NC.

However we are aware of an important limitation in the INN dynamics, as for example in the case of the nitrogen reserve available in perennial organs (e.g. grapevine, illustrated in Figure 3.17.). Consequently we propose an alternative stress variable corresponding to the nitrogen input flux relative to the critical one as proposed by Devienne-Barret *et al.* (2000). It is a kind of instantaneous INN named INNI (eq. 3.33) relying on the daily accumulations of nitrogen (VABSN) and nitrogen dependent biomass (DELTABSO), whose calculation is explained in § 8.6.

eq. 3.33

$$INN(I) = \frac{CNPLANTE(I)}{NC(I)}$$

$$INNI(I) = \frac{VABSN(I)}{DELTABSO(I)\left(\dfrac{dNC}{dMASECABSO}\right)(I)}$$

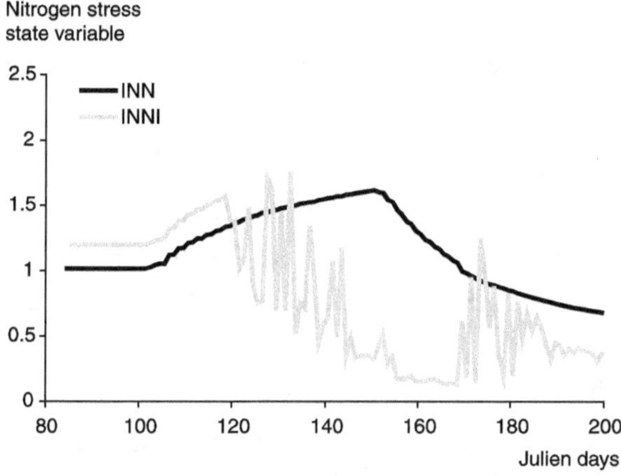

Figure 3.17. Comparison between INN and INNI for a grapevine crop with nitrogen reserve at the beginning of the cycle.

All nitrogen stress indices accept INNMIN$_p$ or INNIMIN$_p$ as the floor value for respectively the "INN" and the "INNI" options. By definition the INNS index corresponds to the INN between INNMIN$_p$ and 1. The INNLAI and INNSENES indices (Figure 3.18) are defined by point [1,1] and by points [INNMIN$_p$, INNTURGMIN$_p$] and [INNMIN$_p$, INNSEN$_p$], respectively.

Such a parameterisation allows the effect of nitrogen deficiency on photosynthesis to be differentiated from that on leaf expansion. In practice it seems that these two functions react very similarly and INNTURGMIN$_p$ is similar to INNMIN$_p$, while INNSEN$_p$ is greater, indicating that the plants accelerate their senescence later than their growth decrease, just as for water stress. A commonly accepted value for INNMIN$_p$ is 0.3 and INNIMIN$_p$ is 0.0. The functions of these three stress indices are summarised in Table 3.2.

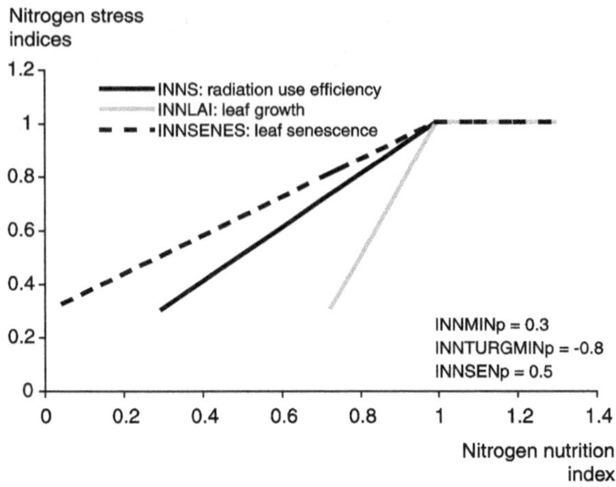

Figure 3.18. Nitrogen stress indices (INNLAI, INNS, INNSENES: leaf senescence) as a function of the nitrogen nutrition index (INN).

Table 3.2. Effect of nitrogen stress on physiological functions through the various nitrogen stress indices.

Physiological function	Nitrogen stress index
Development (delay)	INNLAI
Leaf growth (slowing)	INNLAI
Leaf senescence (acceleration)	INNSENES
Radiation use efficiency (decrease)	INNS

3.4.3 Trophic stress

The trophic stress indices only concern crops simulated as indeterminate. The stress variable (SPLAI) is the ratio of the trophic sources to the sinks, SOURCEPUITS. (Its calculation is explained in § 3.3.4). The SPLAI and SPFRUIT options are defined by the $SPLAIMIN_p$, $SPLAIMAX_p$, $SPFRMIN_p$ and $SPFRMAX_p$ parameters. The various trophic stress indices cannot be considered as equivalent to coefficients of biomass allocation because they are not all applied to biomass. Consequently the relative position of the functions SOURCEPUITS and SPLAI does not indicate any priority between fruit and leaves: the priority needs to be calculated in terms of biomass and depends largely on the relative sink strengths of the organs.

Table 3.3. Effect of trophic stress on physiological functions through the various trophic stress indices.

Physiological function	Trophic stress index
Fruit growth (decrease)	SOURCEPUITS
Leaf growth (slowing)	SPLAI
Fruit number (decrease)	SPFRUIT

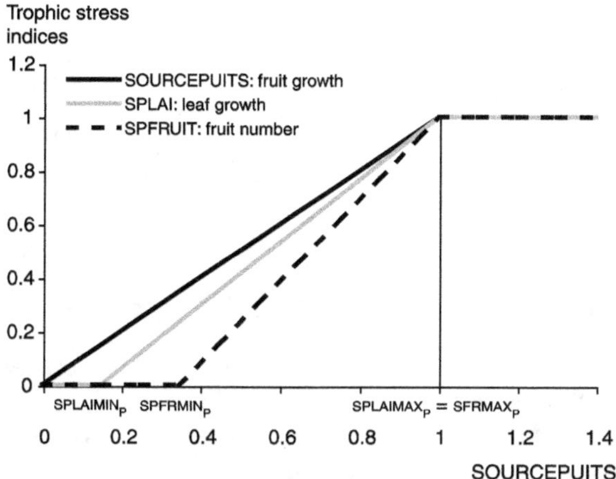

Figure 3.19. Trophic stress indices (SOURCEPUITS, SPLAI, SPFRUIT) as a function of the source/sink ratio (SOURCEPUITS) using the parameters SPLAIMIN$_p$, SPLAIMAX$_p$, SPFRMIN$_p$ and SPFRMAX$_p$.

3.4.4 Temperature stresses

3.4.4.a Frost

The stress variable is the minimum crop temperature, TCULTMIN (see § 6.6.2 on crop temperature for its calculation). The frost stress indices correspond to frost damage (1 for no frost and 0 for lethal frost) and the response to temperature as well as the damage varies as a function of the developmental stage.

Each response is defined by four parameters (Figure 3.20), two of which are independent of the developmental stage: TDEBGEL$_p$ (temperature at the beginning of frost action) and TLETALE$_p$ (lethal temperature); and two others are stage-dependent: TGEL...10$_p$ (temperature corresponding to 10% frost damage) and TGEL...90$_p$ (temperature corresponding to 90% frost damage). For the plantlet phase, it is the TGELLEV10$_p$ and TGELLEV90$_p$ parameters which act on plant density through the index FGELLEV; for the juvenile phase (up to IAMF stage) it is the TGELJUV10$_p$ and TGELJUV90$_p$ parameters which act on foliage (acceleration of senescence: see eq. 3.9) through the index FGELJUV. After the IAMF stage, the TGELVEG10$_p$ and TGELVEG90$_p$ parameters are also active on foliage through the index FGELVEG. For frost affecting flowers and fruits, the TGELFLO10$_p$ and TGELFLO90$_p$ parameters define the dynamics of the FGELFLO index.

Table 3.4. Impact of frost stress on physiological functions through the various frost stress indices.

Physiological function	Frost stress index
Plant density	FGELLEV
Leaf senescence before IAMF (acceleration)	FGELJUV
Leaf senescence after IAMF (acceleration)	FGELVEG
Fruit number	FGELFLO

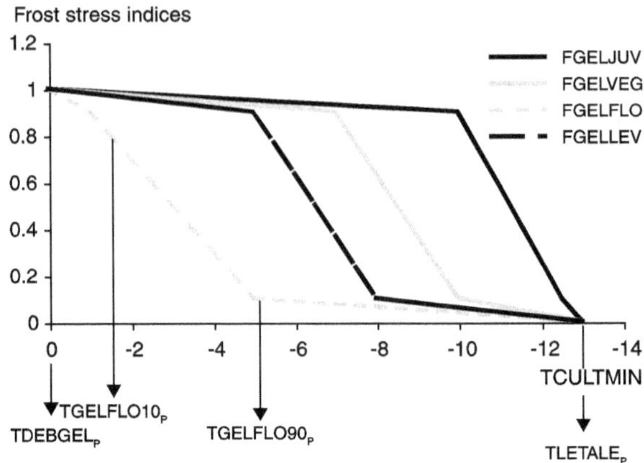

Figure 3.20. Frost stress indices (FGELLEV, FGELJUV, FGELVEG, FGELFLO) as a function of minimal crop temperature (TCULTMIN) using four cardinal temperatures for each index exemplified for FGELFLO.

3.4.4.b Suboptimal temperatures

Stresses linked to temperatures which are too high or too low (without attaining frost thresholds) are included in the temperature effect functions. Temperature usually plays a driving role (development, growth and senescence of LAI, growth and senescence of roots) and the functions concerned accept thermal thresholds (minimum and maximum for functioning). It may also act to reduce activity and be used as a stress variable. This is the case for biomass growth and the filling of storage organs (Figure 3.21). Depending on the vital functions affected and the options chosen, the thermal stress variable changes (average or extreme temperatures, crop, air or soil). The smooth shape of the radiation use efficiency dependency on crop temperature (eq. 3.34, Figure 3.21) is quite classical (Ritchie and Otter, 1984) and comes from the combined responses of photosynthesis and respiration to temperature. Yet the values of the cardinal temperature are highly dependent on the time step used: in our case daily average crop temperatures. As far as fruit filling is concerned, the response in the model is of a yes/no type.

eq. 3.34

$$if\ TCULT(I) < TEOPT_P \quad FTEMP(I) = 1 - \left[\frac{TCULT(I) - TEOPT_P}{TEMIN_P - TEOPT_P} \right]^2$$

$$if\ TEOPT_P \leq TCULT(I) \leq TEOPTBIS_P \quad FTEMP(I) = 1.0$$

$$if\ TCULT(I) > TEOPTBIS_P \quad FTEMP(I) = 1 - \left[\frac{TCULT(I) - TEOPTBIS_P}{TEMAX_P - TEOPTBIS_P} \right]^2$$

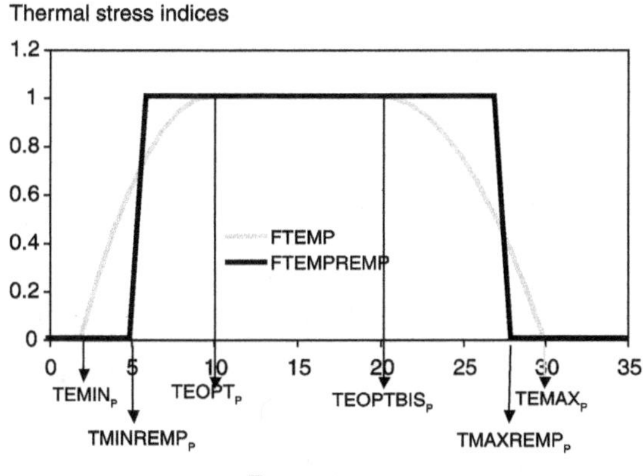

Figure 3.21. Thermal stress indices (FTEMP and FTEMPREMP) as a function of temperature (see Table 3.5) using cardinal temperatures (TEMIN$_p$, TEOPT$_p$, TEOPTBIS$_p$, TEMAX$_p$, TMINREMP$_p$, TMAXREMP$_p$).

Table 3.5. Summary of the role of temperature on the various physiological functions, some of which are of a pilot type and the others of a stress type.

Physiological function	Temperature	Role	Function and thermal stress index
Emergence	Daily average soil temperature	pilot	eq. 2.1 and 2.4
Aboveground development	Daily average crop temperature	pilot	eq. 2.10
Vernalisation and dormancy	Daily average crop temperature	stress	eq. 2.12, eq. 2.13 and eq. 2.14
Leaf growth and senescence	Daily average crop temperature	pilot	eq. 3.3 and 3.10
Root growth and senescence	Daily average soil temperature	pilot	eq. 5.2
Radiation use efficiency (decrease)	Daily average crop temperature	stress	FTEMP
Filling at low temperatures (stop)	Minimum crop temperature	stress	FTEMPREMP
Filling at high temperatures stop)	Maximum crop temperature	stress	FTEMPREMP

3.4.5 Waterlogging

The waterlogging variable is called EXOFAC (eq. 3.35) and corresponds to the proportion of roots flooded, i.e. roots in saturated layers (ANOX=1).

eq. 3.35

$$EXOFAC(I) = \frac{1}{CUMLRACZ(I)} \sum_{Z=PROFSEM_T}^{ZRAC} LRACZ(Z, I) \cdot ANOX(Z, I)$$

The calculation of stress indices (eq. 3.36) is based on the experimental work by Rebière (1996), reviewed in Brisson *et al.* (2002b). IZRAC is the root stress index which limits root growth at an efficient depth and density (see explanations in § 5.2). The index which affects the leaf area index is called EXOLAI and the index affecting radiation use efficiency and transpiration (stomatal effect) is called EXOBIOM.

eq. 3.36

$$IZRAC(I) = 1 - \left[1 - (1.60\ \exp(-0.27\ EXOFAC(I) \cdot 100) - 0.60)\right] \cdot SENSANOX_p$$

$$EXOLAI(I) = 1 - \left[1 - \exp(-0.055 \cdot EXOFAC(I) \cdot 100)\right] \cdot SENSANOX_p$$

$$EXOBIOM(I) = 1 - \left[\frac{1}{1 + \exp[0.14(28.25 - EXOFAC(I) \cdot 100)]}\right] \cdot SENSANOX_p$$

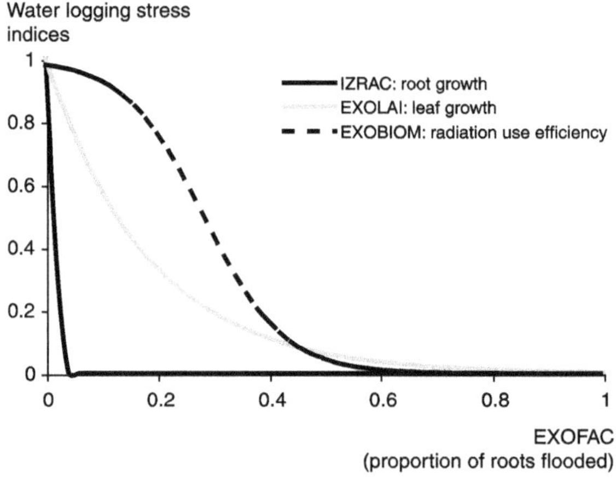

Figure 3.22. Waterlogging stress indices (IZRAC, EXOLAI and EXOBIOM) as a function of the proportion of roots flooded (EXOFAC) for a wheat crop assumed to be of maximal sensitivity (SENSANOX$_p$=1).

The relationships illustrated in Figure 3.22, set up for a wheat crop, are assumed to be relevant for maximum sensitivity to water logging. If the species (or variety) develops resistance mechanisms (e.g. aerenchyma) the effects of excess water will be less pronounced, thanks to the SENSANOX$_p$ sensitivity parameter. If SENSANOX$_p$=1, the sensitivity is maximal and if SENSANOX$_p$=0, the plant is indifferent to excess water (for example, rice).

Table 3.6. Effect of waterlogging stress on physiological functions through the various waterlogging stress indices.

Physiological function	Water logging stress index
Root growth	IZRAC
Leaf growth (slowing)	EXOLAI
Radiation use efficiency (decrease)	EXOBIOM
Transpiration (decrease)	EXOBIOM

3.4.6 Stresses directly linked to the soil structure

At the soil surface, the formation of a crust under certain soil and weather conditions offers a resistance to plant emergence. It can provoke both a delay in emergence dates and a decrease in plant densities (see § 2.2.1).

During the growing period, the soil structure can also limit root soil colonisation because it is either too loose or too compact. The only soil parameter available to describe soil structure is the bulk density (DAF_S) and it is used as a stress variable together with the parameters $DASEUILBAS_G$, $DASEUILHAUT_G$ and $CONTRDAMAX_P$ to calculate a soil structure stress index (see § 5).

3.4.7 Interactions between stresses

How to make the various stresses interact is probably the weakest point of the "limiting factor" approach (Brisson *et al.*, 1997a). In STICS we adopted the principle that stresses are multiplied when their modes of action are assumed independent. When their modes of action interact with each other, the resulting active stress is the more severe, i.e. the one with the lower value (Table 3.7). For instance, water deficiency acts on radiation use efficiency at the stomatal level while the nitrogen deficiency acts on the photosynthesis enzymes: these stresses are assumed to be independent of each other. On the other hand both nitrogen and water stresses limit leaf growth by decreasing the cell membrane expansion and are thus assumed to be mutually dependent. For crop establishment the interactions are more complex, based on the idea of converting a delay in emergence due to stresses into plant mortality.

The trophic stress has a particular status because it does not originate from an environmental resource external to the crop, such as water and nitrogen, but results from the internal crop carbon imbalance. Consequently it already integrates the trophic effects of the primary abiotic stresses, which render unrealistic the hypothesis of stress independence that can lead to an overestimation of stress severity. To cope with such problems of oversimplification of the complex reality, it is required to fit the function parameters using contrasting data sets.

Table 3.7. How the stresses are combined in the model for each of the physiological functions ?*

Physiological function	Combination of stresses (*only for indeterminate crops)
Emergence duration	Water deficiency x Crusting
Plant density establishment	(Water deficiency x Crusting) x Frost
Development	MIN (Water deficiency, Nitrogen deficiency)
Leaf growth	MIN (Water deficiency, Nitrogen deficiency) x Water logging x Trophic*
Senescence	MIN (Water deficiency, Nitrogen deficiency, Frost)
Root growth	Water deficiency x Water logging x Soil structure
Radiation use efficiency	Water deficiency x nitrogen deficiency x Temperature x Water logging
Number of fruits	Nitrogen deficiency x Frost x Trophic*
Fruit growth	Temperature x Trophic*
Transpiration	MIN(Water deficiency, Water logging)

* Means this stress results from internal imbalance.

3.5 Partitioning of biomass in organs

There are models for which allocation of assimilates is critical to the operation of the model (e.g. SUCROS described by Van Ittersum *et al.*, 2003). In STICS this module was added at a late stage, mainly to help dimensioning the reserve pool. For annual plants with determinate growth, the partitioning calculations simply allow the dimensioning of envelopes of harvested organs which may play a trophic role and ensure an input of information for the senescence module. For perennial plants or those with indeterminate growth, those calculations enable the dimensioning of a compartment for reserves which are recycled in the carbon balance. The calculation of root biomass is not directly connected to that of the above-ground biomass.

3.5.1 Organs and compartments identified

The reasons for identifying an organ or a compartment (Figure 3.23 and Table 3.8) are either its internal trophic role within the plant or an external role by participation in the nitrogen balance of the system (such as falling leaves and the recycling of roots). The reserve compartment is not located in a specific organ: it is just a certain quantity of carbon available for the plant growth.

Table 3.8. Various organs identified in STICS for biomass partitioning.

Un-harvested biomass MASECVEG	Leaves MAFEUIL		Green leaves MAFEUILVERTE
		Yellow leaves MAFEUILJAUNE	Remaining attached to the plant
			Falling to the ground and recycled in the nitrogen balance MAFEUILTOMBE
	Stems (only the structural part of stems) MATIGESTRUC		
	Reserves: non-localised compartment which, depending on the plant may be located partly in the stems, roots, or even the leaves RESPERENNE		
Harvested organs	Envelops MAENFRUIT		
	Fruits MAFRUIT		
Roots MSRAC			

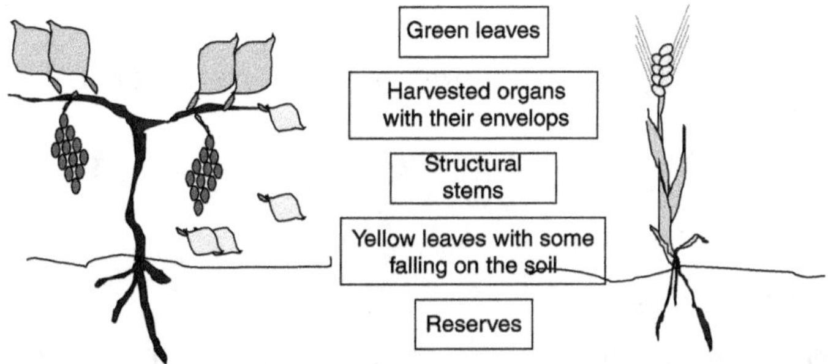

Figure 3.23. Schematisation of the various organs of plants as different as a vine and a wheat plant. For the vine the reserves are perennial ones located in the roots and for the wheat plant they are annual, located in the stem.

3.5.2 Dimensioning of organs

3.5.2.a Leaves

• *Green leaves*

The biomass of green leaves is calculated without accounting for potential reserves that may be stored in the leaves and remobilized later on, which are accounted for in the RESPERENNE non-located reserve pool. The MAFEUILVERTE variable is deducted from the LAI, based on the maximum specific leaf area variable ($SLAMAX_p$). We assume that the difference between the actual SLA and $SLAMAX_p$ corresponds to remobilized leaf carbon.

eq. 3.37

$$MAFEUILVERTE(I) = \frac{LAI(I)}{SLAMAX_P} \, 100$$

• *Yellow leaves*

The biomass of yellow leaves (MAFEUILJAUNE) is calculated in the senescence module. The proportion of leaves in the senescent biomass on a given day (*DLTAMSEN*) is determined using the *PFEUILVERTE* ratio (proportion of green leaves in the non-senescent biomass) on the day of production of this senescent biomass (eq. 3.11).

eq. 3.38

$$MAFEUILJAUNE(I) = \sum_{J=ILEV}^{I} DLTAMSEN(J)$$

Some of these yellow leaves may fall to the ground depending on the $ABSCISSION_p$ parameter (between 0 and 1). The daily falling quantity (DLTAMSTOMBE) is recycled in the nitrogen balance; its cumulative value is MAFEUILTOMBE.

eq. 3.39

$$DLTAMSTOMBE(I) = ABSCISSION_P \cdot DLTAMSEN(I)$$

$$MAFEUILTOMBE(I) = \sum_{J=ILEV}^{I} DLTAMSTOMBE(J)$$

3.5.2.b Stems

This concerns only the structural component of stems (MATIGESTRUC). The non-structural component, if significant, can be included in the reserve compartment (e.g. for cereals) or in the harvested part (sugar cane). The MATIGESTRUC variable is calculated as a constant proportion (TIGEFEUILLE$_p$) of the total mass of foliage (eq. 3.40).

eq. 3.40

$$MATIGESTRUC(I) = TIGEFEUILLE_P \times \left[MAFEUILVERTE(I) + MAFEUILJAUNE(I) \right]$$

For monocotyledonous plants, the stem is secondary and the MATIGESTRUC variable is only incremented from the time when accumulated biomass allows it. It is thus assumed that the first organs to emerge are the leaves. For dicotyledonous plants, it is assumed that the TIGEFEUILLE$_p$ proportionality is always respected. Consequently, if the accumulated biomass and the foliage biomass (calculated from the LAI and SLA) are incompatible with this proportionality, then the SLA (or LAI if the SLA arises from fixed limits) is recalculated.

The MATIGESTRUC variable cannot diminish, except in the case of cutting fodder crops.

3.5.3 Harvested organs

• *Fruits and grains*

The calculation of the number and mass of fruits (indeterminate plants) or seeds (determinate plants) is described in the paragraph on yield formation (see § 4.1 and 4.2)

• *Envelops of harvested organs (pods, raches, etc.)*

The mass corresponding to the envelope is assumed to depend solely upon the number of organs. In any case, it cannot exceed the residual biomass (MASECVEG-MAFEUILVERTE-MAFEUILJAUNE-MATIGESTRUC). The ENVFRUIT$_p$ parameter corresponds to the proportion of membrane related to the maximum weight of the fruit.

eq. 3.41

$$MAENFRUIT(I) = NBFRUIT(I) \cdot ENVFRUIT_P \cdot PGRAINMAXI_P \cdot 10^{-2}$$

If the SEA$_p$ parameter is not zero, then this biomass is transformed into an equivalent leaf surface area, photosynthetically active from the IDRP stage to the IDEBDES stage (eq. 3.13).

3.5.4 Reserves

Reserves (RESPERENNE) are calculated as the difference between the total biomass and the accumulated biomass of leaves, stems and harvested organs (eq. 3.42). For perennial plants, at the beginning of the cropping season, the reserves (carbon) can be initialised at a non-zero value (RESPERENNE0$_I$), so as to represent the role played by root reserves at the resumption of growth.

eq. 3.42

$$RESPERENNE(I) = MASEC(I) - MAFEUIL(I) - MATIGESTRUC(I)$$
$$- MAENFRUIT(I) - MAFRUIT(I)$$

Yet it is assumed that a limit exists to the size of the reserve compartment, parametrized at the plant level by RESPLMAX$_P$. If this limit is reached a "sink on source" effect is simulated.

eq. 3.43

$$if \quad RESPERENNE(I) > 10 \cdot RESPLMAX_P \cdot DENSITE(I) \quad then \quad DLTAMS(I) = 0$$

The use of reserves concerns perennial plants or indeterminate plants. As for determinate annuals, the use of reserves for grain filling is not simulated as such, but taken globally into account when calculating the IRCARB variable (index of progressive harvest: see § 4.1).

The results of the above calculations are illustrated in the case of wheat and grapevine in Figure 3.24.

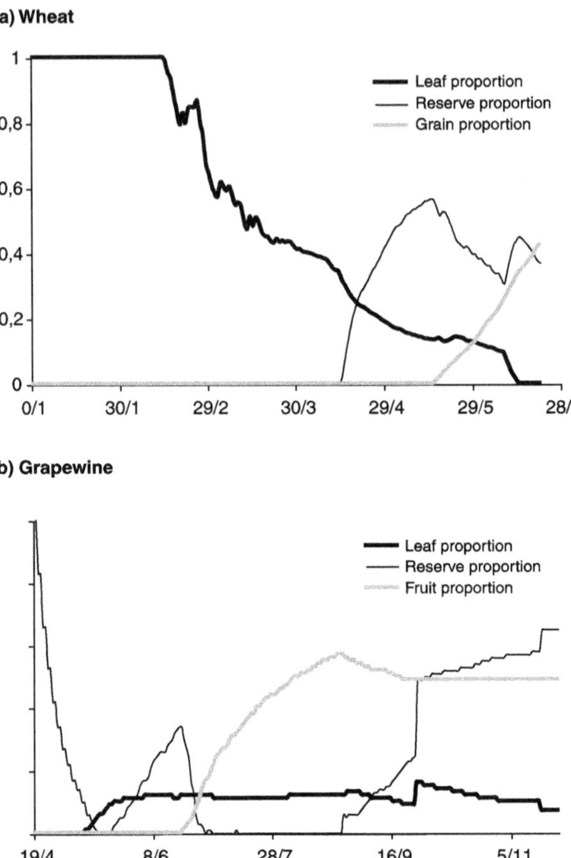

Figure 3.24. Proportion of the shoot biomass allocated to leaves, fruits or grains and to the virtual component of reserves for two different crops: a) wheat and b) grapevine.

The variable RESPERENNE represents the non-structural biomass that can be remobilized.

4 Yield formation

By definition the yield is the weight and the quality of the harvested organs that can be reproductive organs – either grains (dehydrated) or fruits (hydrated), or vegetative storage organs – either stems (sugarcane) or roots (tuber). The determinate or indeterminate character (in the STICS meaning[1]) does not indicate the type of harvested organs. Yet by convention we will call the harvested organs of determinate species "grains" and the harvested organs of indeterminate species "fruit".

Yield prediction is a goal of most crop models. The number of organs harvested is rarely simulated and, if so, is often calculated independently from yield simulation.

In 1972, Warren-Wilson proposed that the plant should be considered as a set of compartments playing the role of sources and/or sinks for assimilates. This concept can be used either for carbon, water, nitrogen or any metabolite of interest. However hereafter we will use it only for carbon, though it is also thoroughly documented for nitrogen (Sinclair and de Wit, 1976; Jeuffroy *et al.*, 2000; Barbotin *et al.*, 2005). The source and sink compartments usually represent organs (e.g. roots, leaves, grains etc.) which can change their function during a cycle: "source then sink" for roots and trunks in perennial plants, or "sink then source" for leaves. Application of this concept to crop models generates self-regulation of the system between the growth of different types of organs. It is particularly well-suited to crops with an indeterminate growth habit and to perennial crops, in which trophic competition exists between growing and storage organs (Jeuffroy and Warembourg, 1991; Munier-Jolain *et al.*, 1998). Source capacity includes both newly-formed assimilates and remobilized resources translocated from vegetative organs. Carbon sink strength i.e. potential growth rate is usually represented by a continuous or discrete function of the physiological age of the organ. The problems with this approach lie in determining the size of the source capacity and remobilized resources, which is difficult to estimate experimentally. Furthermore, it is often necessary to

---

[1] In STICS, "indeterminate" denotes species for which there is significant trophic competition between vegetative organs and harvested organs.

introduce prioritization between organs, thus reproducing the species strategy, and this may be speculative. One alternative is to impose a constant distribution of assimilates by phenological stage, which is frequently applied in determinate crops (Weir *et al.*, 1984). The source-sink approach is used for example by Ritchie and Otter (1984) or Jones *et al.* (2003).

A second alternative, proposed by Spaeth and Sinclair (1985), is to extend the notion of the final harvest index (ratio of grain biomass to total shoot biomass) to the dynamic accumulation of biomass in grains, realizing that a linear variation of the harvest index as a function of time could be assumed. This approach has the advantage of pooling the two sources of assimilates, and is economical in terms of parameters. However, it is important to impose a threshold on this harvest index dynamics, in order to avoid simulating unrealistic remobilization levels or exceeding the maximum filling allowed by the number of organs and the maximum weight of an organ. Apart from cereals, this approach is used for species as different as pea (Lecoeur and Sinclair, 2001) and grapevine (Bindi *et al.*, 1999).

Both these approaches are applied in STICS: the source/sink approach for indeterminate crops and the dynamic harvest index for determinate crops.

4.1 For determinate growing plants

In the case of plants with determinate growth, the hypothesis is made that the number and filling of organs for harvest do not depend on the other organs' growth requirements.

The number of grains is fixed during a phase of variable duration ($NBJGRAIN_P$ in days), which precedes the onset of filling (IDRP). This number depends on the mean growth rate of the canopy during this period (VITMOY in $gm^{-2}d^{-1}$), which in turns depends on dynamics specific to the particular species (eq. 4.1).

eq. 4.1

$$VITMOY(IDRP) = \sum_{J=IDRP-NBJGRAIN_P+1}^{IDRP} \frac{DLTAMS(J)}{NBJGRAIN_P}$$

The number of grains per m^2 (NBGRAINS) is defined at the IDRP stage (eq. 4.2). It depends on the growth variable (VITMOY in $g\ m^{-2}$) that integrates the effect of the prevailing stresses during the period preceding the IDRP stage, on two species-dependent parameters $CGRAIN_P$ (in $g^{-1}\ m^2$) and $NBGRMIN_P$ (grains m^{-2}) and a genetic-dependent parameter $NBGRMAX_V$ (grains m^{-2}). The last two parameters define the limits of variation of NBGRAINS.

eq. 4.2

$$NBGRAINS(IDRP) = CGRAIN_P \cdot VITMOY(IDRP) \cdot NBGRMAX_V$$

$$if\ NBGRAINS(IDRP) > NBGRMAX_V \quad NBGRAINS(IDRP) = NBGRMAX_V$$

$$if\ NBGRAINS(IDRP) < NBGRMIN_P \quad NBGRAINS(IDRP) = NBGRMIN_P$$

According to eq. 4.2, the normalized value NBGRAINS/NBGRMAX$_V$ varies between NBGRMIN$_p$/NBGRMAX$_V$ and 1 and its variability among species (Figure 4.1) expresses the sensitivity of grain onset to growth conditions.

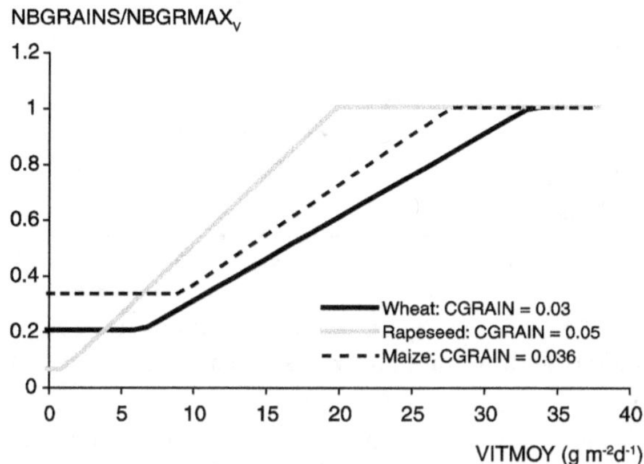

Figure 4.1. Proportion of grain number, for the maximum allowed by the variety (NBGRAINS/ NBGRMAX$_V$), as a function of growth during the pre-grain filling period. Examples of wheat, maize and rapeseed.

After the IDRP stage, the grain number can be reduced in the event of frost (eq. 4.3 and § 3.4) and the daily proportion of grains affected is (1-FGELFLO), whatever their state of growth. The corresponding weight (PGRAINGEL in gm^{-2}) is deducted from the grain weight (eq. 4.6), using the elementary current grain weight (PGRAIN in g) defined in eq. 4.7.

$$eq.\ 4.3$$

$$for\ I > IDRP \quad NBGRAINGEL(I) = NBGRAINS(I-1)(1 - FGELFLO(I))$$

$$and \quad NBGRAINS(I) = NBGRAINS(IDRP) - \sum_{J=IDRP+1}^{I} NBGRAINGEL(J)$$

$$PGRAINGEL(I) = \sum_{J=IDRP+1}^{I} PGRAIN(J-1) \cdot NBGRAINGEL(J)$$

The quantity of dry matter accumulated in grains is calculated by applying a progressive "harvest index" to the dry weight of the plant. This IRCARB index increases linearly with time (VITIRCARB$_p$ in g grain g biomass^{-1} d^{-1}), from the IDRP stage to the IMAT stage and the final harvest index is restricted to the IRMAX$_p$ parameter. The dynamics of IRCARB for various species is depicted in Figure 4.2.

$$eq.\ 4.4$$

$$IRCARB(I) = VITIRCARB_P \cdot (I - IDRP)$$
$$if\ IRCARB(I) > IRMAX_P \quad IRCARB(I) = IRMAX_P$$

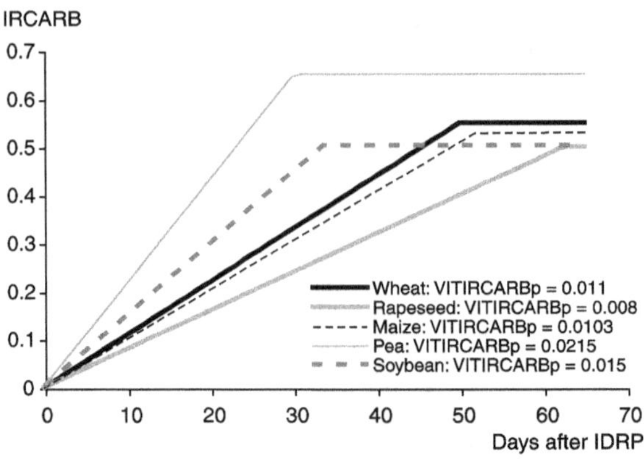

Figure 4.2. Dynamics of the grain to shoot biomass ratio (IRCARB), as a function of time since the stage IDRP, for various species (wheat, maize, rapeseed, pea and soybean).

Yet this dynamics may not be the actual grain filling dynamics since threshold translocation temperatures defining the thermal stress FTEMPREMP (TMINREMP$_P$ and TMAXREMP$_P$; see § 3.4) may stop the carbon filling of harvested organs. Consequently the grain filling is calculated daily (DLTAGS in t ha^{-1}) to allow the effect of the thermal stress (eq. 4.5) and then accumulated within the MAFRUIT (in t ha^{-1}) variable (eq. 4.6).

eq. 4.5

$$DLTAGS(I+1) = [IRCARB(I+1) \cdot MASEC(I+1)$$
$$- IRCARB(I) \cdot MASEC(I)] \, FTEMPREMP(I)$$

eq. 4.6

$$MAFRUIT(I) = \sum_{J=IDRP}^{I} DLTAGS(J) - \frac{PGRAINGEL(I)}{100}$$

$$if \quad MAFRUIT(I) > PGRAINMAXI_V \cdot NBGRAINS(I),$$
$$MAFRUIT(I) = PGRAINMAXI_V \cdot NBGRAINS(I)$$

The mass of each grain is then calculated as the ratio of the mass to the number of grains, although this cannot exceed the genetic PGRAINMAXI$_V$ limit (eq. 4.7).

eq. 4.7

$$PGRAIN(I) = \frac{MAFRUIT(I)}{NBGRAINS(I)} 100$$

$$if, \quad PGRAIN(I) > PGRAINMAXI_V, \quad PGRAIN(I) = PGRAINMAXI_V$$

4.2 For indeterminate growing plants

These species go on growing leaves while producing and growing harvested organs (fruits) during a period of time. There is thus a trophic interaction between the growth of various groups of organs and among successive cohorts of harvested organs that is accounted for in STICS by the source/sink approach using the notion of trophic stress previously defined (see § 3.4). Both processes of organ setting and filling are concerned, assuming that abortion cannot occur during the filling phase.

The simulation technique adopted in STICS was inspired from the "boxcar-train" technique (Goudriaan, 1986) that is used in the TOMGRO model (Jones *et al.*, 1991). During growth, the fruits go through $NBOITE_P$ compartments corresponding to increasing physiological ages. The time fruits spend in a compartment depends on temperature. In each compartment, fruit growth is equal to the product of a "sink strength" function and the source-sink ratio. The fruit sink strength is the derivative of a logistic function that takes the genetic growth potential of a fruit into consideration (Bertin and Gary, 1993).

4.2.1 Fruit setting

Fruits are set between the IDRP stage and the INOU stage (end of setting), defined by the $STDRPNOU_P$ phasic course. If this setting period lasts a long time, then the number of simultaneous compartments (i.e. fruits of different ages) is great which indicates that there must be agreement between the values of $STDRPNOU_P$ and $NBOITE_P$.

During this setting period, on each day, the number of set fruits (NFRUITNOU) depends on $AFRUITSP_V$ (eq. 4.8), a varietal parameter expressed as the potential number of set fruits per inflorescence and per degree.day, the daily development rate (UPVT), the number of inflorescences per plant (NBINFLO), the plant density (DENSITE), the trophic stress index (SPFRUIT) and the frost stress index acting on fruits from flowering (FGELFLO). The introduction of the notion of inflorescence (group of fruits) into the model is only useful when technical or trophic regulation occurs at the inflorescence level (in grapevines for example).

eq. 4.8

$$NFRUITNOU(I) = AFRUITSP_V \times UPVT(I) \times NBINFLO(I)$$
$$\times DENSITE(I) \times SPFRUIT(I) \times FGELFLO(I)$$

If the number of inflorescences is more than 1 (in the case of vines, inflorescences=bunches), it can either be prescribed ($NBINFLO_P$), or calculated as a function of the trophic status of the plant at an early stage (we have chosen IAMF). In the latter case, NBINFLO is calculated using the $PENTINFLORES_P$ and $INFLOMAX_P$ parameters (eq. 4.9). Pruning is not accounted for in this calculation.

eq. 4.9

$$NBINFLO(I) = \frac{PENTINFLORES_P}{DENSITE(IAMF)}[MASEC(IAMF) + RESPERENNE0(IAMF)]$$

$$if,\ NBINFLO(IAMF) > INFLOMAX_P \qquad NBINFLO(IAMF) = INFLOMAX_P$$

RESPERENNE0$_1$ (see § 3.5.4) is the amount of carbon reserves for perennial species coming from the previous cycle.

4.2.2 Fruit filling

The time spent by each fruit in a given compartment is $\dfrac{DUREEFRUIT_V}{NBOITE_P}$, where DUREEFRUIT$_V$ is the total duration of fruit growth expressed in developmental units. In the last box (or age class), the fruits no longer grow and the final dry mass of the fruit has been reached: the fruit is assumed to have reached physiological maturity. A concrete example is shown in Figure 4.3.

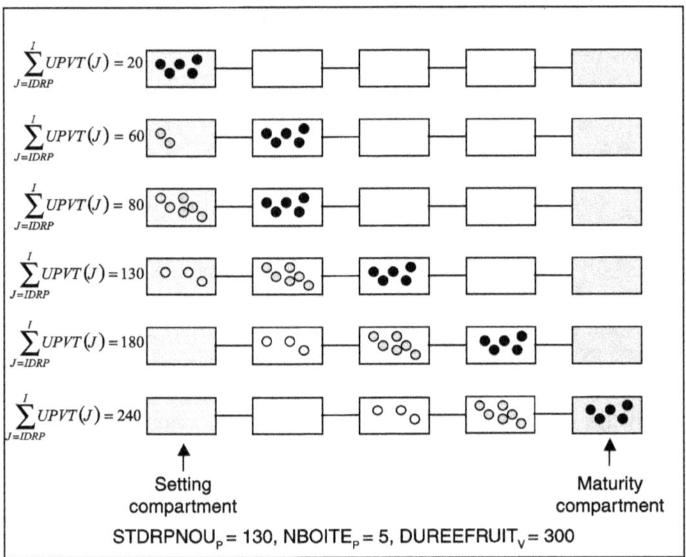

Figure 4.3. Illustration of the dynamics of fruit cohorts using the "boxcartrain" simulation technique.

Each day, in each growth compartment (K), the fruit growth (CROIFRUIT) depends on the number of fruits in the compartment (NFRUIT) multiplied by the growth of each fruit, i.e. the elementary fruit sink strength (FPFT), the trophic stress index (SOURCEPUITS) and the thermal stress index (FTEMPREMP) as given in eq. 4.10.

eq. 4.10

$$CROIFRUIT(I, K) = NFRUIT(I, K) \cdot FPFT(K) \cdot SOURCEPUITS(I) \cdot FTEMPREMP(I)$$

The fruit sink strength function is the derivative of the potential growth of a fruit (POTCROIFRUIT) plotted against the fruit development stage (DFR). There are two successive phases in fruit growth; the first corresponds to a cell division phase while the second is devoted to expansion of the cells already set. In order to account for this double dynamics, the fruit potential cumulative growth is defined as the summation of two functions (eq. 4.11 and Figure 4.4):

• an exponential type function describing the cell division phase (using the parameters $CFPF_p$ and $DFPF_p$)

• a logistic type function describing the cell elongation phase (using the parameters $AFPF_p$ and $BFPF_p$)

eq. 4.11

$$\frac{POTCROIFRUIT(DFR(K))}{PGRAINMAXI_V} = DFPF_P \left(1 - \exp(-CFPF \cdot DFR(K))\right)$$

$$+ \frac{\alpha}{1 + \exp(-BFPF_P (DFR(K) - AFPF_P))} - \beta$$

and α and β values are calculated so as:

$$POTCROIFRUIT(0) = 0$$

$$POTCROIFRUIT(1) = PGRAINMAXI_V$$

$PGRAINMAXI_V$ is the genetic-dependent maximal weight of the fruit and DFR stands for the fruit development stage of each age class, varying between 0 and 1; it is calculated for each age class (K) in a discrete way (eq. 4.12).

eq. 4.12

$$DFR(K) = \frac{K}{NBOITE_P}$$

This double dynamics is particularly interesting for grapevine (Garcia de Cortazar., 2006). In many other cases (tomato, sugar beet, sugarcane) the cell division phase is fast so that the logistic is enough to describe fruit growth (in that case one of the parameters $DFPF_p$ or $CFPF_p$ must be zero).

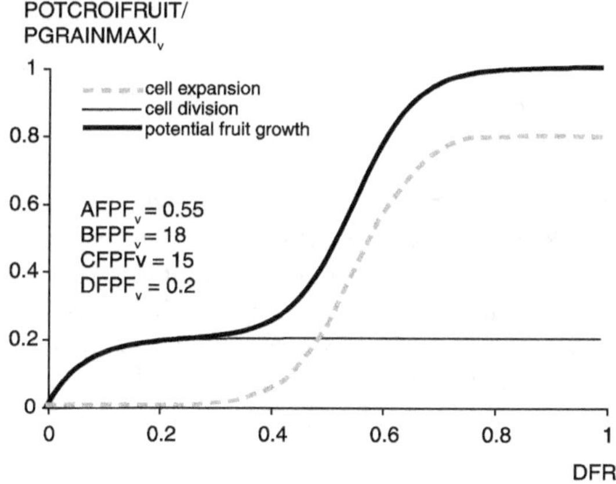

Figure 4.4. Normalized potential fruit growth (POTCROIFRUIT/PGRAINMAXI_v) versus fruit development status (DFR) with its two components: the exponential dynamics representing cell division and the logistic type dynamics representing cell expansion.

If the potential fruit growth is represented by a simple logistic curve, Figure 4.5 shows that when varying the parameters $AFPF_p$ and $BFPF_p$, one can represent various dynamics including the linear one.

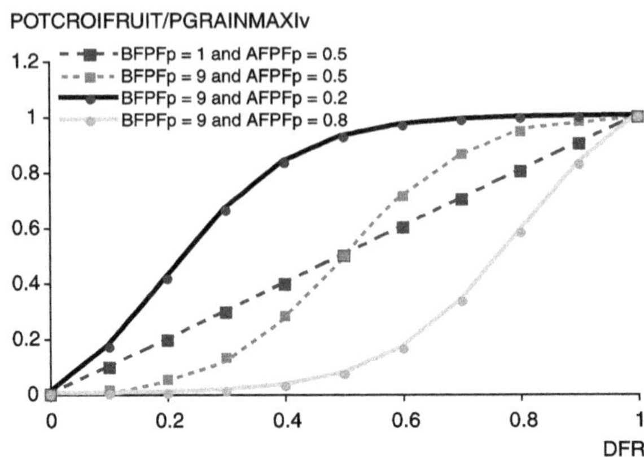

Figure 4.5. Normalized potential logistic fruit growth (POTCROIFRUIT/PGRAINMAXI$_V$) versus fruit development status (DFR) with various parameterizations corresponding to $AFPF_p$ and $BFPF_p$ values.

Then the daily fruit sink strength function (FPFT) is calculated (eq. 4.13) for each age class, accounting for the duration of fruit growth from setting to maturity, expressed in developmental units (DUREEFRUIT$_V$).

eq. 4.13

$$FPFT(I, K) = PGRAINMAXI_V \cdot DEVJOUR(I) \left[DFPF_P \cdot CFPF_P \right.$$

$$\left. \cdot \exp(-CFPF_P \cdot DFR(K)) + \frac{BFPF_P \times \alpha \times Y}{(1+Y)^2} \right]$$

$$\text{with } Y = \exp(-BFPF_P \left(DFR(K) - AFPF_P \right)) \text{ and } DEVJOUR(I) = \frac{TCULT(I) - TDMIN_P}{DUREEFRUIT_V}$$

The sensitivity of the model for subdividing fruit growth into discrete units (NBOITE$_p$ parameter) also depends on the POTCROIFRUIT dynamics, as shown in Figure 4.6. Consequently three elements must be taken into account to give a value to the parameter NBOITEp: the fruit setting duration, the fruit growth dynamics and the location of the IDEBDES stage allowing the fruit water dynamics to be initiated.

If allocation to fruits (ALLOCFRUIT variable calculated in eq. 4.14) exceeds the ALLOCFRMX$_p$ threshold, the SOURCEPUITS variable is reduced in proportion to the ALLOCFRUIT/ALLOCFRMX$_p$ ratio. In the last box, the fruits are ripe and stop growing. The number of fruits present on the plant or fruit load is CHARGEFRUIT. If the CODEFRMUR$_G$ is 1, then the CHARGEFRUIT variable will take account of the fruits in the last box (ripe); if not, it will only take account of the (N-1) first boxes.

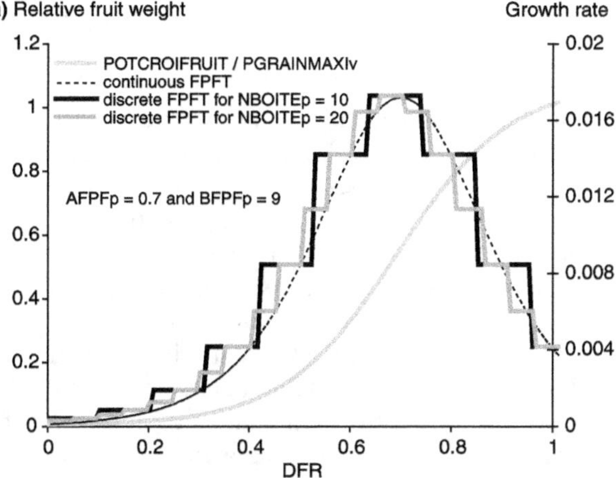

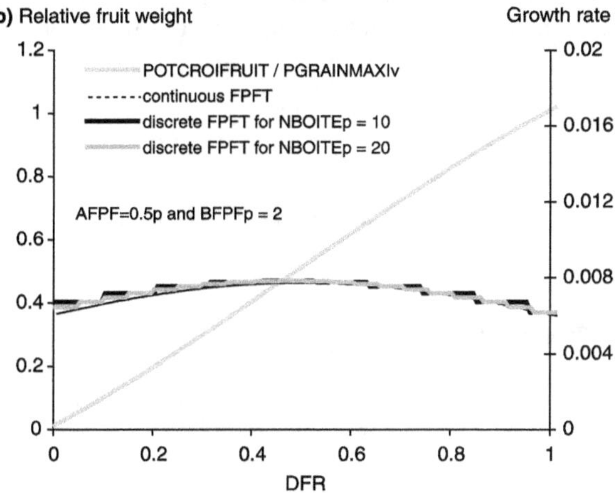

Figure 4.6. Influence of the discretization of fruit growth through the number of boxes parameter (NBOITEp) in relation to the form of the dynamics: "S" shape in a) or nearly linear in b).

eq. 4.14

$$ALLOCFRUIT(I) = \frac{\sum_{K=1}^{NBOITE_p - 1} CROIFRUIT(I, K)}{DLTAMS(I)}$$

4.3 Quality

4.3.1 Water content of organs

For non-harvested organs, the water contents are assumed constant. The corresponding parameters are called $H2OFEUILVERTE_p$, $H2OFEUILJAUNE_p$, $H2OTIGESTRUC_p$ and $H2ORESERVE_p$ for green and dead leaves, stems and reserves respectively: they are expressed in terms of fresh weight (FM), i.e. in g water. g FM^{-1}. They are used to calculate the fresh weight of each organ: MAFRAISFEUILLE (for all green and yellow leaves), MAFRAISTIGE (stems) and MAFRAISRES (reserves).

For harvested organs, it is assumed that the water content is constant (H2OFRVERT$_p$) up to the stage IDEBDES (see chapter 2). This stage may occur before physiological maturity. For indeterminate plants, it does not occur at the same time for all fruit cohorts but it corresponds to one of the age classes. We shall call this stage "onset of fruit water dynamics" that can be hydration or dehydration which results from the concomitant water and dry matter influx into the fruit or grain. As from this stage, we assume that there is a "programmed" time course in the water content of fruits, and this is expressed using the DESHYDBASE$_p$ parameter (g water.g FM^{-1}.d^{-1}), which day after day will modify the fruit water content (TEAUGRAIN) from its initial value H2OFRVERT$_p$. For dehydration DESHYDBASE$_p$ is positive; if the programme evolution tends towards hydration, DESHYDBASE$_p$ is negative. Dehydration may be accelerated (or provoked) by water stress, which is characterised by the difference between the crop and air temperatures. The proportionality coefficient is called TEMPDESHYD$_p$ in g water.g FM^{-1}. °C^{-1}. In summary, the water content (TEAUGRAIN) is the result of eq. 4.15 where the index K (for the box number) is useless for determinate plants.

eq. 4.15

$$TEAUGRAIN(I, K) = H2OFRVERT_P - DESHYDBASE_P (I - IDEBDES(K) + 1)$$

$$- \sum_{J=IDEBDES(K)}^{I} TEMPDESHYD_P (TCULT(J) - TAIR(J))$$

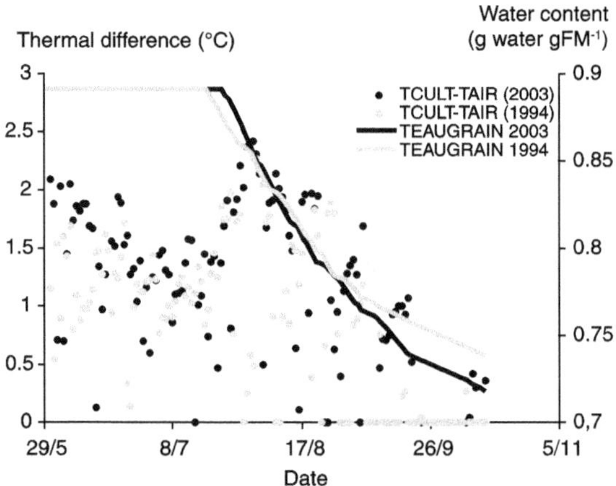

Figure 4.7. Evolution of grape water contents for two different years in Montpellier (France) influenced by the phenological course (the beginning of the dynamics occurs on 20/07 in 1994 and on 26/07 in 2003) and the thermal difference (TCULT-TAIR).

4.3.2 Biochemical composition

The quantity of nitrogen in harvested organs, both for determinate and indeterminate species (QNGRAIN), is an increasing proportion (IRAZO: eq. 4.16) of the quantity of nitrogen in the biomass (QNPLANTE): the concept of the harvest index is extended to nitrogen (Lecoeur and Sinclair, 2001), using the parameter $VITIRAZO_p$. Obviously, as for carbon, the grain/fruit nitrogen filling can be affected by thermal stress which requires a daily calculation (DLTAGN: eq. 4.17). The temperature effect on nitrogen grain filling is assumed to be the same as for carbon. The nitrogen harvest index is assumed to be limited to a value calculated using the carbon parameters ($IRMAX_p$ and $VITIRCARB_p$) as explained in eq. 4.16.

$$\text{eq. 4.16}$$

$$IRAZO(I) = VITIRAZO_P \cdot (I - IDRP)$$

$$if\ IRAZO(I) > IRMAX_P\ \frac{VITIRAZO_P}{VITIRCARB_P} \quad IRAZO(I) = IRMAX_P\ \frac{VITIRAZO_P}{VITIRCARB_P}$$

$$\text{eq. 4.17}$$

$$DLTAGN(I+1) = [IRAZO(I+1) \cdot QNPLANTE(I+1)$$
$$- IRAZO(I) \cdot QNPLANTE(I)] FTEMPREMP(I)$$

To complete the components of the quality of simulated harvested organs, we propose a very simple estimate of the sugar and oil contents. From the beginning of fruit/grain filling until physiological maturity, we assume that there is a gradual increase in the proportions of these two types of components in the dry matter of fruits. This increase is determined using the $VITPROPSUCRE_p$ (see Figure 4.8) and $VITPROPHUILE_p$ parameters expressed in g.g DM^{-1}.degree.day^{-1}. The combination of this evolution and the evolution in the water content in fruits produces contents based on fresh matter, which depends on the development of each crop. For indeterminate crops, the calculation is made for each age category separately, and then combined for all age categories.

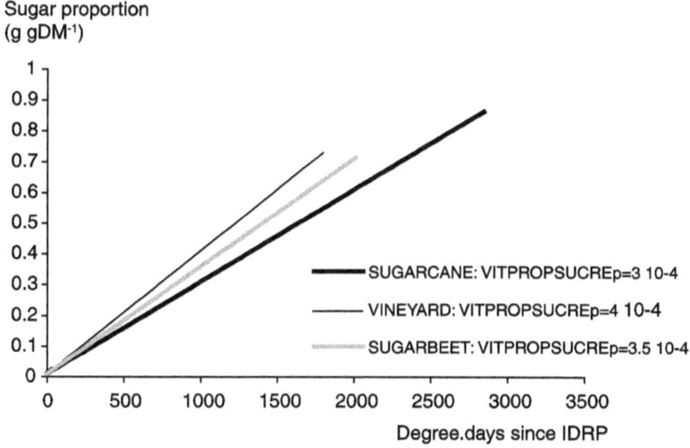

Figure 4.8. Evolution of sugar content in relation to fruit development for sugar cane, sugarbeet and vineyard.

5 Root growth

Apart from plant anchorage, the root system has numerous functions: water and mineral element (mainly N, P, K) uptake, symbiotic fixation (for legumes), rhizodeposition and as a reserve organ, which are variously accounted for in crop models. The root system as a reserve organ can be regarded as a harvested organ (e.g. tubers) or part of the "non-located" reserves (see § 3.5.4). The development of N-fixing nodules and their activity is less dependent on the root system, which plays a support role, than on the physicochemical conditions of the surrounding soil and on the shoot dynamics (Burger, 2001: see § 8.7). Rhizodeposition is accounted for by the recycling of the sloughed roots within the fresh soil organic matter (as a plant residue, see § 6.3.3). The nutritional functions of the root system can be calculated from supply and demand principles, the demand originating from the shoot metabolism while the supply results from the combination of the presence of the elements of interest in the soil and the root's ability to capture those elements.

This ability relies on the efficiency of the root system, which is not simply related to the actual root length profile or to the root biomass and depends very much on the mobility of the element of interest within the soil. For water and nitrate ions, the minimum root length density for unrestricted uptake is 0.5 cm cm^{-3} according to Bonachela (1996), equating to an average soil-root distance of 0.8 cm, which lies within the range proposed by Aura (1996) of 0.5 – 1.0 cm. According to other authors (Kage and Ehlers, 1996; Robertson *et al.*, 1993) it can be lower. This means that the efficient root profile is different from the actual root system, especially in the subsurface layer where roots are more than adequate for nitrate and water uptake, although they are needed for the uptake of less mobile ions.

Moreover this efficiency needs to be dynamically estimated in order to correctly evaluate the supply/demand ratio. Also the effect of the soil (constraints to penetration, sensitivity to anoxia etc.) on the form of the root system (Nicoullaud *et al.*, 1994) must be accounted for. While all these elements are accounted for in architectural root growth modelling approaches (Drouet and Pages, 2003), it is seldom the case in crop models in which roots are not individualized but just layered in the soil.

In crop models, the fact that the soil is regarded in only one dimension requires that growth in depth is treated separately from growth in density. The progression rate of the root front is generally based on degree-days (Giauffret and Derieux, 1991; Hunt and Pararajasingham, 1995) and the root density assumption mostly relies on an exponential decrease of roots with depth (Gerwitz and Page, 1974).

Although we can rely on existing modelling patterns of root/shoot ratio in terms of biomass (Wilson, 1988), the extrapolation to root length is not easy, since the specific root length (length per unit weight) can vary as a function of the phenological stage and experienced stresses in addition to the well-known genetic factor (Bingham, 1995).

This complexity led us to propose optional calculations of root growth in STICS. In the model, roots only act as water and mineral nitrogen absorbers, and are described by their front depth and density profile. The root growth begins at germination (for sown plants) or at planting (for transplanted crops, possibly after a latency phase, see § 2.2.2), and it stops at a given stage of development, depending on the species (STOPRAC$_p$ which can be either LAX, FLO or MAT).

5.1 Root front growth

A first calculation gives the depth of the root front (ZRAC) beginning at the sowing depth (PROFSEM$_T$) for sown crops and at an initial value for transplanted crops (PROFSEM$_T$ + ZRACPLANTULE$_p$) or perennial crops (ZRAC0). The root front growth stops when it reaches the depth of soil or an obstacle that can be physical or chemical (the obstacle depth is defined by the parameter OBSTARAC$_S$) or when the phenological stopping stage has been reached. For indeterminate crops, when trophic competition prevents vegetative growth, the root front growth is stopped (except before the IAMF stage, when root growth is given priority).

The calculation of root front growth rate (DELTAZ in cm.d^{-1}) is broken down in eq.5.1. A first calculation of the front growth rate (DELTAZ$_T$ in cm.d^{-1}) is proportional to temperature with a coefficient depending on the variety (CROIRAC$_V$). This value is then multiplied by the water and bulk density stress indices (DELTAZ$_{stress}$).

eq. 5.1

$$DELTAZ(I) = DELTAZ_T(I) \cdot DELTAZ_{stress}(I)$$

The thermal function relies on crop (eq.5.2) or soil temperature (eq.5.3) according to the root growth dependence on the collar or apex temperature. If the driving temperature is that of the crop, the cardinal temperatures (TCMIN$_p$ and TCMAX$_p$) are the same as those used for the thermal function of the leaf growth rate (see eq. 3.3). If the driving temperature is that of the soil at level ZRAC ($\pm$1cm), the minimum temperature is the base temperature for germination (TGMIN$_p$) but the maximum temperature does not change.

$$\text{eq. 5.2}$$

$$if\ TCULT(I) \leq TCMIN_P \qquad DELTAZ_T(I) = 0.0$$
$$if\ TCMIN_P < TCULT(I) < TCMAX_P \quad DELTAZ_T(I) = (TCULT(I) - TCMIN_P) \cdot CROIRAC_V$$
$$if\ TCULT(I) \geq TCMAX_P \qquad DELTAZ_T(I) = (TCMAX_P - TCMIN_P) \cdot CROIRAC_V$$

$$\text{eq. 5.3}$$

$$if\ TSOL(AP(I), I) \leq TGMIN_P \qquad DELTAZ_T(I) = 0.0$$
$$if\ TGMIN_P < TSOL(AP(I), I) < TCMAX_P$$
$$DELTAZ_T(I) = (TSOL(AP(I), I) - TGMIN_P) \cdot CROIRAC_V$$
$$if\ TSOL(AP(I), I) \geq TCMAX_P \qquad DELTAZ_T(I) = (TCMAX_P - TGMIN_P) \cdot CROIRAC_V$$
$$AP(I) = ZRAC(I) \pm 1cm$$

The water and bulk density stress index ($DELTAZ_{stress}$) is calculated as the product of 3 variables (eq.5.4), depending on soil dryness (HUMIRAC, see eq. 2.3 and Figure 2.3), water logging (IZRAC, see eq. 3.35 and Figure 3.22), and bulk density (EFDA).

$$\text{eq. 5.4}$$

$$DELTAZ_{stress}(I) = HUMIRAC(AP(I), I) \cdot IZRAC(I) \cdot EFDA(ZRAC(I))$$

The HUMIRAC variable, calculated as in eq. 2.3 during emergence, becomes a bilinear variable after emergence (eq.5.5):

$$\text{eq. 5.5}$$

$$if\ HUMSOL(AP(I), I) > HN_S$$
$$then \quad HUMIRAC(AP(I), I) = 1$$
$$if\ HUMSOL(AP(I), I) \leq HN_S$$
$$then \quad HUMIRAC(AP(I), I) = \frac{SENSRSEC_P}{HN_S} HUMSOL(AP(I), I)$$

The EFDA variable constitutes a constraint to penetration in the case of compacted soils, or more rarely a slowing of root penetration linked to a lack of soil cohesiveness. The formalisation proposed by Jones *et al.* (1991) and validated by Rebière (1996), was adapted for STICS (Figure 5.1). Root penetration is not constrained between the bulk density thresholds $DACOHES_G$ and $DASEUILBAS_G$. Above a bulk density threshold $DASEUILHAUT_G$ the effect of bulk density (DA) on root penetration is constant and corresponds to the sensitivity of the plant to the penetration constraint; it is equal to $CONTRDAMAX_P$. $DASEUILBAS_G$ and $DASEUILHAUT_G$ values are 1.4 and 2.0 respectively. The $DACOHES_G$ value is poorly understood and we only provide an order of magnitude. The bulk density is the effective one, taking into account fine earth and pebbles.

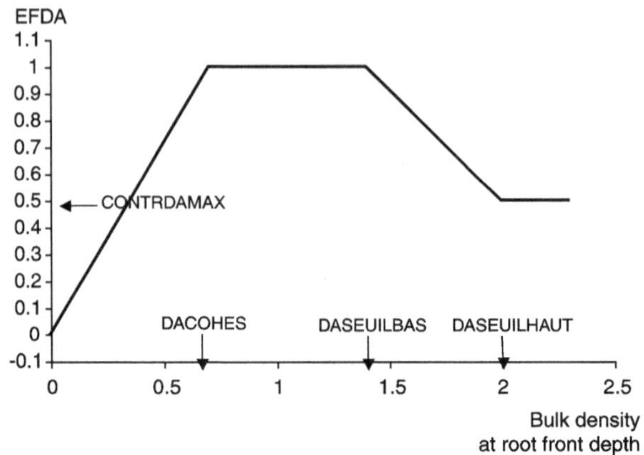

Figure 5.1. Constraint to root penetration (EFDA) as a function of the bulk density (DA).

5.2 Growth in root density

The root density profile is calculated according to two possible options. The 'standard profile' option makes it possible to calculate the root profile that is effective with respect to absorption. The 'true density' option allows the actual root density profile to be estimated, which is more relevant in order to simulate low-density crops, for which root density is never optimal, or in order to take into consideration the effects of constraints imposed by the soil on root distribution.

Whatever the chosen option, roots only play a role as absorbers of water and mineral nitrogen. It is possible to estimate the root mass with the second option and to account for a direct link between shoot and root growing rates. However an indirect link exists in all calculations through temperature, which affects both levels.

5.2.1 Standard profile

This option enables calculation of the root profile which is efficient in terms of absorption. It is defined by the maximum current depth, ZRAC, and a prescribed efficient root density profile, LRACZ (Z). This profile is calculated dynamically as a function of ZRAC and takes a sigmoidal form depending on the $ZLABOUR_p$, $ZPRLIM_p$ and $ZPENTE_p$ parameters (Figure 5.2 and eq.5.6).

These parameters define the form of the reference root profile and are of considerable importance in terms of their interrelationships, but they do not define the final shape of the root system. In this respect, it is the differences between $ZPENTE_p$ and $ZLABOUR_p$, and particularly between $ZPRLIM_p$ and $ZPENTE_p$ which are determinant. $ZLABOUR_p$ corresponds to the depth of the tilled layer, where it is assumed that root proliferation is not limited with respect to water and mineral absorption: root density is optimum at this level ($LVOPT_G$). $ZPENTE_p$ is the depth at which root uptake efficiency is reduced by half, and $ZPRLIM_p$ is the depth of the root front to which this reference profile can be attributed. The value used for the optimum root density threshold, $LVOPT_G$, is 0.5 cm cm^{-3} soil (Brisson, 1998c). In this way, it is possible to represent a root system for various species exhibiting fasciculate or pivotal type root systems (Figure 5.3).

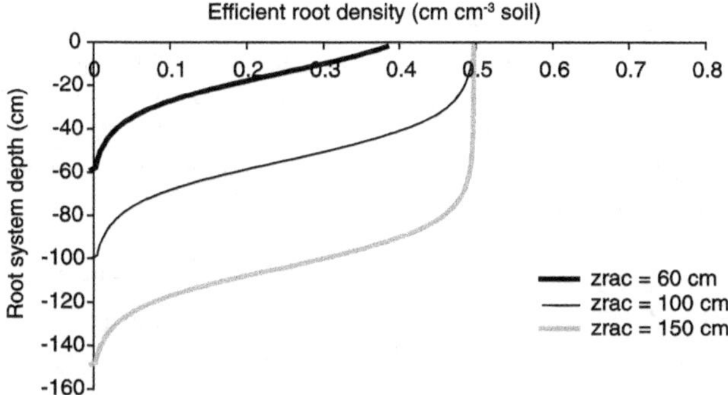

Figure 5.2. Reference root density profile for rapeseed, described by the efficient root density LRACZ(Z) as a function of the root system depth Z and according to the root front depth ZRAC.

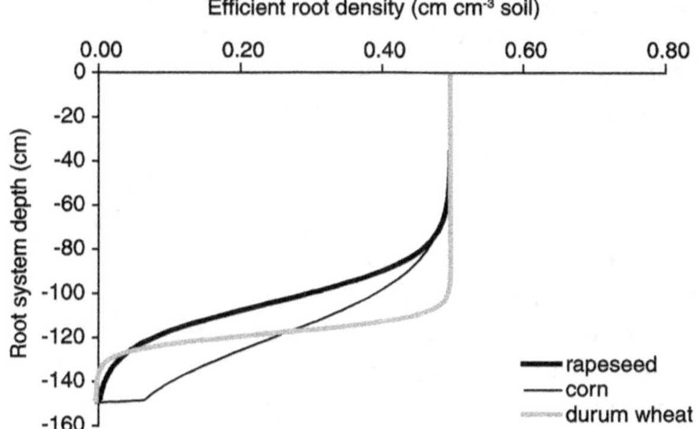

Figure 5.3. Reference root density profile for a root front depth of 150 cm, described by the efficient root density LRACZ (Z) as a function of depth Z, for rapeseed, corn and durum wheat.

$$\text{eq. 5.6}$$

$$LRACZ(Z, I) = \frac{LVOPT_P}{1 + \exp(-S(Z - ZDEMI(Z, I)))}$$

and

$$S = \frac{4.6}{ZLABOUR_P - ZPENTE_P}$$

$$ZDEMI(Z, I) = \max\left(ZRAC(Z, I) - ZPRLIM_P + ZPENTE_P, \frac{1.4}{S}\right)$$

The ZDEMI=1.4/S threshold ensures at least an extraction near the soil surface of 20% of the water available. Roots located in dry layers of soil, with a water content equal to or below the wilting point, are considered as ineffective with respect to water uptake (§ 7.3.3). The total and effective root length throughout the profile is called CUMLRACZ.

Using this method of calculation, any reduction in the root front causes a reduction in density. If the soil contains an obstacle to rooting (calculated as the lesser of the soil depth and an obstacle depth defined by the parameter $OBSTARAC_S$), a fictitious root front (ZNONLI) is calculated until the stage of physiological stoppage $STOPRAC_p$, thus allowing simulation of the course of root proliferation above the obstacle. If the problem is anoxia (inducing a slowing down but not necessarily a cessation of growth), in order to simulate root proliferation above the saturated zone, the $ZPRLIM_p$ parameter continues to grow at a rate reduced by 80% when compared with the rate without waterlogging. This 80% value has been adjusted so as to obtain comparable results between the two root density approaches.

5.2.2 True density

This option enables calculation of a root density profile comparable with measurements. Effectively the hypotheses underlying the "standard profile" formalisation may lead to some problems: i) in the tilled zone, root density is not always optimal with respect to the absorption of water and nitrogen (for woody species in widely-spaced rows, maximum root densities of about 0.2 cm·cm^{-3} are measured, which is lower than the optimum density of 0.5 cm.cm^{-3}(Ozier-Lafontaine *et al.*, 1999) and ii) the influence of constraints imposed by the soil on the distribution of roots in the profile may be far from negligible. Limitations of the "standard profile" formulation could also occur if functions of the root system other than water and nitrogen absorption are considered (e.g. absorption of P and K, supplier of organic matter).

With this option, growth in root length is first calculated, and then distributed to each layer of the soil profile. For sown crops, this calculation begins at emergence: between germination and emergence, it is assumed that only the root front grows. For transplanted or perennial crops, the calculation is initiated with an existing root density profile. After a lifetime characteristic of the species, the roots senesce and enter the mineralization process as crop residue at the end of the crop cycle. Root density above 0.5 cm·cm^{-3} is not taken into account for water and nitrogen absorption.

5.2.2.a Growth in root length

To ensure the robustness of the model, we have chosen to simulate the growth in root length directly, without passing through the root mass, because the specific length (root length/mass ratio) varies depending on the stresses suffered by the plant. Two options are available to calculate the root length. With the first option, we have adopted a formulation similar to that used for the above-ground growth of leaves (Brisson *et al.*, 1998a). With the second, a trophic link between shoot growth and root growth allows increase in root length to be calculated.

• *Self-governing production*

Growth in root length is calculated using a logistic function that is analogous to that of leaves: the calculation of root length growth rate (RLJ in m d^{-1}) is broken down in eq.5.7. A first calculation of the root length growth rate (RLJ$_{dev}$ in m plant^{-1} degree-day^{-1})

describes a logistic curve. This value is then multiplied by the effective crop temperature (RLJ_T in degree-days), the plant density combined with an inter-plant competition factor that is characteristic for the variety (RLJ_{dens} in plant m^{-2}), and the water logging stress index (RLJ_{stress}). Then a second term is added corresponding to the growth at the root front (RLJFRONT), depending on the front growth rate (DELTAZ).

eq. 5.7

$$RLJ(I) = RLJ_{dev}(I) \cdot RLJ_T(I) \cdot RLJ_{dens} \cdot RLJ_{stress}(I) + RLJFRONT(I)$$

The logistic curve describing the root length growth rate RLJ_{dev} (eq.5.8) depends on the maximum root growth parameter $DRACLONG_P$ and on the normalized root development unit URAC, ranging from 1 to 3 (such as ULAI, whose calculation is described in § 3.1.1) and is thermally driven, even when the plant has vernalisation or photoperiod requirements. The plant parameters $PENTLAIMAX_P$ and $VLAIMAX_P$ are the ones already used for the calculation of leaf growth rate (see eq. 3.2).

eq. 5.8

$$RLJ_{dev}(I) = \frac{DRACLONG_P}{1 + \exp(PENTLAIMAX_P(VLAIMAX_P - URAC(I)))}$$

The thermal function RLJ_T relies on crop temperature and cardinal temperatures ($TCMIN_P$ and $TCMAX_P$) which are the same values as for the leaf area growth calculation (eq. 3.3). The inter-plant competition function RLJ_{dens} is the same as the one calculated for the leaf area growth $DELTAI_{dens}$ (eq. 3.4).

Unlike the leaf area index, water and nitrogen deficiencies in the plant do not play any role in root growth, which results in the promotion of root growth relative to above-ground growth in the event of stress. In contrast, anoxia acts via the the water-logging stress index RLJ_{stress}, derived from the IZRAC indicator (eq. 3.35 and eq. 5.9). In view of the difference which may exist between true density and effective density (as much as tenfold), the raw application of IZRAC could have no effect on effective density, which would not accord with experimental results (Rebière, 1996). So when IZRAC is less than 1 (i.e. under water-logging stress conditions), it is multiplied by the ratio between effective (CUMLRACZ) to total (RLTOT) root length ratio before it is applied to the RLJ variable (eq. 5.9).

eq. 5.9

$$RLJ_{stress}(I) = 1 - SENSANOX_P + (IZRAC(I) + SENSANOX_P - 1)\frac{CUMLRACZ(I)}{RLTOT(I)}$$

At the root front, the density is imposed and estimated by the parameter $LVFRONT_P$, and the growth in root length depends directly on the root front growth rate DELTAZ (eq. 5.10):

eq. 5.10

$$RLJFRONT(I) = LVFRONT_P \cdot 10^4 \cdot DELTAZ(I)$$

• *Trophic-linked production*

The root length growth may rely on the daily production of shoot biomass (DLTAMS, eq. 3.26) and on a dynamic underground/total biomass partitioning coefficient (REPRAC) (eq. 5.11 and Figure 5.4). The parameter $LONGSPERAC_p$ is the specific root length/root mass ratio. The plant density effect is not taken into account because it is already integrated in the shoot biomass production. This value can replace calculation by eq 5.7 or just act as a threshold according to the chosing option.

eq. 5.11

$$RLJ(I) = \frac{REPRAC(I)}{1 - REPRAC(I)} \cdot DLTAMS(I) \cdot LONGSPERAC_P \cdot 10^{-2}$$

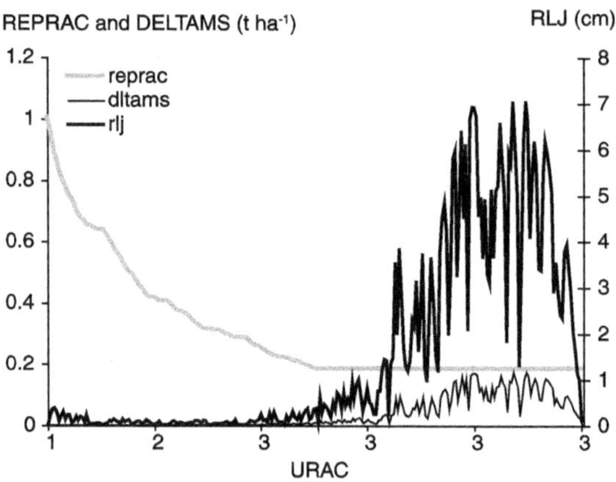

Figure 5.4. Example of the root length growth RLJ as a function of the root development unit URAC, compared to the underground/total biomass portioning coefficient REPRAC and to the daily production of shoot biomass DLTAMS.

The dynamic aboveground / underground partition coefficient (REPRAC) depends on the root development through the normalized root development unit URAC (Baret *et al.*, 1992), and on specific parameters $REPRACMIN_p$, $REPRACMAX_p$ and $KREPRAC_p$ (eq. 5.12 and Figure 5.5).

eq. 5.12

$$REPRAC(I) = (REPRACMAX_P - REPRACMIN_P)$$
$$\cdot \exp(- KREPRAC_P (URAC(I) - 1)) + REPRACMIN_P$$

5.2.2.b Distribution in the profile

The new root length is then distributed in each layer of the soil profile in proportion to the roots present and as a function of the soil constraints.

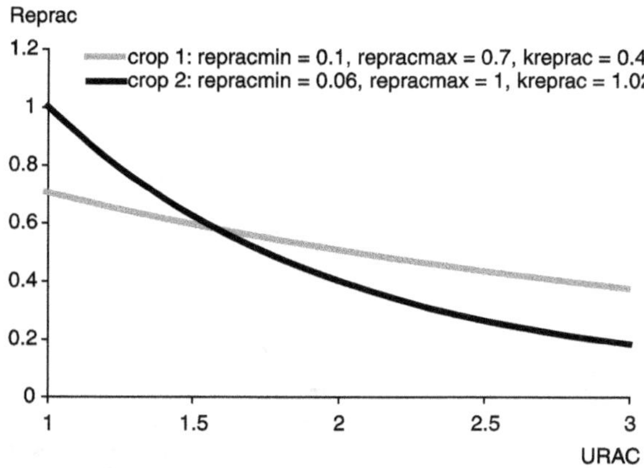

Figure 5.5. Aboveground/underground partition coefficient REPRAC as a function of the root development unit URAC, in the case of two different crops.

A "root sink strength" is defined by the proportion of roots present in the layer. This does not concern the root front, whose growth in density is defined by $LVFRONT_p$ (eq. 5.10). This potential "root sink strength" is then reduced by the soil constraints in each layer. Each constraint is defined at the layer level, in the form of an index between 0 and 1, and assumed to be independent of the others. The resulting index POUSSRAC is the product of elementary indices:

eq. 5.13

$$POUSSRAC(Z, I) = HUMIRAC(Z, I) \cdot EFDA(Z, I)$$
$$\cdot (1 - ANOX(Z, I) \cdot SENSANOX_P) \cdot EFNRAC(Z, I)$$

HUMIRAC (eq. 2.3) defines the effect of soil dryness, taking account of the plant's sensitivity to this effect. EFDA defines the effect of soil compaction through bulk density (§ 5.1 and Figure 5.1). The anoxia index of each soil layer ANOX is assigned the value of 1 if the horizon has reached saturation; it is associated with the sensitivity of the plant to water logging $SENSANOX_p$.

EFNRAC defines the effect of mineral nitrogen, which contributes to the root distribution in the layers with high mineral nitrogen content. It depends on the specific parameters $MINAZORAC_p$, $MAXAZORAC_p$ and $MINEFNRA_p$ which characterize the sensitivity of plant root growth to the mineral nitrogen content in the soil (Figure 5.6). This last constraint is optional and can be inactivated in the model.

5.2.2.c Senescence

A thermal duration in degree days ($STDEBSENRAC_p$) defines the lifespan of roots. Thus, the history of root production per layer is memorized in order to make disappear by senescence the portion of roots $STDEBSENRAC_p$ set earlier. The profile of dead roots is LRACSENZ while the corresponding total amount is LRACSENTOT.

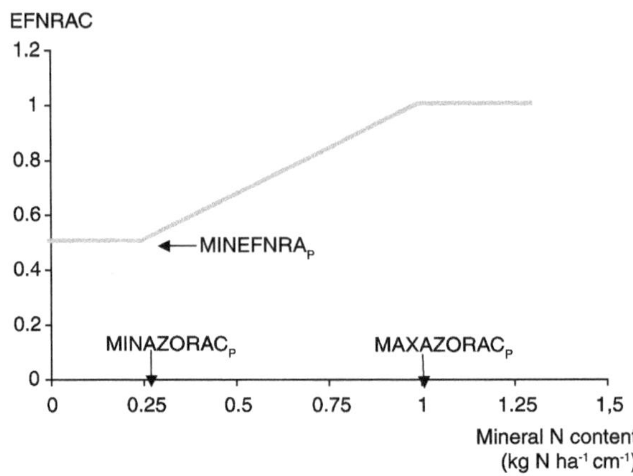

Figure 5.6. Constraint to root distribution (EFNRAC) as a function of the mineral nitrogen content in the soil.

5.2.2.d Root density profiles

The living root density profile is RL, while the total amount is RLTOT. For water and nitrogen absorption, an efficient root length density (LRACZ) is calculated by applying the threshold $LVOPT_G$ (by default equals 0.5 cm cm^{-3}) to the total root length density, RL.

5.2.3 Comparison of the two kinds of density profiles

The differences between the two options in the simulation of the root profiles can be significant (Figure 5.7), but the effect on the simulated water and nitrogen uptakes may not be significant because of the functional root density threshold of 0.5 cm·cm^{-3}.

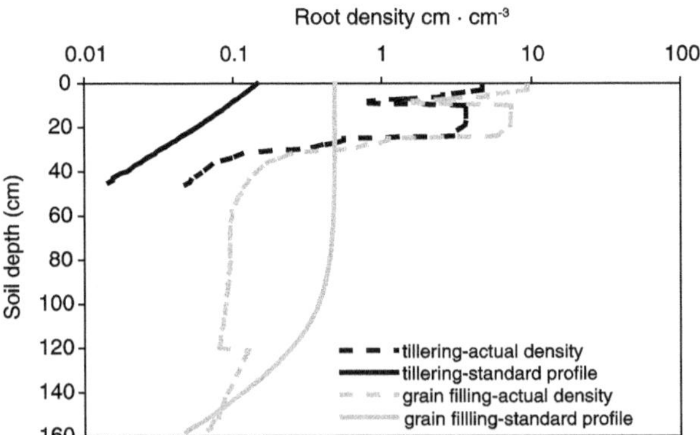

Figure 5.7. Root density profile as a function of the root depth, at tillering and at grain filling, simulated with the standard profile option and the true density option.

6 Management and crop environment

The first models produced (de Wit., 1978; Baker, 1980; Weir *et al.*, 1984) aimed at describing in detail the ecophysiology of crops, often for didactic purposes, but paid little attention to agronomic objectives. Afterwards models started to include farming practices in the inputs (Ritchie and Otter, 1984; Williams *et al.*, 1984) and in particular irrigation and fertilization. Accounting for the techniques requires simulating the appropriate state variable that the technique is supposed to modify, e.g. soil water content for irrigation, organic residue dynamics for manure application or annual wood production for pruning. In STICS, emphasis has been placed on crop management which is important to simulate industrial crops and essential for high value-added crops. However some techniques are not yet accounted for in the model because the corresponding state variable is poorly calculated if at all, for example soil structure or crop sanitary status.

6.1 Effects on plants

For industrial crops the direct effects on plants can be summarized up by the two extreme operations of sowing and harvest, while for high value crops like vegetables or grapevines the farmer also intervenes during the crop cycle to regulate yield. Some of the effects mentioned in the following paragraphs have already been documented in chapters 2 and 3.

6.1.1 Planting design

For annual crops there are two choices, either sowing seeds (industrial crops like wheat, rapeseed, sugarbeet etc.) or transplanting plantlets (lettuce, tomato, strawberry etc.). In the case of sowing the very first stages of the plant occurring beneath the soil are simulated (e.g. germination and underground shoot growth: §2.2.1) depending on sowing depth. In the case of planting, we simply consider a lag time from planting to the start of actual plantlet growth (see §2.2.2); the model requires to initialize the plant status in terms of LAI, biomass, nitrogen status, rooting depth and root density profile.

The assumed precision for sowing depth is 3 cm, since the soil conditions prevailing for emergence are those found in the 3 cm layer above and below the prescribed sowing depth. However this variation in depth does not affect emergence: all the plants that succeed in emerging do so at once.

The model also needs information about the geometrical pattern of the crop, which is very important for radiation interception. It is either homogeneous or in rows. In the latter case, geometrical parameters are required, such as the interrow distance and the row orientation (see §3.2).

A companion crop can also be simulated such as grass in vineyards: in this case the system is simulated as an intercropping system (see §10.3).

6.1.2 Simulation of the decision to sow

It is possible either to prescribe the sowing date ($IPLT_T$) or to calculate it (IPLT) from rules to do with the weather and soil water status.

In the case of calculation, a period when sowing is allowed is defined as the interval [$IPLT_T$, $IPLT_T$ + $NBJMAXAPRESSEMIS_T$]. Three criteria are then taken into account to postpone sowing within the previously defined sowing period.

• The soil must be wet (above wilting point in the seedbed = $PROSEM_T$ ± 2 cm) and warm enough (TAIR above $TDMIN_P$ for several days to allow significant growth: $NBJSEUILTEMPREF_T$) to avoid germination delays or failure of plant emergence (see § 2.2.1)

• The risk of freezing must be low: TMIN above $TDEBGEL_P$ for $NBJSEUILTEMPREF_T$ days

• The soil must be dry enough to avoid compaction risks: the soil water status is considered as damaging if it is above $HUMSEUILTASSSEM_T$ x HUCC in the zone between the surface and $PROFHUMSEM_T$ (see § 6.5.2).

6.1.3 Yield regulation

Yield regulation is generally used for high value production such as tomatoes or grapevines. It can be done either by foliage regulation or by fruit removal.

6.1.3.a Foliage regulation by topping or leaf removal

If the plant exhibits indeterminate growth, a trellis system may be required, which can be simulated by imposing a maximal height and width: $HAUTMAXTEC_T$ and $LARTEC_T$.

Topping only concerns crops having a row structure and consists in restricting growth in terms of height ($HAUTROGNE_T$) and width ($LARGROGNE_T$) of the structure. In order to ensure the efficiency of this technique, a minimum topped shoot biomass threshold must be observed ($BIOROGNEM_T$). The topped biomass and the corresponding LAI are subtracted from the biomass and LAI of the plant. The calculation of this topped

LAI (LAIROGNECUM) and biomass relies on the foliage density DFOL (eq. 3.24), the specific surface area of biomass, SBV (eq. 6.1) using the variable SLA (eq. 3.15):

eq. 6.1

$$SBV(I) = \frac{SLA(I)}{1 + TIGEFEUILLE_P}$$

Topped biomass is recycled in the soil nitrogen balance. Two topping calculations may be employed. With automatic calculation, topping occurs as soon as the plant height exceeds HAUTROGNE$_T$+MARGEROGNE$_T$. The other possible calculation is done at an imposed date, JULROGNE$_T$.

Unlike topping, which is characterised as a function of the canopy geometry, leaf removal (LAIEFFCUM) is expressed directly by reducing the leaf area index, also according to two possible methods. With the automatic calculation, a constant proportion (EFFEUIL$_T$) of the new foliage generated each day $EFFEUIL_T \cdot DELTAI(I)$ is removed as soon as the LAI reaches a threshold value (LAIDEBEFF$_T$). The other possible calculation is done on only one occasion, on day JULEFFEUIL$_T$ and the quantity LAIEFFEUIL$_T$ is removed. The corresponding biomass is calculated from the specific leaf area (SLA) and deducted from the biomass of the plant. Another option concerns the location of leaf removal: the top or bottom of the canopy, which affects the radiation and water balances of crops in rows.

6.1.3.b Fruit removal

Fruit removal occurs on day JULECLAIR$_T$ and the prescribed parameter is the number of fruits or inflorescences removed per plant (NBINFLOECL$_T$). For mono-inflorescence plants the removed fruits are the younger ones (taken from the first "boxes") while for multi-inflorescence plants, the removed fruits are taken from the "boxes" (see §4.2).

6.1.4 Harvest

6.1.4.a Harvest policy

There are two methods of harvest for both types of plant: either cutting (the entire plant is cut and removed or incorporated into the soil) or picking (only the fruits are picked). There may be several cuts (e.g. forage crops) or pickings (e.g. fruit crops with a spread of maturity).

6.1.4.b The particular case of forage crops

Forage crops can be cut using one of the three following methods.

• With automatic calculation, as soon as the crop reaches the stage defined by the STADECOUPEDF$_T$ parameter, it is cut at the cutting height corresponding to HAUTCOUPEDEFAUT$_T$, transformed into biomass using the COEFMSHAUT$_P$ conversion parameter.

• With imposed dates, a table of different cutting dates is entered, associated with the following elements: HAUTCOUPE$_T$ (cutting height) or LAIRESIDUEL$_T$ and

MSRESIDUEL$_T$ and ANITCOUPE$_T$ (LAI, biomass and fertilisation at cutting respectively). If these data cannot be supplied, they are calculated based on the cutting height using the COEFMSHAUT$_P$ conversion coefficient and the height/LAI ratio using eq. 3.17.

• A similar calculation can be made at imposed physiological dates, the cutting dates being defined by cumulative development units.

6.1.4.c The particular case of protracted picking

Protracted maturity occurs for indeterminate crops with a long period of fruit setting (parameter STDRPNOU$_P$) and can lead to a spread of harvest (e.g. tomatoes). The first harvest starts at physiological maturity of the first fruit cohort (passage into the last box) and the data used for summary outputs are those of the last ripe fruits. The number of cuttings and the spread of the harvest depend on the rate of picking (CADENCEREC$_T$) as a number of days between two successive pickings. If the rate is too rapid with respect to the rate of fruit growth, then the harvest is delayed until ripe fruits appear again (in other words, fill the last box of the growth-development period of fruits).

6.1.4.d Simulation of the decision to harvest

The decision to harvest can be taken as a function of crop maturity status but it can also rely on other considerations such as soil water status, sanitary or even economic considerations.

The crop maturity-dependent harvest date depends on one of the following criteria:
– physiological maturity (end of growth-development period)
– maximum water content in fruit which exhibit dehydration dynamics (H2OGRAINMAX$_T$) or minimum water content in fruit that exhibit hydration dynamics (H2OGRAINMIN$_T$) from the onset of water dynamics (IDEBDES stage)
– minimum sugar content in fruit (SUCREREC$_T$)
– minimum nitrogen content in fruit (CNGRAINREC$_T$)
– minimum oil content in fruit (HUILREC$_T$)

If the soil is too wet at this date, it is possible to postpone harvest to avoid compaction. In that case a period (in number of days) after the crop-dependent harvest date is defined (NBJMAXAPRESRECOLTE$_T$) during which the average soil water over the depth affected by the harvesting machinery (PROFHUMREC$_T$) is tested. This soil water status is considered as damaging if it is above HUMSEUILTASSREC$_T$ x HUCC in the zone between the surface and PROFHUMREC$_T$ (see § 6.5.2). Yet this delay cannot exceed IRECBUTOIR$_T$ which is the latest date for harvesting. The reasons for IRECBUTOIR$_T$ are various: risks of sanitary problems, necessity to free the field for the following crop or economical constraints.

6.1.5 Pruning

Winter pruning is used for perennial woody crops like grapevine. On the prescribed day of winter pruning (JULTAILLE$_T$), the structural mass of stems plus the mass of leaves still on the plant are allocated to the MABOIS variable and removed from the plant so that the following cycle starts with the reserves only.

In the model there is no relationship between pruning and the inflorescence or fruit number that develop the following spring ($NBINFLO_P$ or $AFRUITSP_V$), which are predicted independently. In reality pruning is also a technique to regulate yield through the number of remaining buds.

6.2 Soil water supply

The quantity of water reaching the soil is attributable to rain or irrigation, after passage through vegetation and losses by surface runoff. Rain plus irrigation which penetrates into the soil is called PRECIP.

6.2.1 Irrigation

The amounts of water applied can be entered from an irrigation calendar or calculated by the model.

In the latter case, the model automatically calculates water inputs so as to satisfy water requirements at the level of the $RATIOL_T$ parameter: the model triggers irrigation each time the stomatal stress index (SWFAC) is less than $RATIOL_T$. Irrigation amounts (AIRG) are then calculated so as to replenish the soil water reserve (HUR) to field capacity (HUCC) down to the rooting front (ZRAC) without exceeding the maximum dose authorised by the irrigation system ($DOSIMX_T$). At the time of sowing, whatever the soil reserve status, a fixed value of about 20 mm ($IRRLEV_G$ parameter) is supplied to the crop if it has not rained, to enable germination.

eq. 6.2

$$\text{if } SWFAC(I) < RATIOL_T, \; AIRG(I) = EFFIRR_T \sum_{IZ=1}^{ZRAC(I)} HUCC(IZ) - HUR(IZ)$$

$$\text{if } AIRG(I) > DOSIMX_T, \; AIRG(I) = DOSIMX_T$$

Depending on the irrigation system used, water may be applied above or below the foliage or in the soil at a given depth ($LOCIRRIG_T$) intended to mimic drip irrigation. In the case of irrigation below the foliage, water supply is not submitted to the mechanism of rainfall interception by the foliage. In the case of underground irrigation, water supply is also withdrawn from the soil evaporation calculation.

$EFFIRR_T$ is a proportion parameter standing for irrigation efficiency, which makes it possible to empirically account for water losses during irrigation. It is applied as a multiplier to each irrigation amount.

6.2.2 Interception of water by foliage

Interception of water by foliage concerns rainfall (TRR) and irrigation above foliage (AIRG) systems: if irrigation water is provided by drip irrigation or micro-irrigation under the plant canopy, this mechanism does not occur. The persistence of water on the foliage, directly subjected to the evaporative demand of the surrounding atmosphere, may, as it evaporates, significantly reduce the saturation deficit within the canopy and thus affect the water requirements of the crop. In humid, tropical environments, the frequency of rainfall combined with a high evaporative demand (mainly radiative) means that this phenomenon has a marked effect on the water balance (Brisson *et al.*, 1998b). Similar reasoning can be applied in summer to irrigated crops in temperate and Mediterranean climates.

The importance of runoff down stems, or STEMFLOW needs to be evaluated so as to not overestimate the retention of water on foliage. Based on the work by Bussiere (1995), stemflow is considered as a priority and is estimated in proportion to incident rainfall (TRR+AIRG) with a maximum given by STEMFLOWMAX$_p$, as an increasing function of leaf area index (eq. 6.3).

eq. 6.3

$$STEMFLOW(I) = STEMFLOWMAX_P$$
$$[1 - \exp(-KSTEMFLOW_P \cdot LAI(I))] \cdot [TRR(I) + AIRG(I)]$$

The STEMFLOWMAX$_p$ parameter may vary from 0.2 to 0.5, depending on species. The KSTEMFLOW$_p$ parameter is less well known: it can initially be taken to equal the solar radiation extinction coefficient (EXTIN$_p$).

Water which does not flow away via stemflow is partly retained on the foliage (MOUILL), up to a maximum value which is proportional to the LAI. The parameter for the proportionality, or leaves wettability, is called MOUILLABIL$_p$ (in mm LAI^{-1}). It depends on leaf surface properties: shape, texture, pilosity. It is available either by direct measurement or indirectly by solving the water balance equation (examples of values are given in Table 6.1). This water is then evaporated like free water (flux EMPD explained in § 7.2).

Table 6.1. Values of wettability for various plants.

Plant	Forage grass	Maize	Sorghum	Gliricidia	Banana
Indirect estimate	0.27	0.27	0.28	0.23	0.68
Direct measurement	–	–	–	0.17 ($\pm$ 0.03)	–

6.3 Net nitrogen supply

The inorganic N pool in soil can be replenished by the addition of synthetic fertilizers (called 'mineral fertilizers'), by organic fertilizers which contain significant amounts of mineral N (for example: pig slurry, distillery vinasse, etc.), by rainfall or irrigation water.

The N inputs derived from rain and irrigation are summated in the variables AMMSURF (inputs of NH_4^+-N) and PRECIPN (inputs of NO_3^--N). The N inputs derived from mineral fertilizers (NH_4^+ + NO_3^-)-N and from the inorganic fraction of organic fertilizers are summated in the variable TOTAPN.

6.3.1 N inputs from rain and irrigation

The N inputs by rainfall (PLUIEN, in kg ha^{-1}) are the product of the amount of rainfall (TRR, in mm) and its mean concentration in mineral N (CONCRR$_G$, in kg ha^{-1} mm^{-1}). A mean concentration of 1 mg L^{-1} corresponds to 0.01 kg ha^{-1} mm^{-1}. The N input from rainfall occurs at the soil surface and is assumed to consist in 50% of NH$_4^+$ and 50% of NO$_3^-$.

The N inputs due to irrigation water (IRRIGN, in kg ha^{-1}) are also the product of the amounts of water (AIRG, in mm) and its mean concentration, defined in the technical file (CONCIRR$_T$, in kg ha^{-1} mm^{-1}). The N input is located either at the soil surface or at the depth LOCIRRIG$_T$ if the option 'localised irrigation' is activated (CODLOCIRRIG$_T$ = 3). The mineral N in the irrigation water is assumed to be exclusively in the form of NO$_3^-$.

6.3.2 N inputs from mineral fertilisers

The N inputs from mineral fertilizers can be applied either at the soil surface or at a given depth (LOCFERTI$_T$) if the option 'localized fertilization' is activated (CODLOCFERTI$_T$ = 2).

We consider 8 different types of mineral fertilizers. As a simplification, urea is treated as an ammonium fertilizer since its hydrolysis to ammonium carbonate is a very fast process (e.g. Recous et al, 1988; Hah, 2000). The main characteristics of these fertilizers are given in Table 11.12. The fraction of ammonium (or ammonium formed from urea) contained in the fertilizer (ENGAMM$_T$) is used when the option 'nitrification' is activated (in this case, the NH$_4^+$ and NO$_3^-$ forms are distinguished; CODENITRIF$_G$ = 1), which is justified in acid soils (pH < 5.5). The other variables are defined in the following paragraph.

6.3.2.a Nitrogen use efficiency

The (potential) nitrogen use efficiency (EFFN), i.e. the fraction of fertilizer N available for plant uptake, can be either imposed or calculated by the model. If EFFN is fixed, the mineral fertilizer type 8 must be chosen and its nitrogen use efficiency must be defined in the general parameter file. Part of the fertilizer is considered to be unavailable for the plant because it is either immobilized in soil by microbial activity, denitrified or volatilized. The efficiency EFFN is the complement of these 'losses' to 1. It must be noticed that nitrate leaching is not included in these losses since it is simulated directly by the nitrate transfer module.

The nitrogen use efficiency can be measured either by the difference in plant uptake between a fertilized and an unfertilized treatment, relative to the fertilizer rate, or by the ^{15}N method (which gives the recovery of a ^{15}N-labelled fertilizer in the crop directly). The first method often gives higher values than the second; the difference is mainly attributed to substitution effects occurring between soil and fertilizer-N (Recous *et al.*, 1997). In the STICS model the efficiency is intermediate between the two methods because it considers all sources of losses except fertilizer leaching.

The calculation of losses is based on the concept of competition between the soil and the crop. Indeed Limaux *et al.* (1999) have shown that the nitrogen use efficiency depends on the crop growth rate at the time of fertilizer application. The greater the growth rate, the higher is the N use efficiency. Since nitrate leaching from fertilizer is usually negligible, the higher efficiency is attributed to smaller gaseous losses (denitrification and volatilization) from the fertilizer.

In STICS these losses are assumed to depend on nitrogen uptake rate immediately before fertilizer application (VABSMOY, in kg N ha^{-1} day^{-1}). The parameters DENENG$_G$ (f) and VOLENG$_G$ (f) characterize the maximal amounts of N losses for each fertilizer type 'f', by denitrification and volatilization, respectively (Table 11.12). The potential gaseous losses (denitrification and volatilization) are assumed to be proportional to the N fertilizer rate ANIT (kg N ha^{-1}). The actual losses depend on the nitrogen uptake rate (VABSMOY) recorded in the five days before fertilizer application, through a hyperbolic relationship. The N loss through denitrification (N$_2$+N$_2$O) is NDENENG (eq. 6.4).

eq. 6.4

$$NDENENG(I) = DENENG_G(f) \cdot \frac{VABS2_G}{VABS2_G + VABSMOY(I)} \cdot ANIT(I)$$

The parameter VABS2$_G$ corresponds to the crop uptake rate (kg N ha^{-1} day^{-1}) at which losses reach 50% of their maximum.

The N loss through NH$_3$ volatilization (NVOLENG) is calculated similarly, but it also depends on soil pH: it increases linearly when the pH (PH$_S$) increases from PHMINVOL$_G$ to PHMAXVOL$_G$ (eq. 6.5 and eq. 6.6):

eq. 6.5

$$NVOLENG(I) = VOLENG_G(f) \cdot \frac{VABS2_G}{VABS2_G + VABSMOY(I)} \cdot ANIT(I) \cdot FPH$$

with

eq. 6.6

$$\begin{aligned}
&if\ PH_S < PHMINVOL_G \quad FPH = 0.0 \\
&if\ PH_S > PHMAXVOL_G \quad FPH = 1.0 \\
&if\ PHMINVOL_G \leq PH_S \leq PHMAXVOL_G \quad FPH = \frac{PH_S - PHMINVOL_G}{PHMAXVOL_G - PHMINVOL_G}
\end{aligned}$$

Concerning N immobilization, studies made with ^{15}N-labelled fertilizers have shown that the microbial immobilization of N derived from fertilizer depends mainly on the N rate and the type of fertilizer (Powlson *et al.*, 1986; Bronson *et al.*, 1991; Recous *et al.*, 1992; Recous and Machet, 1999; Limaux *et al.*, 1999). Using these published data, we have derived a quadratic relationship between the amount of N immobilized (NORGENG, in kg N ha^{-1}) and the fertilizer N rate (eq. 6.7):

eq. 6.7

$$NORGENG(I) = \frac{ORGENG_G(f)}{XORGMAX_G{}^2} ANIT(I)(2\,XORGMAX_G - ANIT(I))$$

$$if\ ANIT(I) \leq XORGMAX_G$$

$$NORGENG(I) = ORGENG_G(f)$$

$$if\ ANIT(I) \geq XORGMAX_G$$

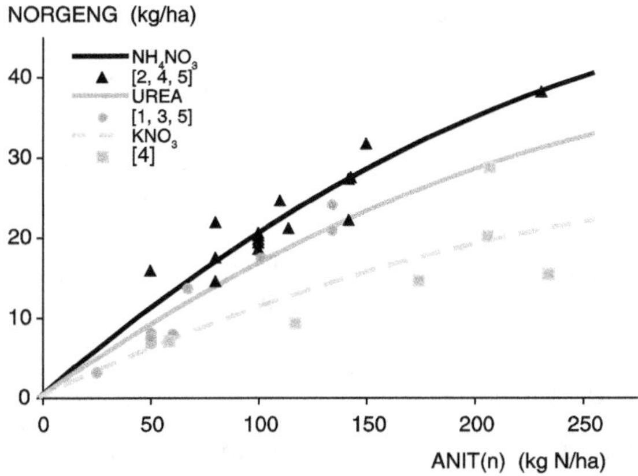

Figure 6.1. Relationship between N immobilised in soil at the expense of N fertilizer and the amount of fertilizer-N added, for three types of ^{15}N labelled mineral fertilizers. References: [1] Recous & Machet (1999); [2] Limaux *et al.* (1999); [3] Bronson *et al.* (1991); [4] Powlson *et al.* (1992); [5] Recous *et al.* (1992).

The parameter $ORGENG_G$ (f) represents the maximal amount of microbial immobilized N from the fertilizer type f and $XORGMAX_G$ is the N rate at which this maximum is reached. Both are expressed in kg N ha^{-1} (Table 11.12).

An example of overall N balance predicted by the model versus N uptake rate is shown in Figure 6.2.

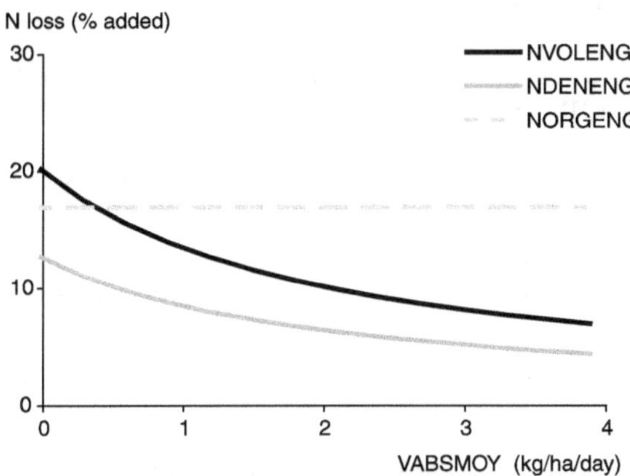

Figure 6.2. Predicted fate of fertilizer-N in the soil-plant-atmosphere system versus crop uptake rate at the time of fertilizer application. The example applies to UAN fertilizer added at the rate of 100 kg N ha^{-1} in a soil with a pH of 7.5.

The fertilizer N losses through immobilization and volatilisation are always calculated as indicated above. However the N losses through denitrification (from soil and fertilizer) can be calculated more mechanistically by activating the option $CODEDENIT_G$. In this case, denitrification is calculated according to NEMIS model (see § 8.5).

Finally, it is also possible to impose a fixed efficiency by choosing fertilizer type 8 (Table 11.12). In that case, the microbial immobilization, the volatilization and the denitrification are fixed and expressed in % of fertilizer-N. The efficiency is the complement of these values to 1.

Whatever the chosen options, the final N use efficiency is calculated from eq. 6.18:

eq. 6.8

$$EFFN(I) = 1 - \frac{NVOLENG(I) + NDENENG(I) + NORGENG(I)}{ANIT(I)}$$

6.3.2.b Fertilisation calendar

Similarly to water applications, fertilizer N applications can be either prescribed (option generally used) or calculated by the model.

• Using the prescribed fertilization option, the model accounts for the fertilization calendar given by the user as a technical input: date of application, N rate and type of fertilizer.

• Using the calculated fertilization option; the model calculates the N applications required to maintain the nitrogen nutrition index (INN) above a given threshold (RATIOLN$_T$). Two other conditions must be fulfilled:

1. The N uptake by the crop needs to be a limiting factor, i.e. the soil supply (CUMOFFRN) must be lower than the plant demand (DEMANDE). This condition is essential because INN characterizes a plant status which can remain N deficient for a long time even though root uptake is maximal. It is no use applying N that the plant cannot absorb when its uptake rate is maximal.

2. The soil must be wet enough in order to allow water and nitrate transport towards the roots. Two technical options are proposed to fulfil this condition: either a test on rainfall (PRECIP > PLNMIN$_G$,) or a test on water availability in the upper soil layer (HUR (1) ≥ HUCC (1)).

Since INN can be greater than 1, the threshold RATIOLN$_T$ can be set at a high value, for example 1.4 or 1.8, in order to mimic early applications of fertilizer which occur in favourable conditions for plant uptake. The calculated N rate (eq. 6.19) is the difference between the maximal amount of N in the crop (QNPLMAX, calculated from the maximal dilution curve, see § 8.6) and the actual amount of N, divided by the N use efficiency EFFN. It is limited by a maximal N rate (DOSIMXN$_T$).

eq. 6.9

$$ANIT(I) = \frac{QNPLMAX(I) - QNPLANTE(I)}{EFFN(I)}$$

$$if \quad ANIT(I) > DOSIMXN_T \quad then \quad ANIT(I) = DOSIMXN_T$$

6.3.3 N inputs from organic residues

The N inputs from organic residues arrive onto the soil either under mineral form (mainly as NH_4^+) or under organic form. The mineral fraction enters the soil mineral pool and is submitted to NH_3 volatilization, nitrification, leaching and plant uptake. The organic fraction decomposes more or less rapidly and mineralizes its C and N according to the decomposition module (see § 8.2). The module is generic and can simulate most types of organic residues. Eight categories are considered: 1) mature crop residues (straw, roots), 2) catch crop residues (young plants), 3) farmyard manures, 4) composts, 5) sewage sludges, 6) distillery vinasses, 7) animal horn and 8) others.

The net mineralization (positive or negative) due to the addition of these residues depends on the category and the C/N ratio of the residue. The characteristics of each organic residue are defined in the technical file: category, depth of incorporation in soil, amount of fresh matter added, carbon content, C/N ratio, water content and mineral N content. Default values are proposed (see Table 11.13).

In the case of chained simulations (see § 10.2), the characteristics of the crop residues returning to the soil are simulated by the model and are taken into account automatically in the next simulation (see § 6.3.4).

Leaves falling onto the soil (the proportion of senescent leaves falling is $ABSCISSION_p$) during crop growth are taken into account by the model as this phenomenon can be important (e.g. rapeseed due to winter frost). Their decomposition at the soil surface is simulated by the decomposition module (category 2, residues of young plants). The C/N ratio of leaves when they fall off is calculated from the nitrogen nutrition index of the whole crop (eq. 6.20) and relies on a plant parameter ($PARAZOMORTE_p$), as proposed by Dorsainvil (2002):

eq. 6.10

$$CNRESIDU(I) = \frac{PARAZOFMORTE_P}{INN(I)}$$

Organic matter decomposition is also affected by the soil tillage operations. The effects of tillage are twofold: i) mixing the newly added organic residues and remixing the previous ones which are decomposing; ii) modifying the environmental conditions of decomposition: temperature, soil water content and particularly mineral N availability, which may have a feedback effect on decomposition.

6.3.4 Crop residues for the following crop

The calculation of crop residues returning to the soil for the following crop, in terms of quantity (QRESSUITE) and quality (CSURNRESSUITE), relies on the parameter $RESSUITE_T$, defining four possible management practices according to the plant fraction remaining in the field and then incorporated into the soil:

• roots ($RESSUITE_T$="RACINES") when harvesting, for example, lettuce or textile flax,

• straw and fine roots ($RESSUITE_T$="PAILLES") when harvesting cereal grains or sugar-beet taproots or potatoes,

• stubble and fine roots ($RESSUITE_T$="CHAUMES") when harvesting cereal grains and straw together or silage maize or cutting meadow,

• whole crop ($RESSUITE_T$="CULTURE") corresponding to catch crops, green manure or crop volunteers.

The mineralization parameters (see Table 11.13) correspond to the first type of residue ($CODERES_T$ =1) except for the last practice ($CODERES_T$ =2). Plant residues are assumed to remain at the soil surface until being buried by the following soil tillage, except for pure roots which are assumed to be located between the surface and the $PROFHUM_S$ depth.

In all cases the fine root biomass (MSRAC) calculation is required, which is done differently according to the option chosen for root profile (see § 5). For the standard profile (see § 5.2.1), root biomass is assumed to be a fixed proportion of shoot biomass (parameter $PROPRAC_G$ affected to 0.20 by default) in eq. 6.11.

eq. 6.11

$$MSRAC(I) = MASEC(I) \cdot PROPRAC_G$$

For the true density profile (see § 5.2.2), root biomass is calculated according to the total root length (living and dead) in eq. 6.12:

eq. 6.12

$$MSRAC(I) = 100 \cdot \frac{RLTOT(I) + LRACSENTOT(I)}{LONGSPERAC_G}$$

The quantities of residues (QRESSUITE, in t DM ha^{-1}) left on the soil at harvest are calculated according to eq. 6.13:

eq. 6.13

$$\textit{if RESSUITE} = "RACINES" \quad \textit{then} \quad QRESSUITE(I) = MSRAC(I)$$
$$\textit{if RESSUITE} = "PAILLES" \quad \textit{then} \quad QRESSUITE(I) =$$
$$MASEC(I) - MAGRAIN(I) + MSRAC(I)$$
$$\textit{if RESSUITE} \quad "CHAUMES" \quad \textit{then} \quad QRESSUITE(I) =$$
$$0.35 \left(MASEC(I) - MAGRAIN(I) \right) + MSRAC(I)$$
$$\textit{if RESSUITE} = "CULTURE" \quad \textit{then} \quad QRESSUITE(I) = MASEC(I) + MSRAC(I)$$

The nitrogen content of the returned residues is CNPLANTE for the whole plant (option RESSUITE='CULTURE') and CNPAILLRAC for the other options (eq. 6.14):

eq. 6.14

$$CNPAILLRAC(I) = \frac{QNPLANTE(I) - QNGRAIN(I)}{MASEC(I) - MAGRAIN(I) + MSRAC(I)}$$

6.4 Physical soil surface conditions

This section is devoted to the characterisation of the soil surface conditions in order to predict their effects on the water and heat balances of the soil-crop system. Those effects will be integrated to the calculations of water requirements, water and heat transfers in the § 7 and 9. However in order to make it easier to understand the formalisations used in these processes, we explain below the modifications induced by taking account of soil surface conditions.

Soil surface conditions are characterised by soil and technical parameters. One of them is the soil albedo under dry conditions ($ALBEDO_S$). Another is the run-off coefficient (RUISOLNU$_S$), giving the proportion of rainfall submitted to run-off which occurs when the soil is bare and when rainfall exceeds a given threshold (PMINRUIS$_G$). These parameters summarize the effects of soil slope and roughness on surface run-off, which are assumed to be constant throughout the simulation. The model also needs to know whether a plant or plastic mulch is present. In the case of plant cover, the user is requested to specify the amount of plant mulch supplied (QMULCH0$_T$), the day it was applied (JULAPPLMULCH$_T$) and the type of mulch (1 = maize stalks, 2= others which have not yet been parameterized). This mulch typology refers to a set of general parameters defining the mulch water retention and decomposition dynamics. In the case of a plastic cover, the user is requested to specify the albedo of the plastic cover, which is related to its colour (ALBEDOMULCH$_G$), and the proportion of soil cover (COUVERMULCH$_T$).

There are six processes to model, the first three of which are devoted to plant mulching, which is the theme of this section:

1. the dynamics of plant mulch and proportion of soil cover,

2. the modification of surface run-off due to the presence of obstacles located on the soil surface,

3. the water interception by the mulch and its direct evaporation (in relation to the energy balance calculations in § 6.6.1),

4. the decrease in soil evaporation induced by the presence of mulch (see § 7.1),

5. the effects of these modified fluxes on the plant's water requirements (see § 7.2) ,

6. the modifications to crop temperature linked to changes in the fluxes and albedo of the soil surface (see § 6.6.1).

In order to give an order of magnitude to the above-mentioned effects of mulch, described in detail later, an example is given in Table 6.2. The increase in mineralization is simply due to the increase in soil temperature.

Table 6.2. Simulated effects of the presence of a mulch on the various processes involved in the water and nitrogen balances, and their consequences on yield. The case study is a sugar cane crop growing in Guadeloupe on a vertisol (1330 mm of rainfall during the season) with the maize mulch parameters proposed by Scopel *et al.* (1998).

	Bare soil	Plant mulch 0.5 t/ha	Plant mulch 5 t/ha	Black plastic mulch
Yield (t/ha)	25	35	40	31
Plant transpiration (mm)	540	839	967	800
Soil evaporation (mm)	382	317	171	99
Mulch evaporation (mm)	0	14	135	0
Drainage (mm)	98	120	212	108
Surface run-off (mm)	492	217	25	492
Mineralisation (kg N/ha)	139	171	182	172

The link between the physical role of the plant mulch accounted for here and its chemical role of carbon and nitrogen mineralisation is not programmed yet. To account for this chemical role, it is essential to also consider the mulch as a crop residue left on the soil surface, and to define its chemical composition with appropriate parameters (see § 6.3.3).

6.4.1 Quantity of plant mulch and proportion of soil cover

Mulch decomposition dynamics relies on the work by Scopel *et al.* (1998) and is a decreasing function of time (eq. 6.15) using the decomposition parameter $DECOMPOSMULCH_G$ (in day^{-1}), one of the typological mulch parameters.

eq. 6.15

$$QMULCH(I) = QMULCH0_T \quad e^{-DECOMPOSMULCH_G\,(I-JULAPPMULCH)}$$

The proportion of soil covered by mulch (COUVERMULCH) is also exponentially related to the quantity of mulch (QMULCH, in t ha^{-1}), using the parameter $KCOUVMLCH_G$ (eq. 6.16). Scopel *et al.* (1998) gave $DECOMPOSMULCH_G$=7 10^{-3} day^{-1} and $KCOUVMULCH_G$ ranging from 0.092 to 0.367, depending on the type of plant residue (entire plant, fresh or decomposed, stalks). This parameterization indicates that the type of plant residue affects both the proportion of soil cover and the rate of decomposition (Figure 6.3). As for plastic mulching, COUVERMULCH is constant and treated as a technical parameter.

eq. 6.16

$$\text{For plant mulching} \quad COUVERMULCH(I) = 1 - e^{-KCOUVMLCH_G \times QMULCH(I)}$$
$$\text{For plastic mulching} \quad COUVERMULCH(I) = COUVERMULCH_T$$

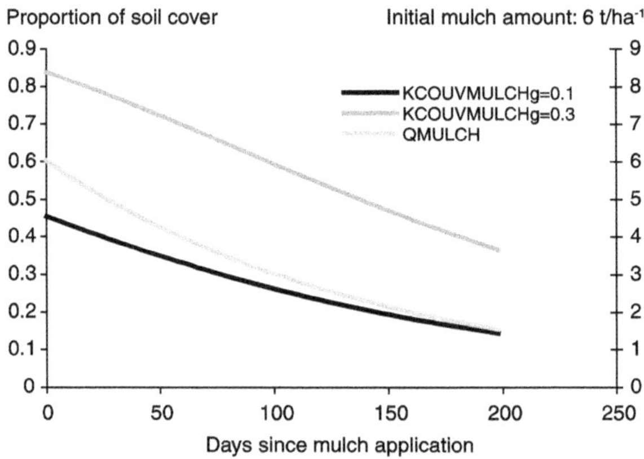

Figure 6.3. Variation in the proportion of the soil covered by a plant mulch, whose quantity (QMULCH in t DM ha^{-1}) decreases as a function of the type of the crop residue given by the parameter $KCOUVMULCH_G$ (a high value for entire fresh plants and a low value for cut stalks)

6.4.2 Surface run-off

We separate "surface" run-off associated with soil surface conditions (RUISSELSURF) and the run-off associated with a lack of soil infiltrability; the latter is simulated by the water and nitrogen transfer (see § 9.2). We calculate the FRUIS variable (eq. 6.18), which is the proportion of run-off water above the activation threshold (PMINRUIS$_G$) as given in eq. 6.17.

108

$$\text{eq. 6.17}$$

$$\text{if} \quad PRECIP(I) - STEMFLOW(I) > PMINRUIS_G$$
$$RUISSELSURF(I) = FRUIS(I)[PRECIP(I) - STEMFLOW(I) - PMINRUIS_G]$$
$$\text{if} \quad PRECIP(I) - STEMFLOW(I) \leq PMINRUIS_G$$
$$RUISSELSURF(I) = 0$$

For values between QMULCH = 0.1 and QMULCH=QMULCHRUIS0$_G$ (another typological mulch parameter), we use the relationship established by Scopel *et al.* (1998) to calculate FRUIS: above QMULCHRUIS0$_G$, FRUIS is zero, and below 0.1 we take the value of RUISOLNU$_S$ (see Figure 6.4).

$$\text{eq. 6.18}$$

$$\text{if} \quad QMULCH(I) < 0.1 \quad FRUIS(I) = RUISOLNU_S$$
$$\text{if} \ 0.1 < QMULCH(I) < QMULCHRUIS0_G \ \ FRUIS(I) = 0.33[QMULCHRUIS0_G - QMULCH(I)]$$
$$\text{if} \quad QMULCH(I) > QMULCHRUIS0_G \ \ FRUIS(I) = 0$$

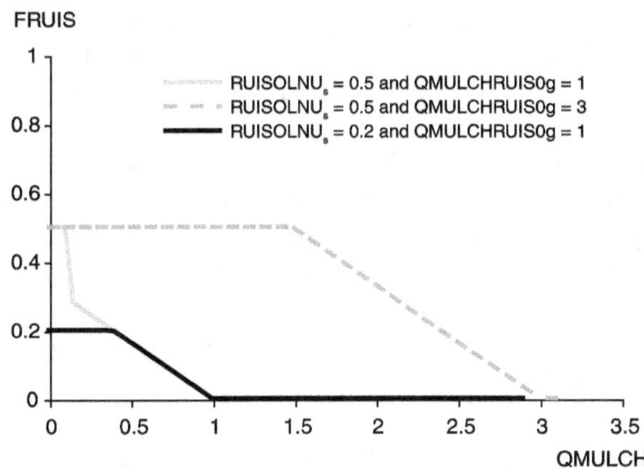

Figure 6.4. Proportion of runoff water as a function of the soil surface (RUISOLNU$_S$) and the plant mulch (QMULCHRUIS0 and QMULCH)

The effect of the presence of vegetation above the soil (LAI) is taken into account via mechanisms for the flow of water along stems (STEMFLOW) as the FRUIS proportion only applies to the amount of water not involved in STEMFLOW.

6.4.3 Modification to water balance induced by the mulch

6.4.3.a Interception of water by mulch

The maximum water reserve of the plant mulch (MOUILLMULCH) is defined (eq. 6.19) as being proportional to its quantity (QMULCH), involving the mulch-dependent parameter of wettability (MOUILLABILMULCH$_G$) that can range between 0.22 and 0.38 mm t^{-1} ha (Scopel *et al.*, 1998). The amount of water retained is limited by the incident rainfall minus the surface run-off.

eq. 6.19

$$MOUILLMULCH(I) = MOUILLABILMULCH_G \times QMULCH(I)$$

The amount of water directly evaporated from the mulch (EMULCH) can be calculated in two ways, using either the reference evapotranspiration (eq. 6.20) given in the weather input file (TETP intended to be the Penman value) or the raw weather variables including wind speed and air humidity. EMULCH is limited by MOUILLMULCH:

eq. 6.20

$$EMULCH(I) = TETP(I) \times COUVERMULCH(I) \ e^{-(EXTIN_P - 0.2) \times LAI(I)}$$

or if cover rate is used instead of LAI

$$EMULCH(I) = COUVERMULCH(I) \cdot TETP(I) \cdot (1 - TAUXCOUV(I))$$

and

$$EMULCH(I) \leq MOUILLMULCH(I)$$

In the first case, it is assumed that the water contained in mulch evaporates in the same way as from a grass canopy, according to a resistance/height compensation phenomenon. This last concept corresponds to the "extinction of energy at the soil level" by the vegetation (as is the case for soil). If the $EXTIN_p$ parameter is not active (because the radiation intercepted by the canopy is calculated with the radiation transfer model and not by the Beer law approach as explained in § 3.2), the value is recalculated and varies depending on the crop geometry and the quality of radiation.

In the second case the Shuttleworth and Wallace formalisation is applied as explained in § 7.2.2, and EMULCH evaporates in the same way as free water located at the soil level and receiving energy. It takes account of the proportion of soil covered by the mulch (COUVERMULCH).

In both cases EMULCH is limited by the amount of water intercepted by the vegetal mulch, MOUILLMULCH

6.4.3.b Modification to soil evaporation due to the presence of mulch

Incident energy at the soil level under the mulch is linearly related to the proportion of soil covered, which considerably reduces direct soil evaporation. In the option using the reference evapotranspiration as an input, the relationship is directly applied to the potential soil evaporation (EOS) as given in eq. 6.21, while it is applied to the radiation balance in the option using Shuttleworth and Wallace formalisation (see § 7.1).

eq. 6.21

$$EOS(I) = TETP(I) \times (1 - COUVERMULCH(I)) \ e^{-(EXTIN_P - 0.2) \times LAI(I)}$$

6.4.3.c Modification to crop requirements due to the presence of a mulch

If the Shuttleworth and Wallace formalisation is used, EMULCH contributes to a reduction in the saturation deficit (DOS) in the same way as direct soil evaporation and the evaporation of water intercepted by the foliage (see § 7). If the "reference potential evapotranspiration" approach is adopted, EMULCH also reduces the evaporative demand according to an empirical formula given in § 7.2.

6.4.4 Modification of crop and soil temperatures by the presence of a mulch

First the mulch influences the temperature regime through the modification of the soil surface albedo as defined in eq. 6.24, using the parameter $ALBEDOMULCH_G$. Secondly the total evapotranspiration from the soil-plant system (evaporation + transpiration), to which is applied the energy balance, accounts for evaporation from the mulch. Taken together, these two elements modify the crop temperature. This modification is of particular importance in the case of plastic mulch.

6.4.5 Influence of soil crusting on emergence

The presence of a crust at the soil surface can hinder emergence: it is taken into account by decreasing the emerged plant density and increasing the emergence time (see § 2.2.1).

6.5 Soil structure modification

6.5.1 The soil structure in STICS

Only very recently has the soil structure been considered as a possible impermanent soil character in STICS (Richard *et al.*, 2007). The parameters accounting for the structure of each soil layer are the bulk density (DAF_S in g cm^{-3}) and the infiltrability ($INFIL_S$ in mm day^{-1}) at the base of the layer (see Table 11.7). We also define the structural porosity (or macroporosity) as the complement of textural porosity (or microporosity) in the total porosity, assuming that field capacity defines the microporosity. If the assumption of the invariability of structure parameters in deep horizons is relevant, it is not the same for those layers whose structure is affected by ploughing, compaction by the farmers' machines or weather.

Table 6.3. Values of parameters linked to the bulk density of the ploughed layer of soils of the Paris Basin, established from the data taken from Mumen (2006) and Viloingt (2005).

DAS in en g cm^{-3}	INFILS in mm d^{-1}	ZESX$_S$ in cm	Q0$_S$ in mm	FMIN1$_G$ in %
1.1	5.2	19	5.4	100
1.2	4.5	–	–	–
1.3	3.5	–	–	–
1.4	2.8	–	–	–
1.5	2.0	–	–	–
1.6	1.3	35	1.5	80

Moreover it was demonstrated for a loamy soil of the Parisian Basin that other parameters are not independent from the structural parameters (Richard *et al.*, 2007). This is the case for parameters driving evaporation: $ZESX_S$ and $Q0_S$ and for the potential mineralization rate $FMIN1_G$ (Table 6.3).

6.5.2 Compaction as influenced by sowing and harvesting machines

It was assumed that the machines likely to cause severe compaction are only those involved in sowing and harvesting operations. The parameters involved, for sowing and harvesting separately, are the average soil water content above which compaction occurs ($HUMSEUILTASSSEM_T$ and $HUMSEUILTASSREC_T$ in proportion to the field capacity), the soil depth affected ($PROFHUMSEM_T$ and $PROFHUMREC_T$ in cm) and the resulting bulk density ($DASEM_T$ and $DAREC_T$ in g cm^{-3}). The maximum effect of compaction is in the two top soil layers. The relationships between these parameters and the nature of the soil machinery could be linked in the future with more mechanistic knowledge of soil mechanics (Defossez *et al.*, 2003).

In the model, compaction results in an increase in bulk density, a decrease in layer thickness and a decrease in their infiltrability. This last effect relies on the data given in Table 6.3. In the absence of more consolidated data, the effects on $ZESX_S$ and $Q0_S$ were not introduced. Also the soil surface roughness ($Z0SOLNU_S$) is assumed not to be affected.

The modification of the soil geometry has repercussions on water and nitrogen profiles: a conservation of the intra-layer amounts is assumed with uniform partitioning within each layer.

6.5.3 Fragmentation under the effects of soil tillage implements

Soil tillage implements, whether or not they invert the soil, fragment it, leading to a decrease in bulk density. Depending on the type of tool, this fragmentation concerns either a superficial layer (e.g. a surface tillage after harvesting) or the whole tilled layer (e.g. a mouldboard plough or a subsoiler). Consequently the soil description in layers should be in agreement with the various soil tillage operations carried out. For each tool, the resulting bulk density and roughness are defined as technical parameters. For instance for a chisel and a plough those parameters are $DACHISEL_T=1.1$ g cm^{-3}, $DALABOUR_T=1.3$ g cm^{-3}, $RUGOCHISEL_T = 0.001$ m and $RUGOLABOUR_T = 0.01$ m. The modification in bulk density affects infiltrability, water and nitrogen profiles, following the rules previously defined. So ploughing tends to increase soil evaporation by increasing its roughness, but the water balance generally remains positive due to the increase in water storage as a consequence of greater infiltrability. The effect of soil tillage on the incorporation of crop residues is described in § 6.3.3.

Of course secondary effects of these management techniques appear on waterlogging, denitrification, nitrate leaching, root growth and water stress, which require careful validation.

6.6 Microclimate

The system microclimate, i.e. its temperature and humidity, drives many processes taking place within the plant canopy: phasic development, photosynthesis, evapotranspiration etc. Moreover it provides the boundary conditions for the calculation of soil temperature and hence influences processes occurring within the soil, such as organic matter mineralization, plant germination etc. Hence the soil colour and dryness, through the soil albedo, can play a significant role on the speed of crop establishment, especially during the spring.

Yet most crop models do not go into these details, using the standard measured weather variables as the driving variables (Brisson *et al.*, 2006). Among the original components of STICS there is the calculation of the temperature and air humidity within the canopy from a daily energy balance, allowing the combined effects of weather and water balance to be accounted for.

The calculations of the energy balance with a daily time step, although questionable physically speaking, have already been done in the framework of an operational estimation of the water requirements of crops (Smith *et al.*, 1996). The daily crop temperature is assumed to be the arithmetic mean of the maximum and minimum crop temperature. Two calculation methods are proposed (depending on the availability of wind and air humidity input data): by using an empirical relationship from Seguin and Itier (1983) or by solving the energy balance. Both methods rely on the calculations of the daily sum of evaporative fluxes and net radiation.

6.6.1 Calculation of net radiation

Net radiation (eq. 6.22) takes account of the surface albedo (ALBEDOLAI) applied to solar radiation (TRG) and long wave radiation (RGLO).

eq. 6.22

$$RNET(I) = (1 - ALBEDOLAI(I)) \, TRG(I) + RGLO(I)$$

6.6.1.a Albedo

The albedo of the surface (ALBEDOLAI) varies between the soil value (ALBSOL) and the vegetation value ($ALBVEG_G$) which is equal to $= 0.23$ (Ritchie, 1985).

eq. 6.23

$$ALBEDOLAI(I) = [ALBVEG_G - (ALBVEG_G - ALBSOL(I))\exp(-0.75 LAI(I))]$$

The soil albedo (ALBSOL) varies as a function of soil type ($ALBEDO_S$ of dry soil), moisture in the surface layer, and the presence of any plastic or plant cover (see § 6.4.4). It decreases linearly with the water content of the surface layer (HUR) according to a relationship established from experimental results obtained for different types of soil (HUCC and HUMIN being the water content at field capacity and wilting point respectively).

113

$$eq.\ 6.24$$

$$ALBSOL(I) = ALBEDO_S \left[1 - 0.517\,\frac{HUR(I,1) - HUMIN(1)}{HUCC(1) - HUMIN(1)}\right]\left[1 - COUVERMULCH(I)\right]$$

$$+ ALBEDOMULCH_G \cdot COUVERMULCH(I)$$

$$and$$

$$if \quad HUR(I,1) \le HUMIN(1) \quad HUR(I,1) = HUMIN(1)$$

$$if \quad HUR(I,1) \ge HUCC(1) \quad HUR(I,1) = HUCC(1)$$

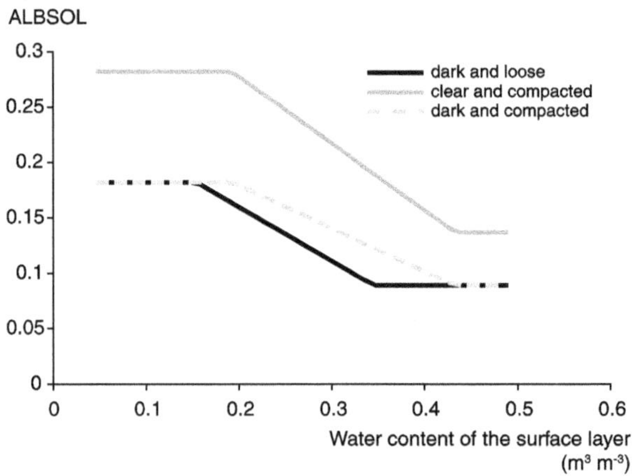

Figure 6.5. Variation in a loam-sandy soil albedo as a function of its surface characteristics: water content, colour (dry albedo of 0.18 for a dark soil or 0.28 for a clear soil) and bulk density (1.2 for a loose soil or 1.5 for a compacted soil).

6.6.1.b Long wave radiation

Two formula are proposed to calculate long wave radiation (RGLO in MJ m^{-2}) based on crop temperature (TCULT in °C), the insolation fraction (FRACINSOL) and the vapour pressure (TPM in mbars). Brunt's formula (1932), given in eq. 6.27, is used in many applications in particular in Penman's potential evapotranspiration formula (1948), while Brutsaert's formula (1982), given in eq. 6.28, is supposed to be more precise (Guyot, 1997). It illustrates clearly the soil and atmospheric components of RGLO using the Stefan-Boltzman law and the emissivity of the atmosphere (EMISSA).

The insolation fraction is estimated using Angström's formula (eq. 6.26), the parameters of which are AANGST$_G$= 0.18 and BANGST$_G$= 0.62. Extraterrestrial radiation (RGEX) is calculated using standard astronomic formulae (Grebet, 1993). If the vapour pressure is not available, it is estimated as the saturated vapour pressure at the temperature TDEW=TMIN − CORRECTROSEE$_C$. The saturated vapour pressure/temperature function (TVAR: eq. 6.25) is represented in Figure 6.6. The order of magnitude of the parameter CORRECTROSEE$_C$ is of a few degrees, from 0 for the wettest locations to 3°C for the driest ones.

eq. 6.25

$$TVAR(TDEW) = 6.1070 \cdot \left(1 + \sqrt{2}\, sin\frac{0.017453293 \cdot TDEW}{3}\right)^{8.827}$$

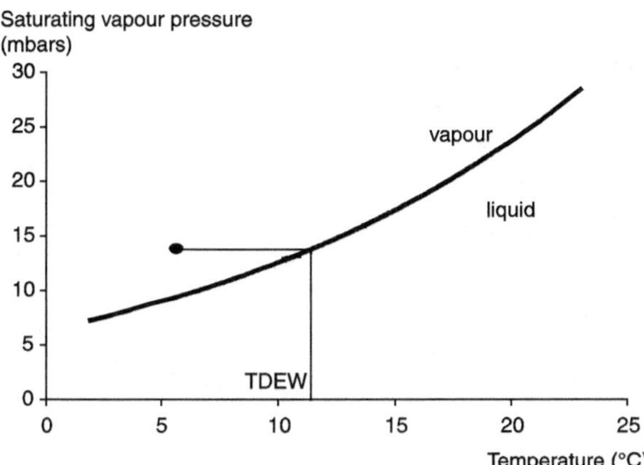

Figure 6.6. Variation in the saturated vapour pressure as a function of temperature according to Alt (1978) referred to by Guyot (1997). The water status in the air is vapour represented by the point and the temperature corresponding to the same pressure on the curve is the dew temperature (TDEW).

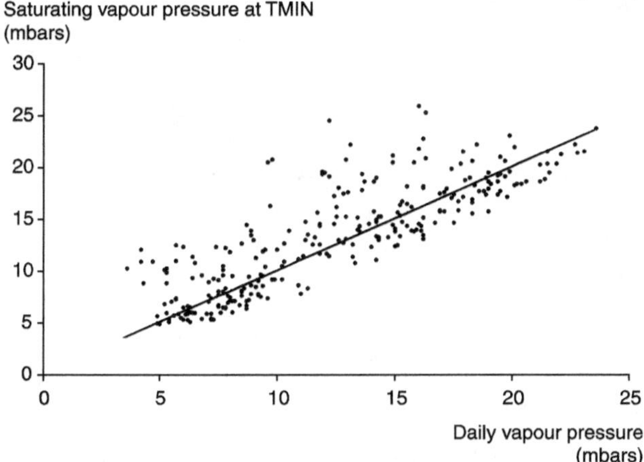

Figure 6.7. Visual evaluation of the estimate of the actual vapour pressure by the hypothesis TDEW=TMIN in Avignon .

eq. 6.26

$$FRACINSOL(I) = \frac{\dfrac{TRG(I)}{RGEX(I)} - AANGST_G}{BANGST_G}$$

eq. 6.27

$$RGLO(I) = -4.9e-9 \quad [TCULT(I)+273.16]^4 [0.1+0.9FRACINSOL(I)]$$
$$[0.56-0.08\sqrt{TPM(I)}]$$

eq. 6.28

$$RGLO(I) = RATM(I) - RSOL(I) = 5.6710^{-8}$$
$$(TCULT(I)+273.15)^4 (1-EMISSA(I))\,3600 \cdot 24 \cdot 10^{-6}$$

and

$$EMISSA(I) = EABRUT(I) + (1-FRACINSOL(I))\,(1-EABRUT(I))\left(1-4\frac{11}{TCULT(I)+273.15}\right)$$

and

$$EABRUT(I) = 1.24\left(\frac{TPM(I)}{(TCULT(I)+273.15)}\right)^{\frac{1}{7}}$$

In both calculations, compared in Figure 6.8, the crop temperature is subjected to an iterative convergence procedure (explained below), meaning that these calculations need to be performed several times in succession.

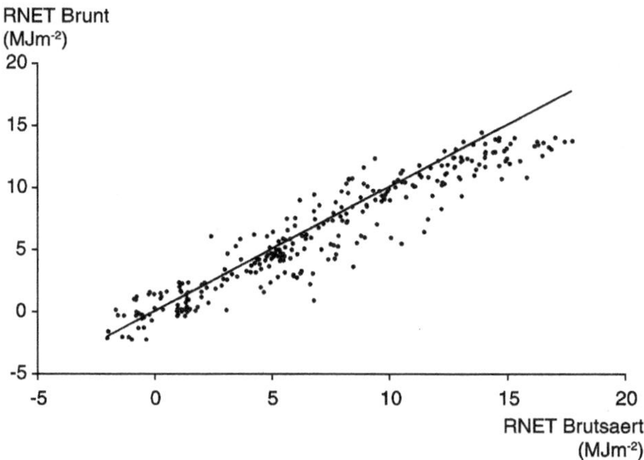

Figure 6.8. Comparison of Brunt's and Brutsaert's formulae for the calculation of net radiation in Avignon.

6.6.2 Calculation of crop temperature

TCULT is assumed to be the arithmetic mean of the maximum crop temperature (TCULTMAX) and the minimum crop temperature (TCULTMIN). Two calculation methods are proposed, depending on the availability of weather data, using either an empirical approach or the energy balance.

6.6.2.a Empirical approach

This method must be used when neither wind speed nor air humidity data are available. It is based on a relationship between midday surface temperature and daily evaporation (Seguin and Itier, 1983), and allows the calculation of TCULTMAX (eq. 6.29) taking in account the parameterization from Riou *et al.* (1988).

eq. 6.29

$$TCULTMAX(I) = TMAX(I) + \left(\frac{RNET(I)}{2.46} - ET(I) - 1.27 \right) \bigg/ \frac{1.68}{\ln \dfrac{1}{Z0(I)}}$$

$$Z0(I) = 0.13 \cdot HAUTEUR(I) \quad and \quad if \; Z0(I) \leq 0.001, Z0(I) = 0.001$$

RNET is the net daily radiation in MJ m^{-2}, ET the daily evapotranspiration in mm and HAUTEUR the canopy height (see § 3.2.1). TCULTMAX cannot be lower than TMAX. In this approach, we assume that TCULTMIN=TMIN.

6.6.2.b Energy balance

Two instantaneous energy balances are calculated to estimate TCULTMAX and TCULTMIN, assumed to occur at midday and at the end of the night, respectively.

eq. 6.30

$$TCULTMAX(I) = TMAX(I) + \frac{RNETMAX(I) - GMAX(I) - ETMAX(I)}{1200} RAAMAX(I)$$

$$TCULTMIN(I) = TMIN(I) + \frac{RNETMIN(I) - GMIN(I) - ETMIN(I)}{1200} RAAMIN(I)$$

In eq. 6.30 appear the minimum and maximum values of the various fluxes: net radiation (RNETMIN and RNETMAX), soil heat (GMIN and GMAX) and evapotranspiration (ETMIN and ETMAX) as well as the minimum and maximum values of the aerodynamic resistance (RAAMIN and RAAMAX).

To calculate long wave radiation i) atmospheric radiation is assumed to remain constant throughout the day, estimated using the Brutsaert formula (eq. 6.28), ii) soil radiation is calculated using TCULTMAX and TCULTMIN, requiring the iterative convergence procedure. At the end of the night, ETMIN and RGMIN are zero, while RGMAX and ETMAX are estimated assuming sinusoidal changes during the day. GMIN is calculated as an empirical function of the wind speed under the cover (Cellier *et al.*,1996). GMAX is taken to be 25% of the maximum net radiation below the cover. In addition to the canopy height (HAUTEUR) and the bare soil roughness (Z0SOLNU$_s$), the calculation of RAAMAX and RAAMIN requires wind speed values (see §7.2.2): the night-time wind speed is assumed to be equal to 50% of the daily mean wind speed, and the daytime wind speed is assumed to be 150% of the daily mean wind speed.

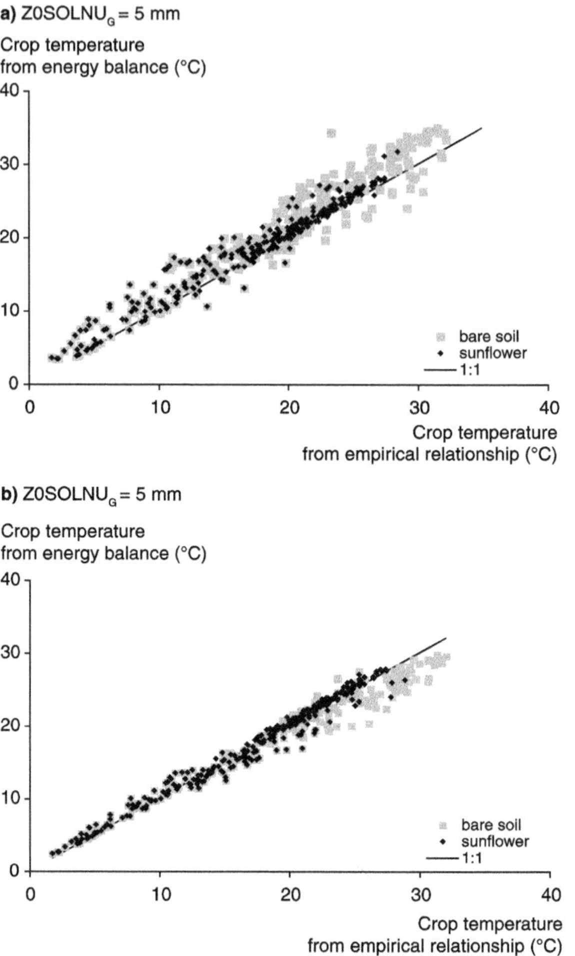

Figure 6.9. Comparison between the empirical relationship and the energy balance for the calculation of crop temperature for two different surfaces (a bare soil and a sunflower crop) in Avignon, using two bare soil roughness factors (1 and 5 mm).

Figure 6.9 shows the impact of surface type and soil roughness on the calculation of the temperature. The rougher the soil, the greater is the soil evaporation. Meanwhile Figure 6.10 shows that the energy balance method, for the minimum temperature, produces results which are identical to the driving hypothesis of the empirical method (TCULTMIN=TMIN).

6.6.2.c Iterative calculation of TCULT

We have seen that TCULT is involved in the calculation of net radiation, which in turn is used to calculate energy balances. In the previous version of STICS, the air temperature was used for calculation of long wave radiation to avoid numerical calculations. This

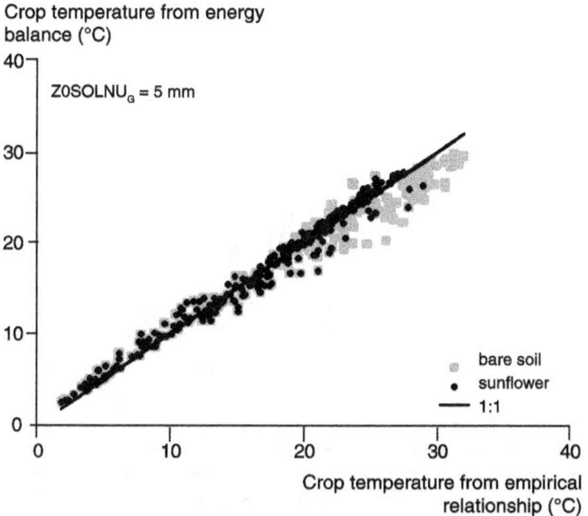

Figure 6.10. Evaluation of the assumption TCULTMIN=TMIN by running the energy balance in Avignon.

hypothesis has demonstrated its limitations (TCULT sometimes greater than 60°C !); this led us to introduce an iterative calculation process based on a difference of 0.5° between two iterations. In the option using Shuttleworth and Wallace the iteration also concerns estimates of water requirements, while in the option using the reference evapotranspiration as an input, the iterative process is only used to calculate net radiation (Figure 6.11).

6.6.3 Calculation of the canopy moisture

6.6.3.a Daily average

The calculation of the saturation deficit within the canopy (DOS in mbars: eq. 6.31) is possible using the Shuttleworth and Wallace formula (1985), and using the sum of evaporation fluxes (evaporation from soil, mulch, free water on leaves and transpiration).

eq. 6.31

$$DOS(I) = DSAT(I) + [DELTAT(I).RNET(I) - (DELTAT(I) + GAMMA) \cdot L(I) \cdot EVAP(I)]\frac{RAA(I)}{105.03}$$

$$with$$

$$DSAT(I) = TVAR(TAIR(I)) - TPM(I)$$

$$DELTAT(I) = TVAR(TAIR(I) + 0.5) - TVAR(TAIR(I) - 0.5)$$

$$EVAP(I) = ESOL(I) + EMULCH(I) + EMPD(I) + EP(I)$$

$$L(I) = [2500840 - 2358.6 \cdot TAIR(I)]10^{-6}$$

$$GAMMA = 0.65\frac{PATM_C}{1000} \quad and \quad PATM_C \ in \ mbars$$

119

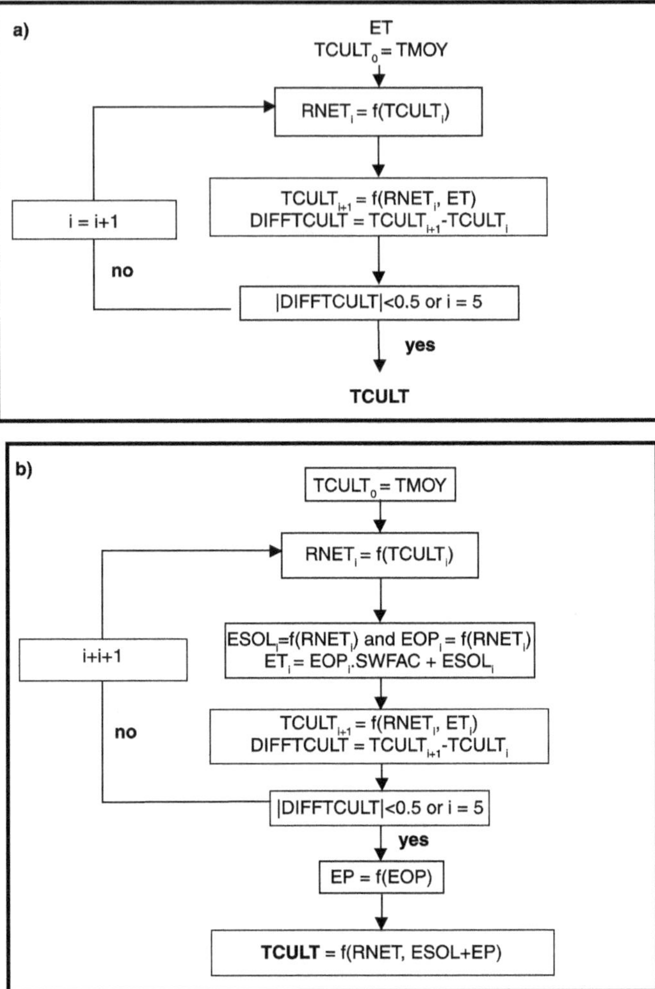

Figure 6.11. Diagrams representing the iteration loop of TCULT calculations for each option: a) reference evapotranspiration and b) Shuttleworth and Wallace. EOP is the maximal plant transpiration, EP the actual plant transpiration, SWFAC the EP/EOP ratio from the previous day and ESOL the actual soil evaporation.

where DELTAT is the gradient of the relationship between saturation vapour pressure and temperature, TAIR (°C) is the average daily temperature, RNET (MJ m^{-2}) is the net daily radiation, L is the latent heat of vaporisation (MJ kg^{-1}), GAMMA is the psychometric constant (mbar °C^{-1}) depending on atmospheric pressure PATM$_C$, DSAT is the air saturation deficit (mbar), TVAR is the saturated vapour pressure as a function of temperature (mbar) (see § 6.6.1.b), RAA is the aerodynamic resistance between the canopy and the reference height of weather measurements (ZR$_C$ generally 2 m) calculated from the canopy height and wind speed (see § 7.2), ESOL, EMULCH and EMPD are evaporation from soil, mulch and free water on leaves respectively (mm) and EP is plant transpiration (mm).

The average daily moisture (HUMIDITE) is then calculated with reference to the crop temperature (eq. 6.32):

$$eq.\ 6.32$$

$$HUMIDITE(I) = \frac{TVAR(TCULT(I)) - DOS(I)}{TVAR(TCULT(I))}$$

If the weather variable "wind speed" is not available, a default value of RAA is used (RA_G). If air humidity is not available the same assumption is made as before, using the parameter $CORRECTROSEE_C$ (see § 6.6.1 b). In this way, the moisture variable can be calculated in the absence of actual weather data.

6.6.3.b Reconstitution of hourly variables

To enable coupling with plant disease models, an hourly reconstitution of micro-climate state variables (crop temperature and air moisture) is made according to the following principle:

• The maximum crop temperature is assumed to occur at 14h00 TU and the minimum at sunrise. Between these two dates, linear interpolations make it possible to reconstitute hourly temperatures.

• The dew point temperature is calculated from TCULT and HUMIDITE by reversing the TVAR function (eq. 6.25). An hourly reconstitution similar to that used for the crop temperature is made by applying recurrent hypotheses to the minimum value of the dewpoint temperature, until there is convergence at the level of average daily moisture levels (Figure 6.12):

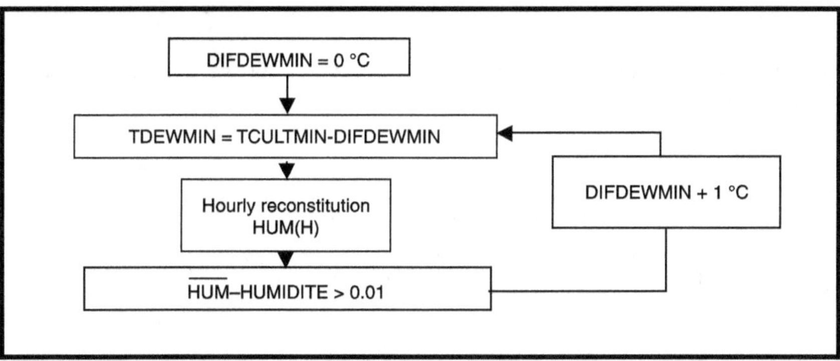

Figure 6.12. Diagrams representing the iteration loop of hourly humidity (HUM (H)) calculations based on the convergence between the averaged hourly values and the daily value, the fitted variable being the minimum value of the dewpoint.

Convergence is generally achieved in less than five iterations, and the comparison between mean daily values and daily values for crop moisture and temperature is satisfactory (no bias and $r^2 > 0.99$: Figure 6.13). The dynamics over a few days are presented in Figure 6.14.

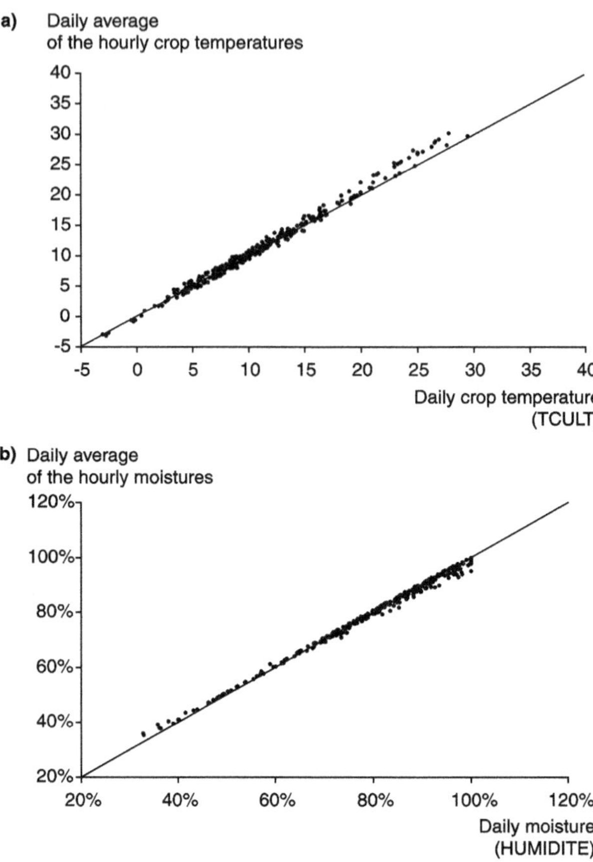

Figure 6.13. Comparisons between average hourly values and daily values of a) crop temperature and b) canopy humidity

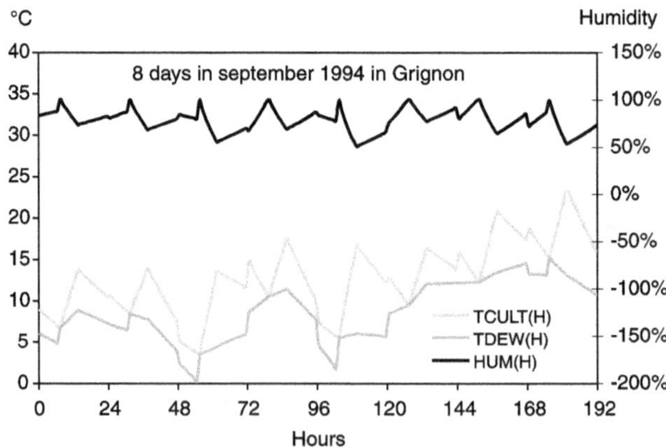

Figure 6.14. Hourly dynamics of the microclimatic variables over 8 days: crop temperature, dew point and humidity.

6.6.4 Estimation of microclimate under shelter

The incoming radiation above the crop (TRG) grown under a shelter is less than the outside radiation (TRGEXT) and the proportionality coefficient between the two (eq. 6.33) is the transmission coefficient (TRANSPLASTIC$_G$), whose value depends on the plastic used for the structure.

eq. 6.33

$$TRG(I) = TRGEXT(I) \cdot TRANSPLASTIC_G$$

In the case of an unheated ("cold") shelter, water requirements are estimated using the reference evapotranspiration approach. The potential evapotranspiration is simply estimated (eq. 6.34) using a multiplicative coefficient of radiation, COEFDEVIL$_G$ (de Villèle, 1974).

eq. 6.34

$$TETP(I) = TRG(I) \cdot COEFDEVIL_G$$

Rainfall is assumed to be zero and thus the crops must be watered by irrigation.

Temperature variations under a cold shelter are estimated using an energy balance based on the work by Boulard and Wang (2000). On a daily time step, the heat flux in the soil is ignored, assuming that the losses and gains balance out. The difference in mean daily temperature inside and outside (DELTATEMP) is thus expressed in eq. 6.35.

eq. 6.35

$$DELTATEMP(I) =$$
$$\frac{1}{(KH(I) + KS(I))24 \cdot 3600 \cdot 10^{-6}}(COEFRNET_G \times TRGEXT(I) - L(I) \times ESTIMET(I))$$

where: KH is the coefficient of heat transfer (W m^{-2} K^{-1}), KS is the coefficient of energy losses between the outside and inside of the shelter (W m^{-2} K^{-1}), COEFRNET$_G$ is a synthetic coefficient which converts external global radiation into net interior radiation (with a standard value of 0.59), L is the latent heat of vaporisation and ESTIMET is the evapotranspiration estimated from the water balance for the previous day and the evaporative demand for the day (eq. 6.36).

eq. 6.36

$$ESTIMET(I) = \frac{ET(I-1)}{TETP(I-1)} TETP(I)$$

KS (eq. 6.37) increases with the external wind speed using the parameters AKS$_G$ and BKS$_G$, equal to 6.0 and 0.5 respectively. KH (eq. 6.37) depends on the proportion of vents related to the total surface area of the greenhouse (SURFOUVRE$_T$) and the wind speed. The values of the constants CVENT$_G$ and PHIV0$_G$ are 0.16 and 4.10^{-3} respectively. SURFOUVRE$_T$ can take three values during the growth cycle.

eq. 6.37

$$KS(I) = AKS_G + BKS_G \cdot TVENT(I)$$

and

$$KH(I) = 1215.6 \left[\frac{SURFOUVRE_T(I)}{2} \cdot CVENT_G \cdot TVENT(I) + PHIV0_G \right]$$

These calculations enable an estimation of the mean elevation of temperature under shelter by comparison with the mean external temperature. This difference is entirely allocated to the maximum temperature (eq. 6.38).

eq. 6.38

$$TMOY(I) = TMOYEXT(I) + DELTATEMP(I)$$
$$TMIN(I) = TMINEXT(I)$$
$$TMAX(I) = 2 \cdot TMOY(I) - TMIN(I)$$

7 Water Balance

The water balance in crop models (Brisson *et al.*, 2005) has a dual purpose: to estimate soil water content and fluxes (which, for example, drives nitrogen mineralization and leaching in the soil) and water stress indices (which drive the behaviour of the plant). The latter objective differentiates crop models from those dedicated to irrigation management, and also forces a distinct separation between evaporation and transpiration.

This separation is usually applied at the level of the crop potential demand, based on its partitioning into potential plant transpiration and potential soil evaporation using a type of Beer's law. The crop potential demand comprises both crop and weather components. However, this variable differs from the classical maximal evapotranspiration variable, as defined for example by Itier *et al.* (1997) because it assumes that all surfaces (soil and foliage) are saturated with water.

As for the weather component, the problems of obtaining meteorological data usually determine the choice of calculation. Yet when compared to standard well-watered grass measurements there appear some differences. Allen *et al.* (1998) showed that the Penman FAO24 predicted too severe water deficit compared to the Penman-Monteith FAO56, and Sau *et al.* (2004) showed that the Priestley-Taylor function (1972), while giving good results for conditions of moderate evaporative demand, tended to over-predict for cool regions.

The crop component is usually linked to the LAI, which represents the increase in crop height and its roughness during growth (with reference to the standard grass evaporation), and affects the degree of the convective component of evapotranspiration. Convection under the plant canopy, which affects maximum transpiration, may be poorly reproduced by this optical analogy, particularly for row crops; this may justify applying a calculation of the energy balance (optional in STICS).

To calculate the quantity of water actually transpired by the crop, most models are based on a concept which includes the quantity of water physically available in the soil and the capacity of the plant to extract this water, due to its root characteristics. This is the fraction of transpirable soil water (Sinclair, 1986 , Lacape *et al.*, 1998, Pellegrino *et al.*, 2002), which also corresponds to the notion of the maximum available water content (amount of water between field capacity and wilting point). This approach does not permit a precise localization of root absorption in the soil layer (on a daily time step, all models assume that transpiration equals absorption), but has the advantage of implicitly taking account of capillary rise within the root zone. However, the threshold of sensitivity may vary over time. This global estimate of transpiration is used in STICS, while in other models (e.g. CERES) the calculation of uptake is differentiated in terms of the soil layer, because of the need to simulate capillary rise. In that alternative approach, water uptake per unit root length is based on the radial flow equation to roots.

7.1 Soil evaporation

Soil evaporation is calculated in two steps: potential evaporation related to the energy available at the soil level and then actual evaporation related to water availability. It is then distributed over the soil profile.

There are two methods for calculating potential evaporation related to plant cover above the soil, using either LAI or fractional ground cover, and the possible presence of an inert cover placed on the soil (Brisson *et al.*, 1998b). The first corresponds to a Beer's Law equivalent applied to the potential evaporation/reference evapotranspiration ratio (Penman) with a constant extinction coefficient. The second is an energy balance approach.

The calculation of actual evaporation, described in detail in Brisson and Perrier (1991), is based on concepts that resemble those put forward by Ritchie (1972).

7.1.1 Potential evaporation

The two methods calculating evaporation (EOS) involve the plant cover above the soil (LAI) and, if relevant, the presence of any mulch over the soil (COUVERMULCH).

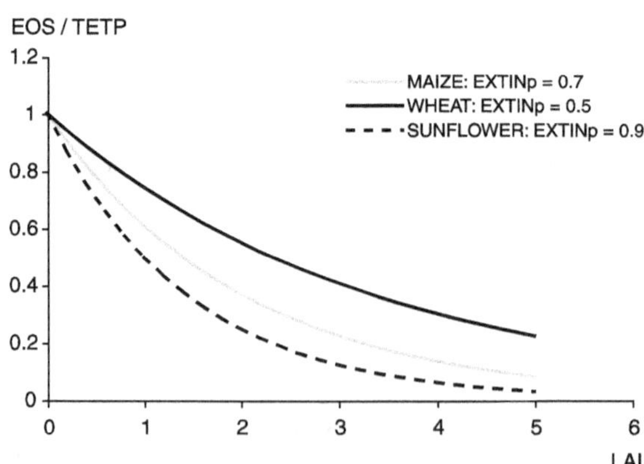

Figure 7.1. Relative potential evaporation as a function of LAI for 3 various crops.

The first method (eq. 7.1) illustrated in Figure 7.1 relies on a Beer's Law equivalent and is linked with the "crop coefficient approach" for the estimation of plant requirements (§ 7.2); it uses the reference potential evapotranspiration (TETP).

eq. 7.1

$$EOS(I) = TETP(I)\,exp\left(-DELTA \cdot LAI(I)\right)\left(1 - COUVERMULCH(I)\right)$$

$$and \qquad DELTA = EXTIN_p - 0.2$$

or if cover rate is used instead of the LAI (see § 3.1.4)

$$EOS(I) = TETP(I)\left(1 - TAUXCOUV(I)\right)\left(1 - COUVERMULCH(I)\right)$$

When using the radiation transfer option the values of EXTIN and DELTA are dynamically recalculated as a function of the canopy geometry and the quality of radiation (direct/diffusive radiation). However for row crops, justifying the use of the radiation transfer calculations, it is highly recommended to use the following energy balance approach.

The second method (eq. 7.2), i.e. the energy balance, is available only if the LAI is explicitly calculated.

eq. 7.2

$$EOS(I) = \frac{DELTAT(I) \cdot RNETS(I) + 105.03 \cdot DOS(I)/RAS(I)}{L(DELTAT(I) + GAMMA)}\left(1 - COUVERMULCH(I)\right)$$

RNETS is the net radiation at soil level (eq. 7.14), DELTAT is the gradient of the relationship between saturation vapour pressure and temperature (eq. 7.3 using the function TVAR explained in eq. 6.21), L is the latent heat of vaporization, GAMMA is the psychrometric constant (eq. 6.27), RAS is the aerodynamic resistance between the soil and the vegetation (eq. 7.16) and DOS is the saturation deficit in the vegetation (eq. 7.4).

eq. 7.3

$$DELTAT(I) = TVAR\left(TAIR(I) + 0.5\right) - TVAR\left(TMOY(I) - 0.5\right)$$

DOS is calculated assuming that, under soil conditions which are kept moist, total evapotranspiration (EPT: soil+canopy) can be written in the form of evaporation according to Priestley and Taylor (Brisson *et al.*, 1998b) in eq. 7.4:

eq. 7.4

$$DOS(I) = DSAT(I) + \left[GAMMA.RNET(I) - (DELTAT(I) + GAMMA) \cdot EPT(I)\right] \times \frac{RAA(I)}{105.03}$$

with

$$EPT(I) = 1.32 \cdot RNETS \cdot \frac{DELTAT(I)}{DELTAT(I) + GAMMA}$$

RAA is the aerodynamic resistance between the vegetation and the reference level (eq. 7.16), and DSAT is the air saturation deficit at the same level.

7.1.2 Actual evaporation

The calculation of actual evaporation relies on a semi-empirical model fully developed and justified in Brisson and Perrier (1991). Following a rain event, soil evaporation is assumed to follow two successive phases, as in Ritchie's (1972) approach, improved by Boesten and Stroosnijder (1986).

During the first phase evaporation is potential until the accumulation of daily evaporation reaches the $Q0_S$ threshold. During the second phase evaporation decreases and this decrease depends on the weather and soil type, through parameter A (eq. 7.5).

eq. 7.5

$$\sum_{\substack{beginning \\ second\ phase}} ESOL\ (I) = \sqrt{2\,A \sum_{\substack{beginning \\ second\ phase}} EOS\ (I) + A^2} - A$$

and

$$A = \frac{1}{2}\,ACLIM_C.(0.63 - HA)^{\frac{5}{3}}\,(HX_S - HA) \quad with \quad HA = \frac{ARGI_S}{1500}$$

$ACLIM_C$ is a weather parameter which depends mainly on the average wind speed. HX_S is the volumetric moisture content at field capacity of the surface layer, and $ARGI_S$ is the clay content used here to estimate the residual moisture, HA. Nevertheless the sensitivity of soil evaporation to these parameters (Figure 7.2) is rather low compared to the sensitivity to $Q0_S$ (Figure 7.3). Although $Q0_S$ depends on the soil texture and structure it is difficult to infer it from soil particle size distribution or bulk density. It generally varies between 0 to 30 mm.

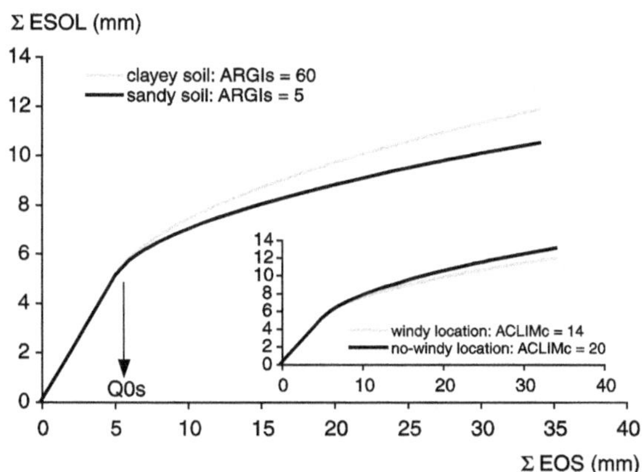

Figure 7.2. Sensitivity of cumulative soil evaporation (ΣESOL) as a function of the cumulative evaporative demand (ΣEOS) for various soil types (clayey soil: $ARGI_S$=60 and HN_S=0.4, sandy soil: $ARGI_S$=5 and HN_S=0.2) and weather conditions (for an average wind speed of 1ms^{-1} $ACLIM_C$=20, for 2 ms^{-1} $ACLIM_C$=14) without varying the $Q0_S$ parameter (5 mm).

The formalisation (eq. 7.5 and eq. 7.6) also provides an estimate of the thickness of the dry layer in the surface (or natural mulch: XMULCH) which is taken into account in the water profile in the soil, in the sense that this layer is supposed not to participate in evaporation.

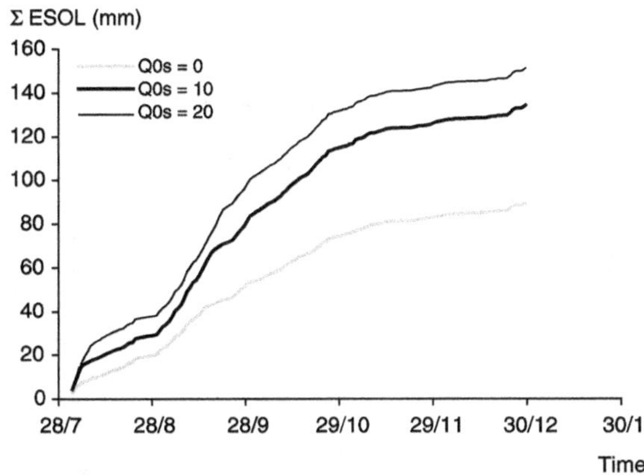

Figure 7.3. Cumulative soil evaporation from August 1st until the end of December in the north of France for three values of $Q0_S$.

7.1.3 Distribution in the soil profile

The method of calculating the distribution of evaporation resembles that of the LIXIM model (Mary *et al.*, 1999). The daily evaporation value ESOL, calculated above, is assumed to affect the layers of soil up from the base of the natural mulch XMULCH (if present) to a maximum depth of $ZESX_S$. Below this depth, there is no evaporation. The contribution of each basic soil layer to evaporation ESZ decreases with depth, according to the following function (eq. 7.6).

$$\text{eq. 7.6}$$

$$\text{for} \quad XMULCH < Z \leq ZESX_S$$

$$\frac{ESZ(Z,I)}{ESOL(I)} = \frac{\left(1 - \dfrac{Z}{ZESX_S}\right)^{CFES_S} K(Z,I)}{\sum \left(1 - \dfrac{Z}{ZESX_S}\right)^{CFES_S} K(Z,I)} \quad \text{and} \quad K(Z,I) = \frac{HUR(Z,I) - HA_S}{HN_S(Z) - HA_S}$$

$CFES_S$ is a slope coefficient, and K is an "evaporative conductance". HUR is the actual volumetric soil water content, HUCC the soil water content at field capacity of layer Z and HA is residual soil water content defined in eq. 7.5. By varying parameters ZESX and CFES, it is possible to take account of differences in hydraulic conductivity from one soil to another. A very high surface moisture gradient during soil drying is correctly represented by a high $CFES_S$ value. The sensitivity of the soil evaporation depth partitioning to the parameters $CFES_S$ and $ZESX_S$ is represented in Figure 7.4. If nothing is known about the soil one can use the standard values proposed: $CFES_S$=5 and $ZESX_S$=60 cm.

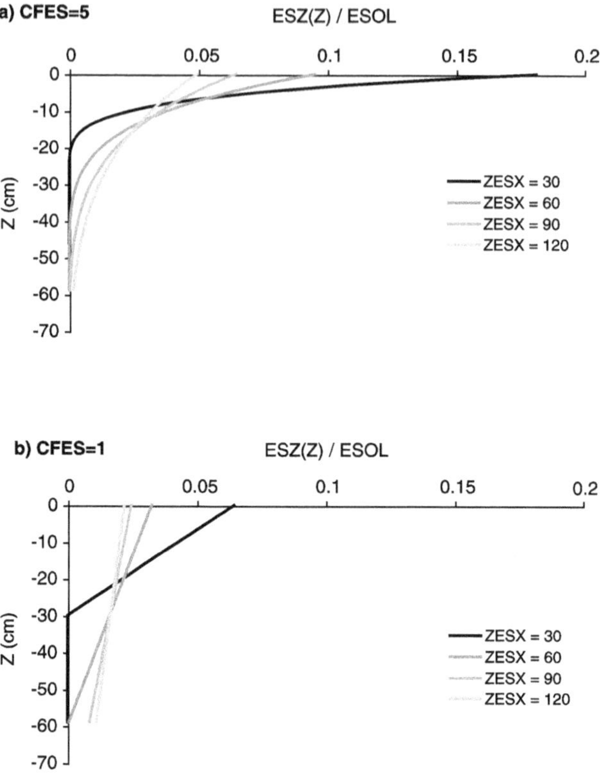

Figure 7.4. Partitioning of soil evaporation with depth as a function of the parameters $ZESX_S$ and $CFES_S$ assuming K=1 and XMULCH=0.

7.2 Crop water requirements

The two approaches described for soil potential evaporation have their equivalent for plant water requirements (or maximum transpiration EOP).

7.2.1 The crop coefficient approach

In the crop coefficient approach, fully documented in Brisson *et al.* (1992b), plant water requirements (maximum transpiration) are calculated in several steps, using the potential evapotranspiration as the driving variable.

First of all, calculation of what the crop evaporation value would be if none of the soil or plant surfaces had limited water (EO). This evaporation is a logistic function of the LAI (or a linear function of the ground cover) which involves the $KMAX_p$ parameter, the maximum crop coefficient of the crop (eq. 7.7 and Figure 7.5). $KMAX_p$ is attained when the LAI is approximately 5 (or TAUXCOUV equals $TCKMAX_p$ generally taken to 1) and depends on the reference evapotranspiration used (Penman, Penman-Monteith or Priestley- Taylor: Penman, 1948, Monteith, 1965, Priestley and Taylor, 1972).

eq. 7.7

$$EO\left(I\right) = TETP\left(I\right)\left[1 + \frac{KMAX_P - 1}{1 + exp\left(-1.5\left(LAI\left(I\right) - 3\right)\right)}\right]$$

or if ground cover is used instead of the LAI (see § 3.1.4)

if $TAUXCOUV\left(I\right) < TCKMAX_P$

$$then\ EO\left(I\right) = TETP(I)\left[\frac{TAUXCOUV\left(I\right)}{TCKMAX_P}\left(KMAX_P - 1\right) + 1\right]$$

if $TAUXCOUV\left(I\right) \geq TCKMAX_P$ *then* $EO\left(I\right) = KMAX_P \cdot TETP\left(I\right)$

If the leaves (MOUILL $\neq$ 0), or the plant mulch laid on the soil surface (MOUILLMULCH $\neq$ 0), have intercepted water (see chapter 6), then this water will evaporate depending on the reference evaporative demand (TETP): EMPD for leaves (eq. 7.8) and EMULCH for mulch (eq. 6.20). Naturally, the EMPD threshold is set by the amount of water retained on the foliage (MOUILL) while the EMULCH threshold is set

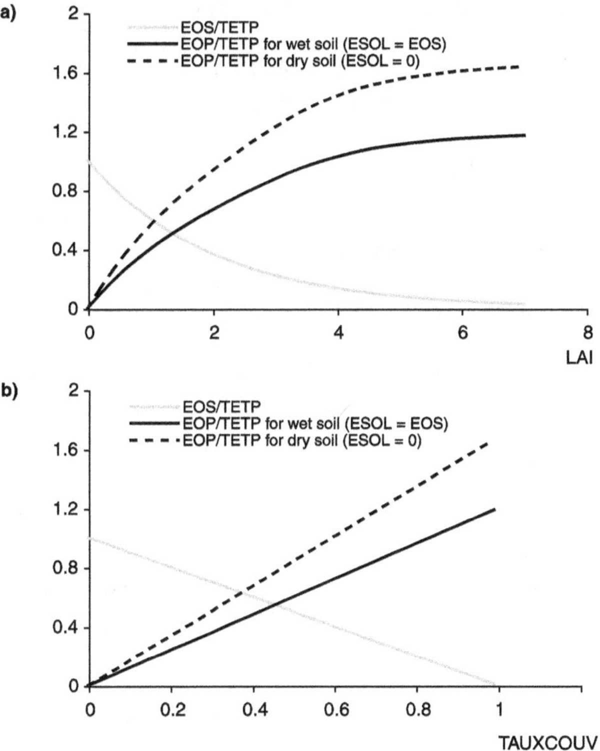

Figure 7.5. Relative evaporative demand applied to soil (EOS/TETP) and plants (EOP/TETP) accounting for the actual soil surface water status (ESOL=EOS or ESOL=0) for canopy qualified in LAI (a) or in ground cover (b).

by the amount of water retained in the mulch (MOUILLMULCH). The evaporated water contributes to reducing evaporative demand at the plant level.

eq. 7.8

$$EMPD\ (I) = EO\ (I) - TETP\ (I) \cdot exp\ (- DELTA \cdot LAI\ (I))$$

or if ground cover is used instead of the LAI (see § 3.1.4)

$$EMPD\ (I) = EO\ (I) - TETP\ (I) \cdot (1 - TAUXCOUV\ (I))$$

$$\text{and } EMPD\ (I) \le MOUILL\ (I)$$

Maximal transpiration depends on the available energy in plants, estimated by subtracting EOS from EO but also on atmospheric conditions in the vegetation. In order to take into consideration the increase in plant demand due to the dryness of the soil below the vegetation, we use the empirical relationship (eq. 7.9) based on the parameter $BETA_G$ deduced from work by Denmead (1973), Ritchie (1985) or Feddes (1987).

eq. 7.9

$$EOP(I) = (EO(I) - EDIRECTM\ (I)) \times \left[BETA_G - (BETA_G - 1) \frac{EDIRECT(I)}{EDIRECTM\ (I)} \right]$$

considering that EDIRECTM corresponds to evaporation of water intercepted by soil, mulch and leaves together, and that EDIRECT corresponds to the actual evaporation of the three together. A value of 1.4 is taken for $BETA_G$. It causes EOP to increase by a maximum of 40 % when the soil is completely dry.

7.2.2 The resistance approach

The "crop coefficient" approach can create problems in cases where it is not possible to apply Beer's law in a straightforward way (see § 3.2), or when the relationship between LAI and canopy height is not stable. Moreover, the previous approach is somewhat unreliable with regard to the "soil evaporation" variable and the microclimatic effect around the plant. We therefore suggest an alternative approach which consists of estimating plant water requirements and soil evaporation using the Shuttleworth and Wallace daily time-step model (Brisson *et al.*, 1998b). This has proved to be effective for explaining the energy budget of canopies (Sene, 1994) provided that appropriate empirical resistance parameters are used (Fisher and Elliott, 1996).

7.2.2.a Theoretical bases

The calculations are based on the resistance diagram shown in Figure 7.6, involving four flows (soil evaporation (ES), maximum plant transpiration (EOP), direct evaporation of water intercepted by the foliage (EMPD) or by mulch (EMULCH)) and two types of resistance (resistance to diffusion between canopy and soil, cover and reference level, respectively: RAS, RAA; surface resistance of canopy, of canopy boundary layer, respectively: RC, RAC). In this case all the fluxes are actual ones except the plant transpiration flux, which is the maximal one.

Each flux is calculated using a formula of the same type as for the potential soil evaporation (eq. 7.2); that leads to write eq. 7.10, eq. 7.11 and eq. 7.12.

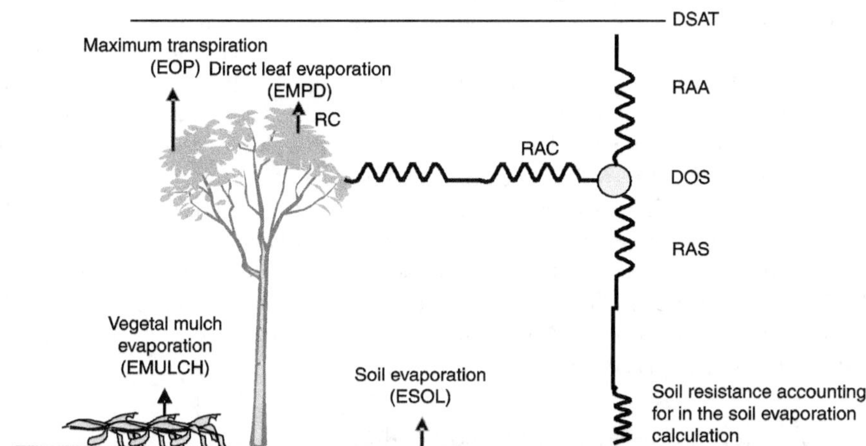

Figure 7.6. Drawing of the resistive diagram applied to the soil-crop system.

eq. 7.10

$$EMULCH(I) = \frac{DELTAT(I) \cdot RNETS(I) + 105.03 \cdot DOS(I)/RAS(I)}{L(DELTAT(I) + GAMMA)} COUVERMULCH(I)$$

limited by $MOUILLMULCH(I)$

eq. 7.11

$$EMPD(I) = \frac{DELTAT(I) \cdot RNETP1(I) + 105.03 \cdot DOS(I)/RAC(I)}{L(DELTAT(I) + GAMMA)}$$

limited by $MOUILL(I)$

eq. 7.12

$$EOP(I) = \frac{DELTAT(I) \cdot RNETP2(I) + 105.03 \cdot DOS(I)/RAC(I)}{L(DELTAT(I) + GAMMA[1 + RC(I)/RAC(I)])}$$

The amount of energy required for the direct evaporation of water on leaves (EMPD) is RNETP1 while this energy used for direct evaporation has to be deducted to evaluate the resulting energy available for transpiration (RNETP2). Energy distribution between bare soil and the soil cover (mulch) depends on COUVERMULCH (eq. 7.10).

DOS, the saturation deficit within the vegetation, is the variable linking all the fluxes. It is calculated (eq. 7.13) by relationship of Shuttleworth and Wallace (1985).

eq. 7.13

$$DOS(I) = DSAT(I) + [DELTAT(I) \cdot RNET(I) - (DELTAT(I) + GAMMA) \cdot L(I) \cdot EVAPO(I)]\frac{RAA(I)}{105.03}$$

with

$$EVAP0(I) = ESOL(I) + EMULCH(I) + EMPD(I) + EOP(I)$$

eq. 7.13 is very similar to eq. 6.31 (qv. for the meaning of the various terms) except that the evaporative term EVAPO implies potential transpiration flux: EVAPO is the accumulation of evaporative fluxes EOP and all the direct evaporation fluxes, i.e. ESOL, EMPD and EMULCH. It is the value of this direct evaporation which affects DOS and can cause the evaporative demand of the plant to fluctuate. The three components of the direct evaporation are calculated from an intermediate value of the saturation deficit DOS based on the hypothesis that, at complete saturation of the surfaces, the evaporation can be treated using a formalisation of the Priestley-Taylor type (Brisson *et al.*, 1998b).

In order to solve the above equations several terms have to be calculated: 1) the distribution of the energy sources between the soil and the plants (RNETS, RNETP1 and RNETP2) the water retention on the foliage and in the mulch (EMPD and EMULCH), 3) the resistances to diffusion (RAA and RAS), 4) the surface resistances (RC and RAC) and 5) soil evaporation (ESOL).

7.2.2.b Available energy and its distribution

The calculation of the whole surface net radiation was described in § 6.6.1. To evaluate the distribution of this available energy between the soil and the plants, we use the fraction of PAR intercepted by the plants (FAPAR) calculated using the RAINT variable (see § 3.2 and eq. 7.14). Thornley (1996) inferred the net radiation extinction coefficient from the extinction coefficient of the total radiation by applying a coefficient of 0.83, which corresponds to the range of measurements under a soybean canopy (Brisson *et al.*, 1998b).

eq. 7.14

$$RNETP1\,(I) = 0.83 \cdot FAPAR\,(I) \cdot RNET\,(I) \quad and \quad FAPAR\,(I) = \frac{RAINT\,(I)}{PARSURRG \cdot TRG\,(I)}$$

$$RNETS\,(I) = RNET\,(I) - RNETP1\,(I)$$

$$RNETP2\,(I) = RNETP1\,(I) - EMPD\,(I)$$

7.2.2.c Calculation of resistances to diffusion

We have adopted the formalisations proposed by Shuttleworth and Wallace (1985) which are described in detail in Brisson *et al.* (1998b) and in the box below. The characteristic lengths, bare soil roughness ($ZOSOLNU_s$) crop roughness (Z0) and displacement height (DH) are usually estimated as a function of the canopy height (HAUTEUR), when plants are present, and as a fixed value for bare soil (eq. 3.15).

eq. 7.15

$$For\ a\ plant\ stand \quad Z0(I) = 0.10\ HAUTEUR(I)$$

$$For\ a\ bare\ soil\ stand\ \ Z0(I) = ZOSOLNU_S$$

$$D(I) = 6.6\ Z0(I)$$

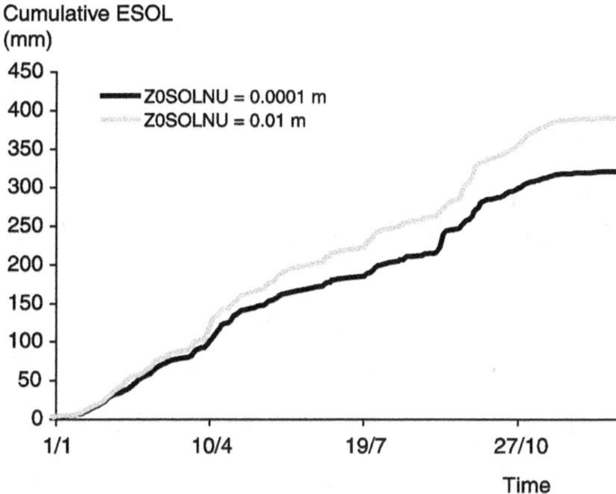

Figure 7.7. Influence of soil roughness (Z0SOLNU$_S$) on the cumulative soil evaporation during one year in the region of Avignon (South of France).

The bare soil value of roughness can vary between 10^{-2} and 10^{-4} m, corresponding to roughness values between 10 cm (after a very rough ploughing) and 1 mm (very flat soil). Figure 7.7 shows the effect of this parameter on soil evaporation: the greater the roughness the higher the soil evaporation, and in the given example the annual difference between extreme values is about 70 mm. The reference height taken for meteorological data is 2m. If the plant canopy height exceeds this threshold, a wind speed value is recalculated at a reference height of over 2 m (parameter ZR$_C$) by applying a logarithmic profile. The other meteorological values are not recalculated. The calculations of diffusive resistances are different for bare soil (eq. 7.16) and covering crops (eq. 7.17 for LAI≥4), while for non-covering crops (LAI<4) a LAI-dependent linear combination of the two first values is used (eq. 7.18). They all depend on wind speed (TVENT).

$$\text{eq. 7.16}$$

$$LAI = 0$$

$$RAS_0(I) = \frac{\ln\left(\dfrac{ZR_C}{ZOSOLNU_S}\right) \cdot \ln\left(\dfrac{Z0(I) + D(I)}{ZOSOLNU_S}\right)}{0.41^2\, TVENT(I)}$$

$$and$$

$$RAA_0(I) = \frac{\left[\ln\left(\dfrac{ZR_C}{ZOSOLNU_S}\right)\right]^2}{0.41^2\, TVENT(I)} - RAS_0(I)$$

$$\text{eq. 7.17}$$

$$LAI \geq 4$$

$$RAS_\infty(I) = \frac{\ln\left(\dfrac{ZR_C - D(I)}{Z0(I)}\right)}{0.41^2 TVENT(I)} - \frac{HAUTEUR(I)}{2.5(HAUTEUR(I) - D(I))}\left[12.18 - exp\left[2.5\left(1 - \frac{D(I) + Z0(I)}{HAUTEUR(I)}\right)\right]\right]$$

$$RAA_\infty(I) = \frac{ln\left(\frac{ZR_C - D(I)}{Z0(I)}\right)}{0.41^2 TVENT(I)}$$

$$\left[ln\left(\frac{ZR_C - D(I)}{HAUTEUR(I)}\right) + \frac{HAUTEUR(I)}{2.5(HAUTEUR(I) - D(I))}\left[exp\left[2.5\left(1 - \frac{D(I) + Z0(I)}{HAUTEUR(I)}\right)\right] - 1\right]\right]$$

eq. 7.18

$$0 \le LAI < 4$$

$$RAS(I) = \frac{1}{4}\left[LAI(I) \cdot RAS_\infty(I) + (4 - LAI(I)) \cdot RAS_0(I)\right]$$

and

$$RAA(I) = \frac{1}{4}\left[LAI(I) \cdot RAA_\infty(I) + (4 - LAI(I)) \cdot RAA_0(I)\right]$$

7.2.2.d Calculation of surface resistances

To simplify calculations, the resistance of the canopy boundary layer *(RAC)* is a function solely of the leaf area index of the canopy (eq. 7.19).

eq. 7.19

$$RAC(I) = \frac{50}{2LAI} \quad \text{with a lower threshold at} \quad RAC(I) = 12.5\ sm^{-1}$$

The canopy resistance *(RC)* is the product of four factors (eq. 7.20).

eq. 7.20

$$RC(I) = RSMIN_P \left(\frac{0.5LAI(I) + 1}{LAI(I)}\right)(0.039\ DSAT(I) + 0.45)\left(\frac{28}{2.5 + TRG(I)}\right)FCO2S$$

$RSMIN_P$ is the minimal stomatal resistance of leaves, DSAT (in mbars) is the saturation deficit, TRG (in MJ m^{-2} s^{-1}) is the global radiation and FCO2S is a $CO2_C$-dependent variable.

Due to the daily time step, the parameter $RSMIN_P$ cannot be inferred from the instantaneous values of measurements but must be derived from a top-down approach (Brisson *et al.*, 1998b Tolk *et al.*, 1996, Baldocchi *et al.*, 1991). Values of 250, 215 and 220 sm^{-1} where found for soybean, maize and sorghum respectively (Brisson *et al.*, 1998b; Brisson *et al.*, 2004). As shown in Figure 7.8, by a 150-250 sm^{-1} range of variation, the physiological characteristic $RSMIN_P$ has a less influence on the transpiration calculation than the morphological one, $HAUTMAX_P$.

The "saturation deficit" and "radiation" components are taken from research by Stockle and Kjelgaard (1996). With regard to the conditions for applying the proposed formulae, the saturation deficit is calculated at the meteorological scale and the radiation is the incident radiation above the crop.

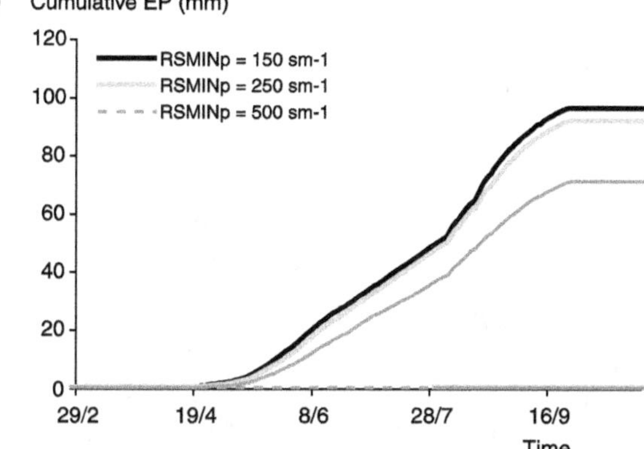

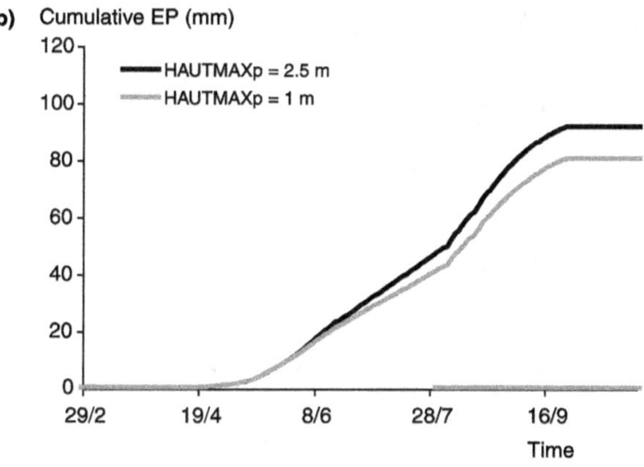

Figure 7.8. Influence of the minimal stomatal resistance of leaves ($RSMIN_p$) and the plant maximal height ($HAUTMAX_p$) on the cumulative transpiration of a rainfed vineyard in the region of Avignon (South of France).

If the atmospheric $CO2_C$ is high, stomatal conductance falls. Idso (1991) demonstrated the existence of proportionality between the CO2 effect on conversion efficiency and the CO2 effect on stomatal conductance, at a ratio of 2.5 for the addition of 300 ppm in the nominal concentration. Furthermore, Stockle *et al.* (1992) proposed a species-dependent formalisation (Figure 7.9). We propose to combine these two approaches to take account of the species and ensure a continual effect of $CO2_C$ (eq. 7.21). FCO2 is the species-dependent CO2 effect on conversion efficiency (see eq. 3.27), which affects the value of the species-dependent factor on stomate closure (FCO2S).

eq. 7.21

$$FCO2S = \cfrac{1}{1 + 0.77\left(1 - \cfrac{FCO2}{2.5}\right)\left(1 - \cfrac{CO2}{330}\right)}$$

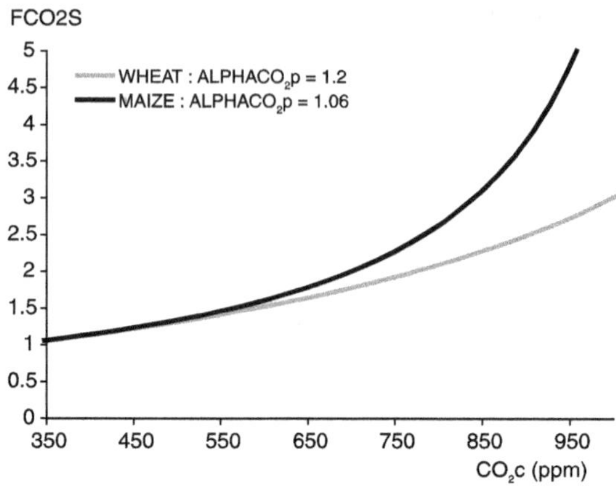

Figure 7.9. Influence of the species on the stomatal resistance CO2-dependent effect.

7.2.2.e Calculation of soil evaporation

Soil evaporation, already described in § 7.1, is calculated from a potential evaporation value obtained from an intermediate value of the saturation deficit based on the hypothesis that, at complete saturation of the surfaces, the evaporation can be approached using a Priestley-Taylor type formalisation.

7.3 Plant transpiration and derived stresses

To calculate actual transpiration we chose to use a relationship linking relative transpiration (ratio of actual to maximal transpiration) to soil water content. Such a simplified mathematical representation was proposed by Van Bavel (1953) for the total evapotranspiration. He suggested a straight-line relationship allowing simple calculations of soil water balance. Subsequent studies have shown that the relationship was more likely to be curvilinear (Denmead and Shaw, 1962; Eagleman, 1971) or exponential (Baier, 1969). Nevertheless a bilinear function may be a good representation of the experimental data (Burch *et al.*, 1978; Meyer and Green, 1981; Rosenthal *et al.*, 1985; Robertson and Fukai, 1993) and was adopted as the driving equation of many simple water balance models (Leenhardt *et al.*, 1995). Such a relationship assumes that a crop is able to take up soil water at a maximal rate to meet atmospheric demand until the soil water content falls below some threshold value. Though in many models, this threshold is assumed to be a constant equal to 30, 40 or 50% of the maximal available water content (Hunt and Pararajasingham, 1995; Robertson and Fukai, 1993; Fisher and Elliott, 1996; Mailhol *et al.*, 1996), it was shown to depend on atmospheric demand, species and time within the crop cycle (Hallaire, 1964; Doorenbos and Kassam, 1979; Cordery and Graham, 1989; Burch *et al.*, 1978; Novak, 1989; Gardner, 1991;Teixera *et al.*, 1996; Palacios and Quevedo, 1996). Relying on work by Slabbers (1980), we proposed an operational formula to calculate this threshold using the above-mentioned variable, derived from basic laws governing water transfer in the soil-plant atmosphere continuum (Brisson, 1998b).

7.3.1 Actual transpiration

On a daily time scale, root uptake can be considered to be equal to leaf transpiration. Root uptake calculated overall is then distributed between the soil layers. Relative transpiration, i.e. the relationship between actual transpiration and maximal transpiration (EP/EOP), is a bilinear function of the available water content in the root zone, TETA (i.e. the water content above the wilting point in cm3 of water/cm3 of dry soil). The EP/EOP ratio is considered as the stomatal water stress (SWFAC, eq. 7.22) which is represented in Figure 3.15.

eq. 7.22

$$SWFAC(I) = \frac{EP(I)}{EOP(I)}$$

The water content threshold separating the maximal transpiration stage and the reduced transpiration stage (TETSTOMATE) depends on root density, the stomatal functioning of the plant, and the evaporative demand; that is formalized in eq. 7.23 according to Brisson, (1998). It was shown that this threshold does not depend on the soil type, for example via the maximal available water content, as is commonly assumed.

eq. 7.23

$$TETSTOMATE(I) =$$

$$\frac{1}{80} \ln\left[\frac{EOP(I)}{2\pi\,CUMLRACZ(I) \cdot PSISTO_P \cdot 10^{-4}} \times \ln\left(\frac{1}{RAYON_G \sqrt{\dfrac{CUMLRACZ(I)}{ZRAC(I)}}} \right) \right]$$

where CUMLRACZ is the summation over the whole rooting depth, ZRAC, of effective root length density LRACZ, $PSISTO_P$ is the critical potential of stomatal closure

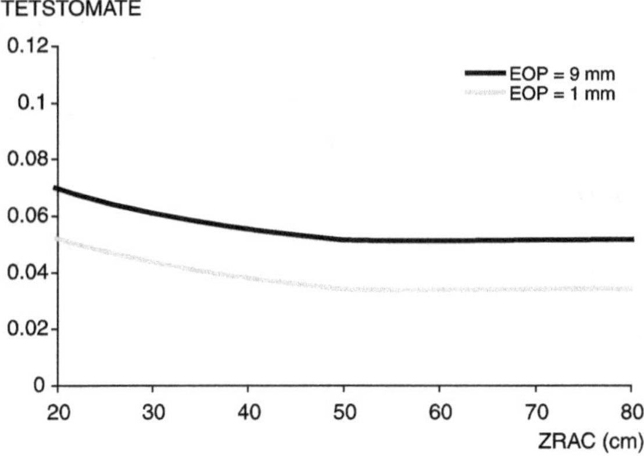

Figure 7.10. Influence of the rooting depth (ZRAC) and the maximal daily transpiration (EOP) on the threshold of soil water content above wilting point (TETSTOMATE) below which the transpiration is reduced.

(positive in bars) and $RAYON_G$ is the mean root radius which is assumed to be equal to 0.02 cm. When using this formula we find that, beyond a certain root depth, the TETSTOMATE threshold tends to be stable (Figure 7.10). The role played by the various factors influencing this value is summarised in Table 7.1. It highlights the dominant effect of evaporative demand, EOP, and the parameter of stomatal closure, $PSISTO_p$.

Table 7.1. Sensitivity analysis of the threshold TETSTOMATE (in cm³ water cm³ soil above the wilting point).

Parameters	Nominal value	TETSTOMATE sensitivity	
Root profile (pivot, ramified)	In-between	0.050	0.068
RAYON (5e-3 to 7e-2 cm)	0.02	0.052	0.060
PSISTO (5 to 25 bars)	15	0.050	0.070
EOP (1 to 9 mm)	4	0.039	0.066

7.3.2 Extrapolation to the water stress turgor index

The EP/EOP ratio is equal to the stomatal stress index, SWFAC. The stress turgor index TURFAC which affects leaf growth comes into play earlier. The method for calculating it is copied from the method used for SWFAC using the critical potential of cell expansion $PSITURG_p$ in eq. 7.23. Since $PSITURG_p$ is lower than $PSISTO_p$, we obtain a higher TETURG threshold. In other words, leaf growth can be inhibited even when transpiration is still at its maximum level (Figure 3.15).

7.3.3 Distribution of root water extraction within the profile

Water absorption EP is distributed in the root zone (EPZ profile) according to two factors, each of them having the same influence: the effective root density profile, LRACZ and the available water content (HUR-HUMIN): eq. 7.24 where CUMLRACZ and HCUM are the cumulative efficient roots and available water over the rooting zone (layers without roots are excluded from HCUM calculation) . The roots are assumed to be effective whenever the soil layer water content is above wilting point. Moreover it is assumed that the water located in the macroporosity does not contribute to transpiration. It is assumed that macroporosity fills up when microporosity is already filled.

$$eq.\ 7.24$$

$$if\ HUR(Z,I) \geq HMIN(Z)\quad then$$

$$EPZ(Z,I) = \frac{EP(Z)}{2} \times \left[\frac{LRACZ(Z,I)}{CUMLRAC(Z)} + \frac{HUR(Z,I) - HMIN(Z,I)}{HCUM(Z)} \right]$$

$$if\ HUR(Z,I) < HMIN(Z)\quad then\quad EPZ(Z,I) = 0$$

8 Nitrogen transformations

The nitrogen balance in the soil-plant system depends on the main processes affecting the mineral nitrogen content of the soil (mineralization, immobilization, nitrification, volatilization, denitrification, leaching) and the source/sink effect of the crop (symbiotic N fixation, absorption of mineral N).

The net N mineralization, i.e. the net production of mineral nitrogen by the soil, is the sum of two components:

• Humus mineralization, which results from the decomposition of stabilized organic matter in soil. This is a permanent process which always leads to a release of mineral N, i.e. a positive net mineralization, called "basal" mineralization.

• The mineralization of organic residues, which is associated with the decomposition of crop residues (straw, roots etc.) or organic wastes added to the soil. It is a process which is very variable in time, linked to the application of organic residues. During a first phase after the addition of residues, the mineralization can be positive or negative (immobilization of soil mineral N). During the second phase, it is positive through the "re-mineralization" process which releases N coming from either the residue or the microbial biomass which has decomposed it.

These processes are very dependent on soil and weather, and may be affected differently by them, particularly soil temperature and water content. The effect of temperature on C or N mineralization in soil is still a matter of controversy, as pointed out by Kirschbaum (2006). We attribute some of the disagreement between authors to the fact that the temperature response differs according to the type of organic matter decomposed. In STICS we use a different function for decomposition of humus and (fresh) organic residues. The effect of soil moisture might also be different for the two processes, but little is known on this aspect. Therefore we use a single function to describe the effect of water content on decomposition and N mineralization.

8.1 Mineralization of soil organic matter

Although the soil below the plough depth and the subsoil may contain important reserves of organic C and N, their decomposition rate appears to be slow or negligible compared to the upper soil layer (e.g. Fontaine *et al.*, 2007). In STICS, mineralization is assumed to occur up to a maximum depth (PROFHUM$_S$, in cm) and be negligible below that depth. The basal mineralization rate (VMINH, in kg N ha^{-1} day^{-1} in eq. 8.1) depends on the amount of active soil organic nitrogen (NHUM, in t ha^{-1} calculated in eq. 8.4), the soil type (its clay and calcium carbonate contents) and environmental factors, namely the water content and temperature in each soil layer (FH and FTH).

eq. 8.1

$$VMINH(I) = K2(I) \cdot NHUM$$

$$K2(I) = K2HUM \cdot \sum_{Z=1}^{PROFHUM} FH(Z) \cdot FTH(Z)$$

In eq. 8.1 K2 is the actual mineralization rate (kg N day^{-1}) and K2HUM is the potential mineralization rate (kg N day^{-1}), i.e. the mineralization rate constant of a soil which contains clay and calcium carbonate and is maintained at constant temperature and moisture content (reference conditions).

The soil water content (HUR) modifies the mineralization rate according to a linear function (eq. 8.2). The maximum value is reached for soil water contents equal or above HOPTM$_G$ (expressed as a proportion of field capacity), whereas mineralization is stopped when the soil water content is below HMINM$_G$ (Figure 8.1). Values of these parameters can be different for temperate and tropical soils (Figure 8.1).

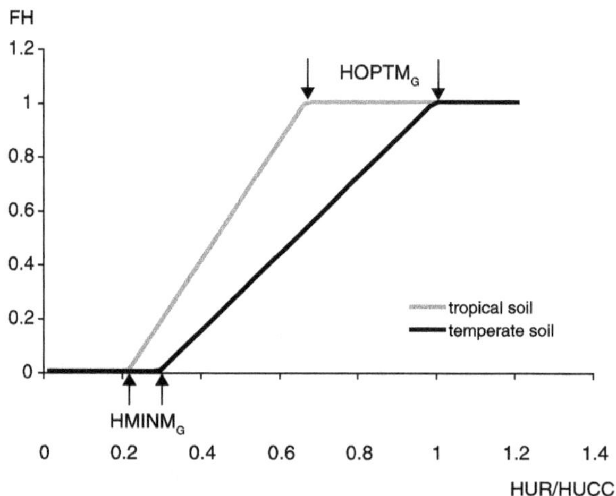

Figure 8.1. Influence of relative soil water content (HUR/HUCC) on decomposition rate of organic matter and N mineralization for a temperate and a tropical soil. Values for the temperate soil are HOPTMG=1.0 and HMINMG= 0.30 (Rodrigo *et al.*, 1997) while for the tropical soil they are HOPTMG=0.67 and HMINMG= 0.22 (Sierra *et al.*, 2003).

eq. 8.2

$$FH(Z,I) = \frac{HUR(Z,I) - HMINM_G \cdot HUCC(Z)}{(HOPTM_G - HMINM_G)HUCC(Z)}$$

$$if \quad FH(Z,I) > 1 \quad then \quad FH(Z,I) = 1$$
$$if \quad FH(Z,I) < 0 \quad then \quad FH(Z,I) = 0$$

The effect of soil temperature on basal mineralization (FTH) can be described either by an Arrhenius or a logistic function (Valé, 2006). We have chosen a logistic function because it makes it possible to simulate the slower increase in mineralization rate at high temperatures when microbial activity slows down (Figure 8.2).

The proposed function is roughly exponential from 0 to 25°C and increases more slowly above this temperature (eq. 8.3). It relies on three parameters including the reference temperature ($TREF_G$ chosen at 15°C). The parameter $FTEMHA_G$ corresponds to the asymptotic value of FTH and has been set to 25. Using these settings, the two parameters defining FTH are $FTEMH_G = 0.120$ K^{-1} and FTEMHB=145.

eq. 8.3

$$FTH(Z,I) = \frac{FTEMHA_G}{1 + FTEMHB \cdot \exp(-FTEMH_G \cdot TSOL(Z,I))}$$

$$if \quad TSOL(Z,I) < 0 \quad then \quad FTH(Z,I) = 0$$
$$with \quad FTEMHB = (FTEMHA_G - 1) \cdot \exp(FTEMH_G \cdot TREF_G)$$

The logistic function thus parameterized makes it possible to simulate adequately C or N mineralization kinetics measured in controlled conditions for several temperate and tropical soils (Balesdent and Recous, 1997; Valé, 2006; Nicolardot *et al.*, 2006). It is very close to an Arrhenius function with an activation energy $E_A = 78$ kJ mol^{-1} K^{-1} between

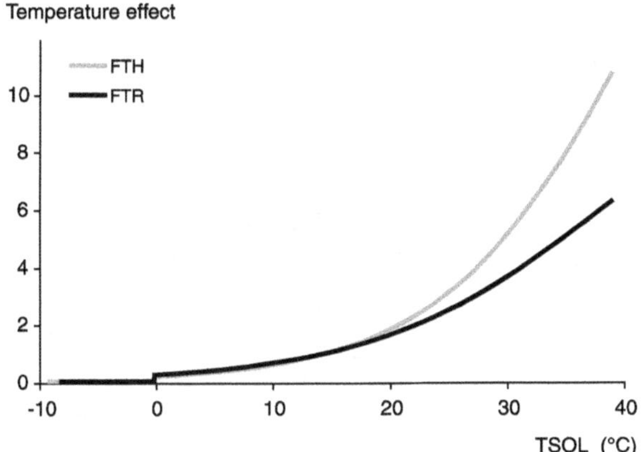

Figure 8.2. Influence of temperature on decomposition rates of organic matter (humus and organic residues).

0 and 35°C. It is also equivalent to a Van't Hoff function between 0 and 25°C with a Q_{10} coefficient equal to 3.15.

The amount of active organic N (NHUM, in t ha^{-1} in eq. 8.4) is the product of the soil organic nitrogen content in the upper layer (NORG$_S$, in %), the proportion of active organic nitrogen (1-FINERT$_G$, the default value for FINERT$_G$ is 0.65), the bulk density of the upper layer (DA (1), in g cm^{-3}) and the equivalent mineralization depth (PROFHUM$_S$, in cm).

eq. 8.4

$$NHUM = NORG_S \cdot (1 - FINERT_G) \cdot DA(1) \cdot PROFHUM_S$$

The effect of soil tillage is not explicitly considered but is accounted for as follows. In a regularly ploughed soil, the nitrogen content is homogeneous in the ploughed layer and corresponds to NORG$_S$. In this case PROFHUM$_S$ must be equal to or greater than the ploughing depth in order to take into account the contribution of lower layers to the total mineralization. However some studies (e.g. Valé, 2006; Oorts *et al.*, 2007) suggest that this contribution is small.

If the soil is no longer ploughed, the organic nitrogen distribution in the old plough layer becomes heterogeneous. The same calculation can be applied but PROFHUM$_S$ corresponds to the old depth of ploughing and NORG$_S$ represents the mean organic N concentration over this depth. If NORG$_S$ is measured over the new (and smaller) depth of soil tillage, PROFHUM$_S$ must correspond to that depth. In both cases, the possible change in bulk density (increase due to the reduction in soil tillage) must be taken into account.

The potential rate of mineralization (K2HUM) is affected by the mineralogic clay content (ARGI$_S$, in %) and the CaCO$_3$ content (CALC$_S$, in %) which reduce mineralization (eq. 8.5). It involves three parameters: FMIN1$_G$, FMIN2$_G$, FMIN3$_G$.

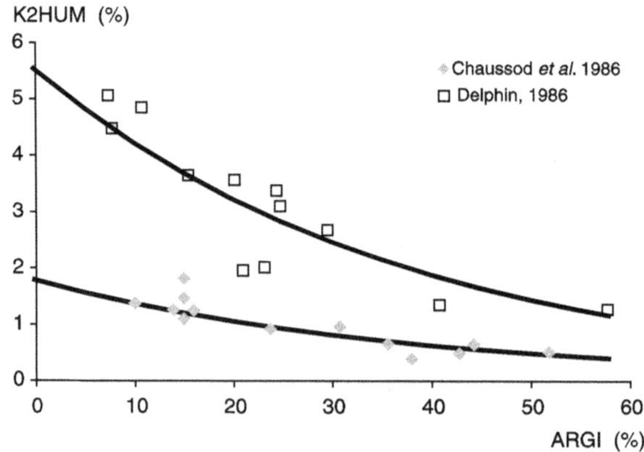

Figure 8.3. Influence of clay content on N mineralization rate from humus measured in laboratory conditions. The continuous lines are the simulated values using eq. 8.5.
Parameter values: FMIN1$_G$ = 6 10^{-4}; FMIN2$_G$ = 0.0272; FMIN3$_G$ = 0.0167.

eq. 8.5

$$K2HUM = FMIN1_G \; \frac{\exp\left(- FMIN2_G \cdot ARGI_S\right)}{1 + FMIN3_G \cdot CALC_S}$$

eq. 8.5 is based on soil incubation data obtained by Delphin (1986) and Chaussod *et al.* (1986) for N mineralization, as shown in Figure 8.3. This relationship was found to be applicable to another set of incubation data (Valé, 2006), although it explained only 29% of the variance. It was validated by Saffih and Mary (2008) for predicting the evolution of soil organic carbon over the long term.

The parameter $FMIN1_G$ was calibrated using the field experiments described by Mary *et al.* (1999): its value is 6.10^{-4} day^{-1}. Using this value, the mean turnover time of the whole soil organic matter (1/K2) lies between 30 and 60 years in temperate soils. The N/C ratio of humified organic matter in soil is assumed constant and equal to WH_G, whose standard value is 0.105.

As indicated in eq. 8.4, the humified organic N in soil (NHUMT, in t ha^{-1}) is composed of 2 pools: an "active" pool (NHUM) and a "stable" pool ($FINERT_G$. NHUMT). The first pool only contributes to mineralization and humification, whereas the second is assumed to be inert on the time scale of a century. Such an inert pool is included in most models simulating the evolution of soil organic matter over the long term. However the models differ greatly in the size of this pool, which may vary from 10% (e.g. in ROTHC; Coleman and Jenkinson, 1996) to 50% (in CENTURY; Parton *et al.*, 1987) and even 60-80% (Ludwig *et al.*, 2003, 2007).

In STICS the initial size of the stable pool has been assigned a nominal value of $FINERT_G$=65%. This value allowed the evolution of soil organic carbon to be simulated in nine long-term experiments (Saffih and Mary, 2008). However this value should depend on the cropping history of the field. It should be smaller if the soil has a recent grassland or forest land use or if it received large amounts of organic manure.

8.2 Mineralization of organic residues

STICS simulates the decomposition of various organic residues and their humification due to microbial activity, as described by Nicolardot *et al.* (2001) for crop residues. Nitrogen mineralization depends on the decomposition rate of organic residues (i.e. carbon fluxes), their C/N ratio ($CSURNRES_T$), the C/N ratio of the zymogeneous biomass (CNBIO) and of the newly formed humified matter (CNHUM).

Eight categories of organic residues have been defined: 1) crop residues from mature crops (e.g. straw), 2) crop residues from young plants (e.g. catch crops), 3) farmyard manures, 4) composts, 5) sewage sludges, 6) vinasses, 7) animal horn and 8) other (any other residue can be included). The fate of residues in each category (r) is followed separately. The carbon and nitrogen flows occurring during the decomposition of the organic residues is given in Figure 8.4.

The model is defined by 6 parameters, most of them being residue-dependent: two decomposition rate constants (KRES and $KBIO_G$, in day^{-1}), two partition parameters ($YRES_G$ and HRES) and two C/N ratios (CNBIO and CNHUM). For a given category,

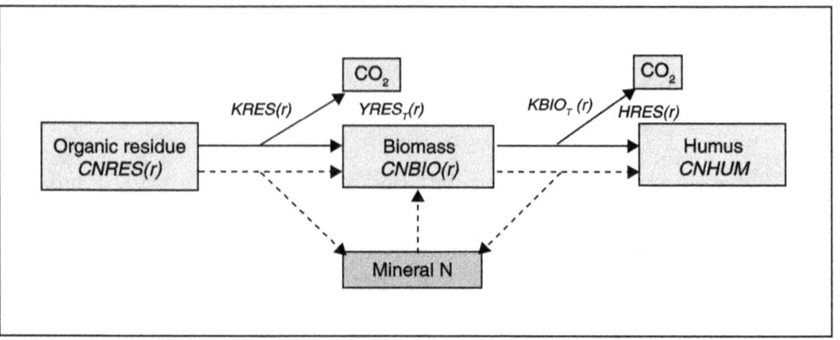

Figure 8.4. Flow diagram of the decomposition of organic residues in soil (from Nicolardot *et al.*, 2001). The continuous lines indicate carbon flows, and the dashed lines nitrogen flows.

the parameters are either constant ($KBIO_G$, $YRES_G$, and $CNHUM=1/WH_G$) or dependent upon the C/N ratio of the organic residue ($CSURNRES_T$), according to the following relationships:

$$\text{eq. 8.6}$$

$$KRES(r)= AKRES_G(r)+ \frac{BKRES_G(r)}{CSURNRES_T(r)}$$

$$CNBIO(r)= AWBS_G(r)+ \frac{BWB_G(r)}{CSURNRES_T(r)}$$

$$HRES(r)= 1-AHRES_G(r)+ \frac{CSURNRES_T(r)}{BHRES_G(r)+CSURNRES_T(r)}$$

The decomposition rate of organic residues (DCRES, in kg C ha^{-1} day^{-1}) is assumed to follow first order kinetics (eq. 8.7) against the amount of decomposable carbon (CRES) and depends on their nature (KRES), on soil temperature (FTH), water content (FH) and the available soil nitrogen in the vicinity of residues (FN).

$$\text{eq. 8.7}$$

$$DCRES(Z,I,r)=-KRES(r)\cdot CRES(Z,I,r)\cdot FTR(Z,I)\cdot FH(Z,I)\cdot FN(Z,I)$$

The change in the associated microbial biomass (DCBIO) and the rate of humus formation (DCHUM), both in kg C ha^{-1} day^{-1}, are given in eq. 8.8 and eq. 8.9.

$$\text{eq. 8.8}$$

$$DCBIO(Z,I,r)=$$
$$-YRES_G(r)\cdot DCRES(Z,I,r)-KBIO_G(r)\cdot CBIO(Z,I,r)\cdot FTR(Z,I)\cdot FH(Z,I)$$

$$\text{eq. 8.9}$$

$$DCHUM(Z,I,r)= KBIO_G(r)\cdot HRES(r)\cdot CBIO(Z,I,r)\cdot FTR(Z,I)\cdot FH(Z,I)$$

The soil moisture content influences decomposition similarly to the decomposition of humified materials (eq. 8.2) whereas the soil temperature has a specific effect on the decomposition rate of organic residues. The thermal effect on residue mineralization FTR is based on the data published by Balesdent and Recous (1997). It is similar to the logistic function defined for humus decomposition (eq. 8.3), but with different parameters. Using the same reference temperature ($TREF_G$), the parameters are: $FTEMH_G = 0.103$ K^{-1} and $FTEMHA_G = 12$. The shape of the curve is shown in Figure 8.2.

A lack of mineral nitrogen reduces both the decomposition rate (factor FN) and the N immobilization rate (Recous *et al.*, 1995; Giacomini *et al.*, 2007). It also reduces and postpones the subsequent remineralization of nitrogen.

The depth of residues incorporation in soil modifies their decomposition since water content and temperature vary with depth and their localization determines the amount of mineral nitrogen available for the microbial biomass. Each tillage operation is assumed to mix the residues uniformly with the soil over the depth defined by a minimal value ($PROFRES_T$) and a maximal value ($PROFTRAV_T$).

The net N mineralization rate (DN, in kg N ha^{-1} day^{-1}, positive or negative) resulting from residue decomposition is calculated as the complement of the variation in the three organic pools (eq. 8.10).

eq. 8.10

$$DN(Z,I) = \sum_{r=1}^{8} -DNRES(Z,I,r) - DNBIO(Z,I,r) - DNHUM(Z,I,r)$$

The changes in the three N pools (residue, microbial biomass, humus) are calculated using the C/N ratio of the three compartments (eq. 8.11)

eq. 8.11

$$DNRES(Z,I,r) = \frac{DCRES(Z,I,r)}{CSURNRES_T(r)}$$

$$DNBIO(Z,I,r) = -\frac{YRES_G(r) \cdot DCRES(Z,I,r)}{CNBIO(r) \cdot FBIO(Z,I)} - KBIO_G \cdot NBIO(Z,I) \cdot FTR(Z,I) \cdot FH(Z,I)$$

$$DNHUM(Z,I,r) = \frac{DCHUM(Z,I,r)}{CNHUM}$$

The factor FBIO is normally equal to 1. It can be greater in the case where the soil mineral N is exhausted and cannot satisfy the needs of the decomposers. In that case, there is a change in the composition of the microbial biomass which requires less N, so that its C/N ratio increases. The factor FBIO becomes greater than 1; it is calculated in order to fit the microbial requirements to the available soil mineral N.

Examples of C and N dynamics predicted by the model are given in Figure 8.5 and Figure 8.6. The variation in organic pools may exceed the amount of N added by the residue when mineral N is immobilized.

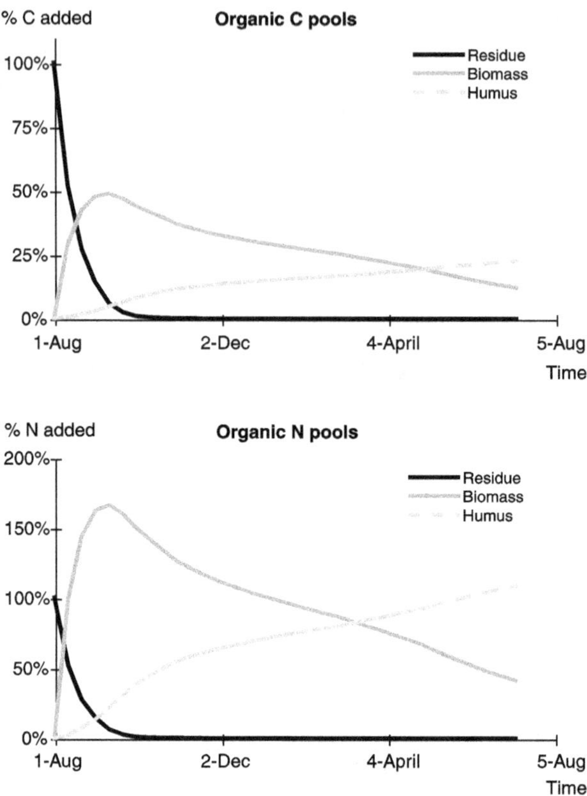

Figure 8.5. Evolution of C and N pools simulated during the decomposition of a crop residue (rapeseed straw) finely mixed in the soil, assuming no limitation by mineral N. The ordinate is expressed in % of C or N added by the residue.

The C and N mineralization kinetics differ according to the type of organic residue (Figure 8.6). Decomposition results in net release of N for residues with a low C/N ratio (vinasse, C/N =7; catch crop C/N =12) and net immobilization with residues poor in N (rapeseed straw, C/N =46).

8.3 Nitrification

Nitrate production in soil results from two successive processes: mineralization (or ammonification) and nitrification. Nitrification is often a rapid process in cultivated soils under temperate climates, which may justify avoiding describing nitrification and equating mineral N to nitrate-N. However in some soil and climatic conditions (acidic, hydromorphic or tropical soils etc.), the nitrification process may be much slower and ammonium ions may persist in soil. Furthermore, the simulation of ammonia volatilization is highly dependent on NH_4^+ concentration and requires a description of nitrification. Therefore the present version of STICS is able to simulate nitrification (in terms of slowing down of mineralization) and the two forms of mineral N separately.

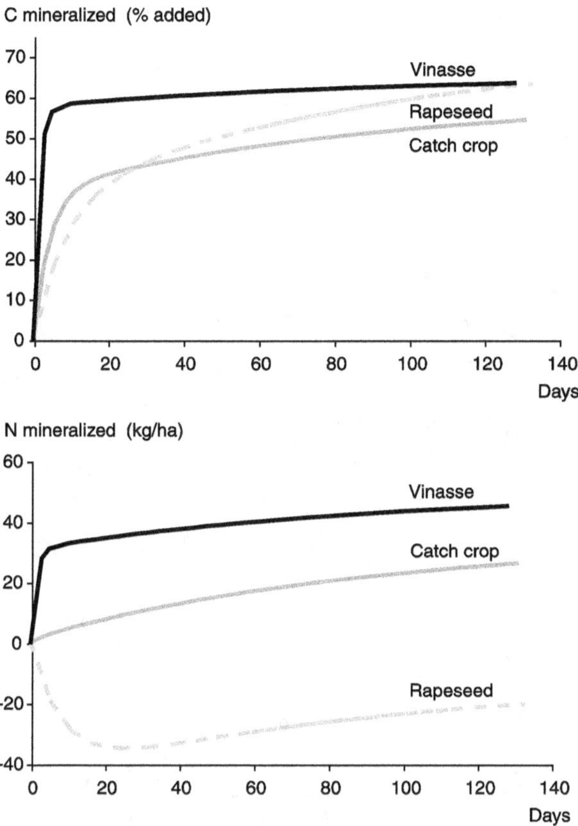

Figure 8.6. Evolution of C and N mineralized due to the decomposition of three types of organic residues (vinasse, catch crop shoots and rapeseed straw). The abscissa represents the normalized time (constant temperature and moisture, no limitation by mineral N). The ordinate is expressed in % of C added or kg N/ha.

Nitrification is assumed to occur in the biologically active layer, i.e. up to the depth PROFHUM$_S$. It is a first order process against NH$_4^+$ concentration, and depends on soil temperature (TSOL), soil water content (HUR) and soil pH (PH$_S$). These factors do not interact with each other. The fraction of NH$_4^+$ transformed into NO$_3^-$ every day in each layer (TNITRIF given in eq. 8.12) cannot exceed the value FNX$_G$ (in day^{-1}). This parameter has been assessed at 0.5 in a tropical soil (Sierra *et al.*, 2003).

eq. 8.12

$$TNITRIF(Z,I) = FNX_G \cdot FPHN \cdot FHN(Z,I) \cdot FTN(Z,I)$$

The effects of soil pH (PH$_S$) and water content (HUR) are linear as described in eq. 8.13 and eq. 8.14 and illustrated in Figure 8.7 and Figure 8.8, involving the parameters PHMINNIT$_G$, PHMAXNIT$_G$, HMINN$_G$ and HOPTN$_G$, whose default values are 3.0, 5.5, 0.67, 1.0 respectively.

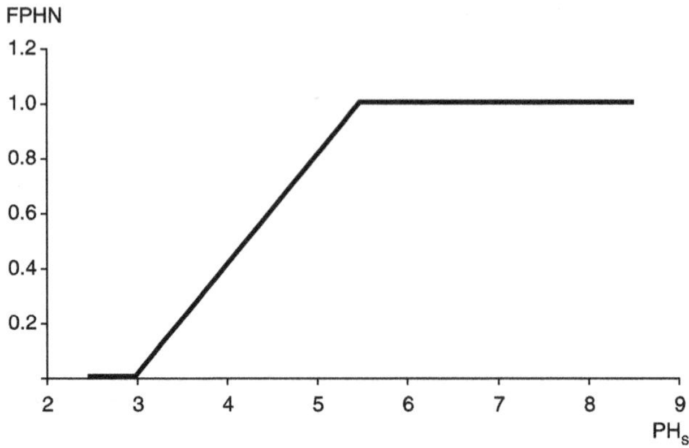

Figure 8.7. Effect of soil pH on nitrification.

eq. 8.13

$$FPHN = \frac{PH_S - PHMINNIT_G}{PHMAXNIT_S - PHMINNIT}$$

$$if \quad PH_S < PHMINNIT_G \quad then \quad FPHN = 0$$
$$if \quad PH_S > PHMAXNIT_G \quad then \quad FPHN = 1$$

eq. 8.14

$$FHN(Z,I) = \frac{HUR(Z,I) - HMINN_G \cdot HUCC(Z)}{(HOPTN_G - HMINN_G)HUCC(Z)}$$

$$if \quad FHN(Z,I) > 1 \quad then \quad FHN(Z,I) = 1$$
$$if \quad FHN(Z,I) < 0 \quad then \quad FHN(Z,I) = 0$$

As shown in Figure 8.8, the optimal water contents for nitrification ($HOPTN_G$) and mineralisation ($HOPTM_G$) are different, which can lead to significant NH_4^+ accumulation in soil.

The temperature function increases linearly from the threshold $TNITMIN_G$ up to the optimum $TNITOPT_G$ and then decreases linearly to $TNITMAX_G$, after which it becomes nil (Figure 8.9 and eq. 8.15).

eq. 8.15

$$if \quad TSOL(Z,I) \leq TNITOPT_G \quad FTN(Z,I) = \frac{TSOL(Z,I) - TNITMIN_G}{TNITOPT_G - TNITMIN_G}$$

$$if \quad TSOL(Z,I) \geq TNITOPT_G \quad FTN(Z,I) = \frac{TSOL(Z,I) - TNITMAX_G}{TNITOPT_G - TNITMAX_G}$$

$$if \quad FTN(Z,I) > 1 \quad then \quad FTN(Z,I) = 1$$
$$if \quad FTN(Z,I) < 0 \quad then \quad FTN(Z,I) = 0$$

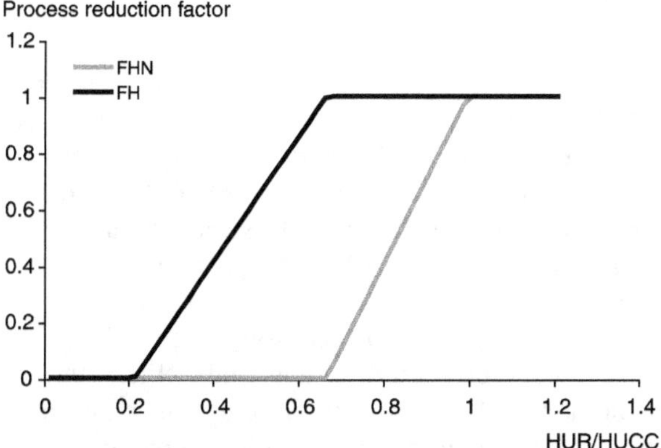

Figure 8.8. Effects of soil water content on nitrification and mineralization in the case of a tropical soil.

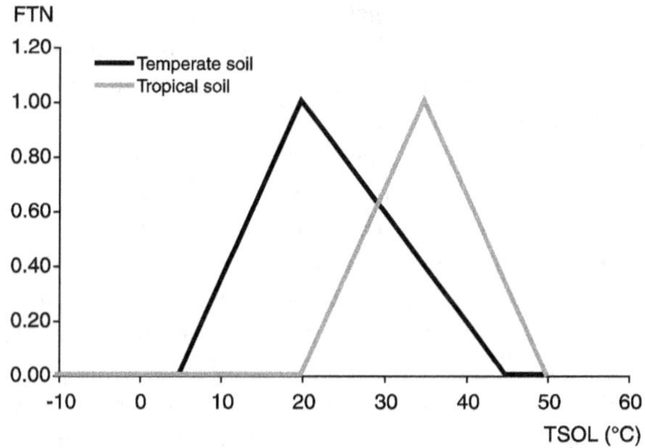

Figure 8.9. Temperature effects on nitrification rate in temperate and tropical soils with their respective cardinal temperatures.

Nitrification is also accompanied by N_2O emissions. Under satisfactory aerobic conditions, it has been shown that the amount of N_2O emitted is a constant proportion of the nitrified NH_4 (Garrido *et al.*, 2002; Khalil *et al.*, 2004), called $RATIONIT_S$. The rate of N_2O emission through nitrification (N2ONIT, in kg N ha^{-1} day^{-1}) is given in eq. 8.16 as the complement of the nitrate production (NITRIF, in kg N ha^{-1} day^{-1}).

eq. 8.16

$$NITRIF(I) = (1 - RATIONIT_S) \sum_{Z=1}^{PROFHUM} TNITRIF(Z, I) \cdot AMM(Z, I)$$

$$N2ONIT(I) = RATIONIT_S \sum_{Z=1}^{PROFHUM} TNITRIF(Z, I) \cdot AMM(Z, I)$$

8.4 Ammonia volatilization

Ammonia volatilization is a purely chemical process which operates on the soil ammonium pool (NH_4^+) and converts it into gaseous ammonia (NH_3). It affects the ammonium derived from mineral fertilizers or from organic fertilizers which contain large amounts of ammonium (such as liquid manure) and/or which have rapid mineralizing potentials (e.g. vinasses). The current STICS version only simulates explicitly the volatilization following an application of liquid organic manure (see § 6.3.2 for volatilization from mineral fertilizers).

In order to simulate volatilization, it is necessary to consider four forms of ammonia compounds which are in equilibrium (Génermont and Cellier, 1997):
- NH_4s: ammonium ions (NH_4^+) adsorbed onto the mineral or organic soil fractions
- NH_4l: ammonium ions in solution in the liquid soil phase
- NH_3l: ammonia molecules (NH_3) in solution in the liquid soil phase
- NH_3g: ammonia molecules in the gaseous soil phase.

All conditions which move these equilibrium towards the last form (e.g. high pH and temperature) stimulate volatilization. Volatilization occurs at the soil surface and depends on the NH_4^+ concentration there: therefore it is affected by fertilizer type, fertilizer rate, soil water content and NH_4^+ movement in soil. The equilibrium between NH_4s and NH_4l forms can be characterized by an adsorption isotherm which depends on soil CEC (itself linked to clay and organic matter contents). NH_4l and NH_3l are linked through a chemical equilibrium which is pH- and temperature- dependent. The solubility equilibrium between NH_3l and NH_3g forms mainly depends on temperature.

The first step consists of defining the volatilizable NH_4^+ immediately after the application. The exchangeable NH_4^+ ($NMINRES_T$, in kg N ha^{-1}) is split into two pools: a pool which remains at the soil surface and which can be volatilized (NVOLATORG, in kg N ha^{-1}) and a pool which infiltrates and is not volatilizable. The proportion of the volatilizable fraction (PROPVOLAT, eq. 8.17) increases with the dry matter content of the manure (Figure 8.10) or its water content ($EAURES_T$). It is also affected by soil tillage: it decreases if the soil has been tilled during the last 7 days before manure spreading

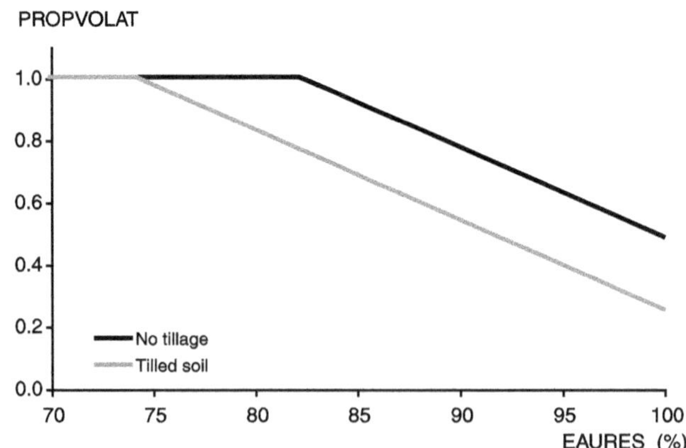

Figure 8.10. Effect of dry matter content of slurry and soil tillage on the volatilizable fraction.

(TRSOLVOLAT=-1) and increases otherwise (TRSOLVOLAT=+1) (Morvan, 1999). The volatilizable NH_4^+ at the time of application (IAP) is:

eq. 8.17

$$NVOLATORG(IAP) = NMINRES_T \cdot PROPVOLAT$$

with

$$PROPVOLAT = 0.37 + 0.029 \cdot (100 - EAURES_T) + 0.117 \cdot TRSOLVOLAT$$

Furthermore, the addition of manure (containing urea type compounds and bicarbonates) is accompanied by a pH increase which is considered in the calculations. Immediately after the manure application, the soil pH at soil surface (PHVOL) increases by a value DPHVOL, which varies with the mineral N level as follows (Figure 8.11):

eq. 8.18

$$if \ \ 7 \leq PH_S \leq PHVOLS_G \ \ then \ DPHVOL = DPHVOLM \cdot \frac{PHVOL_G - PH_S}{PHVOL_G - 7.0}$$

$$if \ PH_S < 7.0 \ \ then \ \ DPHVOL = DPHVOLM$$

$$if \ PH_S > PHVOL_G \ then \ \ DPHVOL = 0$$

with

$$DPHVOLM = ALPHAPH_G \cdot NMINRES_T \ \ and \ \ DPHVOLM \leq DPHVOLMAX_G$$

Using the data given by Morvan (2001) and Chantigny *et al.* (2004), we can propose the following parameter values: $ALPHAPH_S = 0.005$, $DPHVOLMAX_G = 1.0$ and $PHVOLS_G = 8.6$.

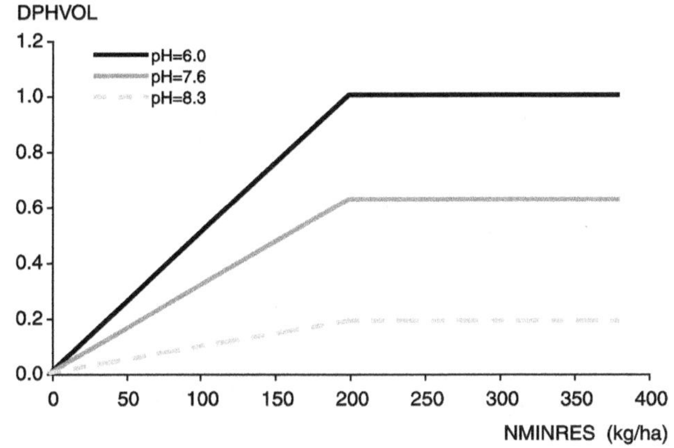

Figure 8.11. Effect of mineral N content of slurry and soil pH on the initial pH increase.

During the following days, the pH at the soil surface (PHVOL) returns to the soil pH value (PH_S), at a rate proportional to the decrease in the volatilizable pool: eq. 8.19.

eq. 8.19

$$PHVOL(I) = PH_S + DPHVOL \times \frac{NVOLATORG(I)}{NVOLATORG(IAP)}$$

The model then calculates the amounts of the four forms: NH4s, NH4l, NH3l and NH3g (in mol m^{-2}), using the acido-basic equilibria equations, Henry solubility equation the transfer equations of Beutier and Renon (1978). These amounts depend on soil temperature, water content, soil porosity, pH at soil surface and wind speed. The ammonia concentration at the soil surface (NH3SURF, in µg N m^{-3}) is:

eq. 8.20

$$NH3SURF(I) = 1.2 \cdot NH3g(I) \cdot 10^9$$

The potential ammonia volatilization rate (FSNH3, in µg N m^{-2} s^{-1}) is:

eq. 8.21

$$FSNH3(I) = \frac{NH3SURF(I) - NH3REF_C}{RAS(I) + RAA(I)}$$

where NH3REF$_C$ is the atmospheric ammonia concentration, which is about 10 µg m^{-3} in cattle production areas and 0 elsewhere; RAS and RAA (in s m^{-1}) are the resistances calculated according to Shuttleworth and Wallace (1985) if the "resistive" option is activated (eq. 7.18); otherwise RAA is the default parameter (RA$_G$=50) and RAS = 0. The calculation of FSNH3 is made hourly because volatilization decreases rapidly, assuming constant weather data throughout the day.

The actual ammonia volatilization rate (NVOLORG, in kg N ha^{-1} day^{-1}) is proportional to FSNH3 through a coefficient 0.036 which is a unit conversion factor (µg m^{-2} s^{-1} into kg ha^{-1} day^{-1}). However it can exceed neither the amount of ammonium at the soil surface (AMM(1), in kg N ha^{-1}) nor the volatilizable pool (NVOLATORG): eq. 8.22.

eq. 8.22

$$NVOLORG(I) = 0.036 \cdot FSNH3(I)$$
and
$$NVOLORG(I) \leq AMM(1)$$
$$NVOLORG(I) \leq NVOLATORG(I)$$

Finally, the volatilizable pool is updated daily (eq. 8.23)

eq. 8.23

$$NVOLATORG(I) = NVOLATORG(I) - NVOLORG(I)$$

8.5 Denitrification

Denitrification and N$_2$O emissions are calculated according to the model proposed by Hénault *et al.* (2005). The actual rate NDENENG (kg N-(N$_2$O+N$_2$) ha^{-1} day^{-1}) is assumed to be affected by soil temperature (FDENT), nitrate content (FDENNO3) and water content (FDENW), as follows:

$$\text{eq. 8.24}$$

$$NDENENG(I) = \frac{VPOTDENIT_S}{PROFDENIT_S} \sum_{Z=1}^{PROFDENIT_S} FDENT(Z,I) \cdot FDENNO3(Z,I) \cdot FDENW(Z,I)$$

In eq. 8.24 $PROFDENIT_S$ is the thickness (cm) of the denitrifying layer and $VPOTDENIT_S$ is the total denitrification potential rate (kg N ha^{-1} day^{-1}) of the soil, assumed to be constant with time. The effect of the three limiting factors (temperature: TSOL, nitrate content: NIT and saturation soil status: WFPS) are detailed in eq. 8.25, eq. 8.26 and eq. 8.27 and illustrated in Figure 8.12, Figure 8.13 and Figure 8.14. Obviously the daily denitrification rate in each layer cannot exceed the amount of nitrate-N in that layer.

$$\text{eq. 8.25}$$

$$if\ TSOL(Z,I) < TDENRE1_G\ then\ FDENT(Z,I) =$$
$$\exp\left[(TSOL(Z,I) - TDENRE1_G) \cdot 0.449 - 0.668\right]$$

$$if\ TSOL(Z,I) \geq TDENRE1_G\ then\ FDENT(Z,I) =$$
$$\exp\left[(TSOL(Z,I) - TDENRE2_G) \cdot 0.0742\right]$$

The default values for ($TDENREF1_G$, $TDENREF2_G$) are (11°C, 20°C) and (20°C, 29°C) for temperate and tropical soils respectively.

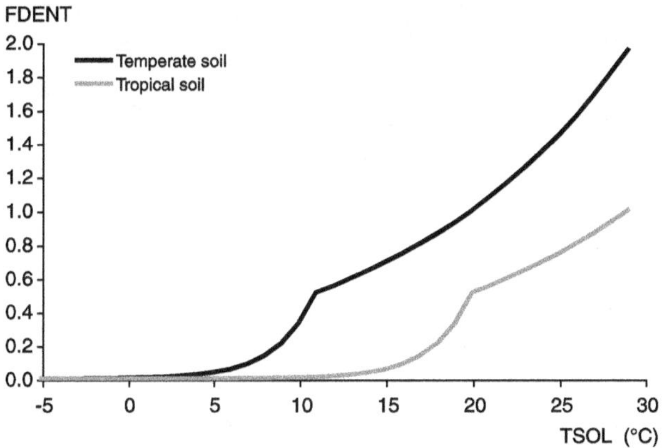

Figure 8.12. Relative effect of temperature on the denitrification rate.

The denitrification rate increases with the nitrate content in soil and depends also on bulk density (eq. 8.26)

$$\text{eq. 8.26}$$

$$FDENNO3(Z,I) = \frac{NIT(Z,I)}{NIT(Z,I) + 2.2\,DA(Z)}$$

In eq. 8.27, the soil water factor is in interaction with mineralisation through the SWRMIN variable using the $TREF_G$ parameter (Sierra, comm. pers.).

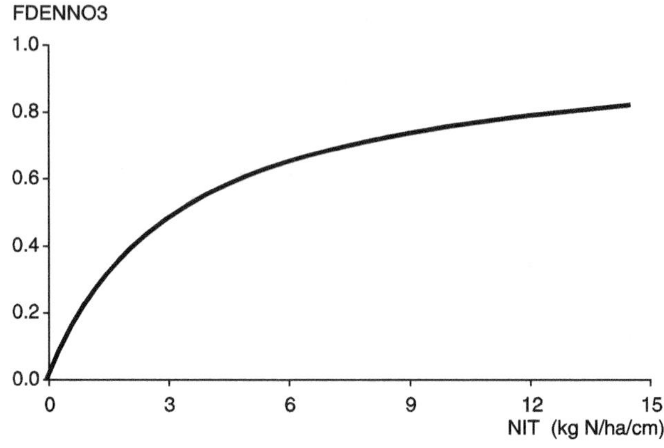

Figure 8.13. Relative effect of nitrate concentration on the denitrification rate.

eq. 8.27

$$FDENW\,(Z,I) = 0 \qquad\qquad\qquad if\ \ WFPS\,(Z,I) < SWRMIN(Z,I)$$

$$FDENW\,(Z,I) = \left[\frac{WFPS\,(Z,I) - SWRMIN\,(Z,I)}{1 - SWRMIN\,(Z,I)}\right]^{1.74} \quad if\ \ WFPS\,(Z,I) \geq SWRMIN(Z,I)$$

with

$$WFPS\,(Z,I) = \frac{HUR\,(Z,I) + SAT\,(Z,I)}{10\left(1 - \dfrac{DA(Z)}{2.66}\right)}$$

$$and\ \ SWRMIN\,(Z,I) = 0.62 - \frac{TSOL\,(Z,I) - TREF_G}{100}$$

Denitrification is highly sensitive to soil properties, both water content at field capacity and bulk density (Figure 8.14).

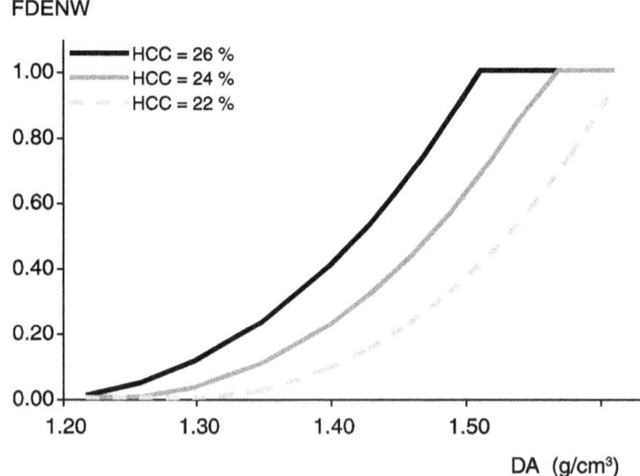

Figure 8.14. Effect of bulk density (DA) and water content at field capacity (HCC) on the relative denitrification rate when soil water content is equal to field capacity.

The N_2O evolved during denitrification (N2ODENIT, in kg N ha^{-1} day^{-1}) is calculated assuming a constant ratio between N_2O-N emissions and total denitrification (i.e. (N_2O+N_2)-N production), called RATIODENIT$_S$. The total amount of N2O emissions is N2ONIT (eq. 8.16) + N2ODENIT (eq. 8.28).

eq. 8.28

$$N2ODENIT(I) = NDENENG(I) \cdot RATIODENIT_S$$

8.6 Nitrogen uptake by plants and plant nitrogen status

Nitrogen uptake by the crop is simulated using the concept of either soil availability or crop demand being the more limiting. The model calculates and compares these two terms every day. The effective N uptake rate is equal to the smaller of these terms.

Lemaire and Gastal (1997) have shown that one can define a nitrogen content in shoots below which the plant metabolism is affected, which is called the 'critical N content'. Its value decreases with time and with plant biomass. Yet the way this decrease occurs is not the same throughout crop life. It depends on 3 factors:

• the plant metabolism requiring more or less nitrogen, illustrated by the difference between C3 and C4 plants,

• the plant's ability to store nitrogen in the form of reserves (proteins, amino-acids etc.) that explains the differences between cereals and proteinaceous plants,

• the inter-plant competitive processes that play a role on senescence and thus on the C/N ratio within the plant. This requires to consider differently isolated plants and plants within a dense canopy.

These three components are not always considered with the same attention. In the first versions of STICS the early crop phase was simply considered as constant in terms of maximal and critical N content. In practice only the critical curve can be found from experiments (Justes *et al.*, 1994), the maximal level being very difficult to ascertain.

8.6.1 The dilution curves

If NMAX is the maximal crop nitrogen content and written as a function of plant biomass (W that can be slightly different from MASEC), the daily N demand (DEMANDE, in kg N ha^{-1} day^{-1}) is the product of the crop growth rate (DLTAMS, in t ha^{-1} day^{-1}) and the derivative of NMAX relative to W:

eq. 8.29

$$DEMANDE(I) = \frac{\Delta NMAX(I)}{\Delta I} = \frac{\Delta NMAX(W)}{\Delta W} \cdot DLTAMS(I)$$

In STICS the expression of NMAX varies as a function of 2 criteria: the density of the canopy and the presence of storage organs; the first one defining the parameters of the NMAX=f(W) curves (according to Lemaire and Gastal, 1997, or Justes *et al.*, 1997) and the second one defining W.

8.6.1.a The dilution curves of N in aboveground biomass

Two curves define the critical response function (NC): one for low biomass corresponding to isolated plants (NI) and one for high biomass with dense canopies (NP). Similarly, two curves can be derived to characterize the N demand of these two populations (NMAXI and NMAXP). These 4 curves can be described by similar power functions:

eq. 8.30

$$NI(I) = ADILI \cdot W(I)^{-BDILI}$$

$$NP(I) = ADIL_P \cdot W(I)^{-BDIL_P}$$

$$NMAXI(I) = ADILMAXI \cdot W(I)^{-BDILMAXI}$$

$$NMAXP(I) = ADILMAX \cdot W(I)^{-BDILMAX}$$

In addition to the prescribed parameters $ADIL_P$ and $BDIL_P$, the other parameters are obtained using the following assumptions:

• There is a value of metabolic-N concentration ($NMETA_P$) corresponding to the plantlet nitrogen content that is composed of functional organs only. This value is a function of species metabolism: 6.47% for C3 crops (e.g. wheat) and 4.8% for C4 crops (e.g. maize) (Justes *et al.*, 1997; Lemaire and Gastal, 1997).

• It is possible to define an arbitrary biomass for this plantlet status ($MASECMETA_G$ = 0.04 t ha^{-1}; Justes *et al.*, 1997).

• It is possible to define experimentally the biomass value at the intersection of the two curves that depends on the form of the canopy ($MASECDIL_P$) and at this point the reserve N content is $NRES_P$.

• The curvature of the maximal curve is the same than that of the critical curve for dense canopy: $BDILMAX = BDIL_P$

We can then calculate the missing parameters of eq. 8.30
eq. 8.31)

eq. 8.31

$$BDILI = \frac{BDIL \cdot \log(MASECDIL_P) - \log\left(\dfrac{ADIL_P}{NMETA_P}\right)}{\log\left(\dfrac{MASECDIL_P}{MASECMETA_G}\right)}$$

$$ADILI = \frac{NMETA_P}{MASECMETA_G^{-BDILI}}$$

$$ADILI = \frac{NMETA_P}{MASECMETA_G^{-BDILI}}$$

$$ADILMAX = ADIL_P + \frac{NRES_P}{MASECDIL_P^{-BDIL_P}}$$

$$BDILMAXI = -\frac{\log\left(\dfrac{ADILMAX \cdot MASECDIL_P^{-BDIL_P}}{NMETA_P}\right)}{\log\left(\dfrac{MASECDIL_P}{MASECMETA_G}\right)}$$

$$ADILMAXI = \frac{NMETA_P}{MASECMETA_G^{-BDILMAXI}}$$

An example of these dilution curves is given in Figure 8.15.

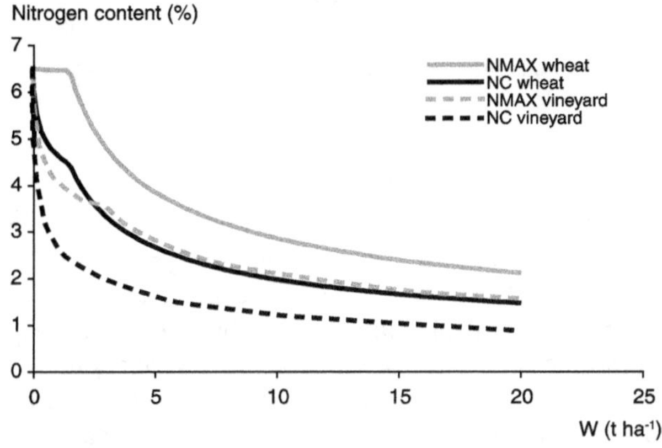

Figure 8.15. Maximal (NMAX) and critical (NC) dilution curves for wheat and vineyard.

8.6.1.b The presence of storage organs

The N demand due to vegetative organs is assumed to follow the maximal dilution curve, whereas the demand associated with the "fruit" (either grains or storage organs) depends on the nitrogen status of the crop through the variable ABSODRP. The biomass (W) used to calculate the N demand from the maximal dilution curve can be reduced using the parameters $INNGRAIN1_P$ and $INNGRAIN2_P$ (eq. 8.32)

eq. 8.32

$$W(I) = MASECVEG(I) + ABSODRP(I) \cdot MAGRAIN(I)$$

and

$$ABSODRP(I) = \frac{INNS(I) - INNGRAIN2_P}{INNGRAIN1_P - INNGRAIN2_P}$$

$$if \quad INNS(I) \le INNGRAIN2_P \quad ABSODRP(I) = 1.0$$

$$if \quad INNS(I) \ge INNGRAIN1_P \quad ABSODRP(I) = 0.0$$

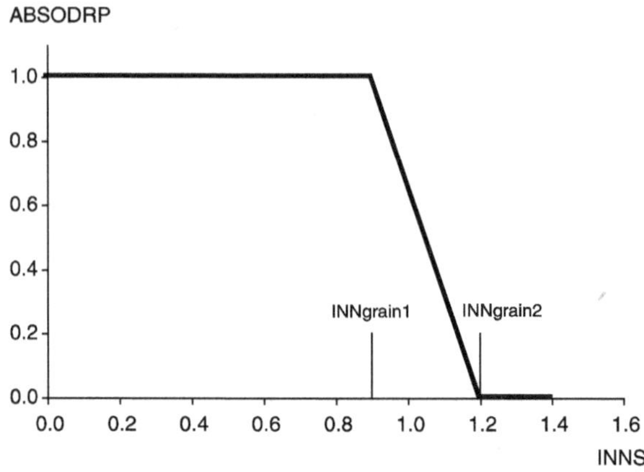

Figure 8.16. ABSODRP versus nitrogen stress index.

8.6.2. The N supply from the soil

The soil supply is the maximum amount of mineral N that the soil can deliver to the surface of roots, for a given status of soil and plant root system. It is calculated for each elementary layer (1 cm thick) from the surface to the maximum rooting depth (ZRAC, in cm). It does not account for possible nitrate upflow by capillary rise (this would require a knowledge of the nitrate concentration in the soil below the rooting depth).

The soil N supply in each soil layer (FLUXSOL, in kg N ha^{-1} day^{-1}) is determined by the transport of mineral N from a given soil location to the nearest root by convection and diffusion (eq. 8.33).

eq. 8.33

$$FLUXSOL(Z,I) = CONV(Z,I) + DIFF(Z,I)$$

The convection flow in each elementary soil layer (CONV, in kg N ha^{-1} day^{-1}) is the product of the water flow (i.e. the transpiration flow EPZ, in mm day^{-1}, see § 7.3.3) and the mean NO_3 concentration (CONCN, in kg N ha^{-1} mm^{-1} water). The exchangeable NH_4 is not included, since it assumed to be immobile. There is no nitrate transport by convection if the transpiration is nil (due to absence of roots or severe water stress):

eq. 8.34

$$CONV(Z,I) = EPZ(Z,I) \cdot CONCN(Z,I)$$

The diffusion flow in each elementary soil layer (DIFF, in kg N ha^{-1} day^{-1}) is the product of the effective diffusion coefficient of mineral N (DIFE, in cm^2 day^{-1}) and the horizontal gradient of mineral N concentration in the soil (in kg N ha^{-1} mm^{-1} water cm^{-1} soil). This gradient is calculated from the effective root density profile (LRACZ), assuming that roots are vertical and equidistant and that mineral N concentration decreases linearly from the middle of two adjacent roots up to root surface (mineral N concentration is nil at the root surface). These assumptions lead to eq. 8.35.

eq. 8.35

$$DIFF(Z,I) = 4\sqrt{\pi} \cdot DIFE(Z,I) \cdot [NIT(Z,I) + AMM(Z,I)] \cdot \sqrt{LRAC(Z,I)}$$

with

$$DIFE(Z,I) = DIFN_G \cdot \frac{HUR(Z,I) - HUMIN(Z)}{HUCC(Z) - HUMIN(Z)}$$

$$if \quad DIFE(Z,I) < 0 \quad then \quad DIFE(Z,I) = 0.0$$

The effective diffusion coefficient is a function of soil water content and bulk density (de Cockborne *et al.*, 1988). Only the moisture effect (which is the main effect) is considered in STICS. The hypothesis of uniform root distribution leads to maximize the diffusive flow. In fact, roots are heterogeneously distributed so that the diffusive flow is smaller. In order to account for this effect, the diffusion coefficient at field capacity ($DIFN_G$) used in STICS (0.018 cm^2 day^{-1}) is lower than the measured values reported in the literature (0.10-0.25 cm^2 day^{-1}) (Barber and Silberbush, 1984; de Cockborne *et al.*, 1988; Kersebaum and Richter, 1991).

8.6.3 The N uptake capacity

The N uptake by the root system is an active physiological process which depends on the intrinsic absorption capacity of the plant, its root density and the nitrate concentration in the soil. The specific absorption capacity VABS (per unit of root area, in μmol N h^{-1} cm^{-1} root) increases with nitrate concentration according to a double Michaëlis-Menten kinetics (Devienne-Barret *et al.*, 2000) (eq. 8.36). These kinetics correspond to two types of transport systems: a high affinity transport system 'HATS' (with low VMAX1$_p$ and KM1$_p$) and a low affinity transport system 'LATS' (with high VMAX2$_p$ and KM2$_p$).

eq. 8.36

$$VABS(Z,I) = \frac{VMAX1_P \cdot CONCN(Z,I)}{KM1_P + CONCN(Z,I)} + \frac{VMAX2_P \cdot CONCN(Z,I)}{KM2_P + CONCN(Z,I)}$$

In eq. 8.36 CONCN is the molar concentration of mineral nitrogen (μmol l^{-1}) and the parameters VMAX are in μmol cm^{-1} h^{-1}. Mineral N is considered as a whole ($NH_4 + NO_3$), so that any selectivity between ammonium and nitrate absorption is not accounted for.

The potential uptake rate in each soil layer is FLUXRAC (kg N ha^{-1} day^{-1}). It is proportional to the effective root density (eq. 8.37) which is limited by the threshold $LVOPT_G$ above which uptake is no longer limited by root density:

eq. 8.37

$$FLUXRAC(Z,I) = 33.6 \cdot VABS(Z,I) \cdot LRACZ(Z,I)$$

The coefficient 33.6 is the ratio of μmol cm^{-2} h^{-1} to kg ha^{-1} day^{-1}. Figure 8.17 shows the dynamics of FLUXRAC versus the nitrate concentration in soil and the contribution of both transport systems to the uptake capacity.

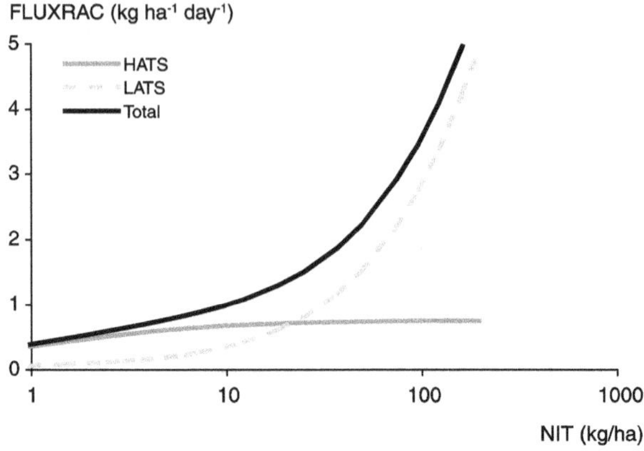

Figure 8.17. N uptake capacity versus nitrate content in soil. Parameter values: VMAX1$_p$=0.0018; KM1$_p$=50; VMAX2$_p$=0.050; KM2$_p$=25000; LRACZ=0.20; ZRAC=60; HUR=0.20.

8.6.4 The actual N uptake

The mineral N available for root uptake in each layer (OFFRN) is equal to the smallest of the three terms (eq. 8.38): soil supply, uptake capacity and available mineral N:

eq. 8.38

$$OFFRN\,(Z,I) = MIN\,(FLUXSOL\,(Z,I),\,FLUXRAC\,(Z,I),\,NIT\,(Z,I)+AMM\,(Z,I))$$

The integration of OFFRN over the whole profile yields CUMOFFRN. In each layer, the N supply can be compared to the crop demand through the ratio PROP (eq. 8.39):

eq. 8.39

$$PROP\,(I) = \frac{DEMANDE\,(I)}{CUMOFFRN\,(I)} \quad with \quad PROP\,(I) \leq 1.0$$

If PROP = 1, the soil N supply is the factor limiting N uptake. In this case the N uptake in each layer is equal to the N supply OFFRN. Conversely, the demand is the factor limiting N uptake if PROP < 1; in this case, the actual N uptake in each layer is smaller than the N supply and proportional to it.

In both cases, the actual N uptake in each soil layer (ABSZ) and the total uptake over the root profile (ABSO) can be written as functions of the PROP variable (eq. 8.40):

eq. 8.40

$$ABSZ\,(Z,I) = OFFRN\,(I) \cdot PROP\,(I)$$
$$and$$
$$ABSO\,(I) = CUMOFFRN\,(I) \cdot PROP\,(I)$$

8.7 Nitrogen fixation by legumes

The influence of crop growth and phenology on the activity of biologic nitrogen fixation (BNF) has been shown experimentally by many authors, as well as the influence of environmental factors, and in particular of soil nitrate availability (Voisin *et al.*, 2002).

Firstly, BNF intensity is known to vary during crop growth. It increases until the early stages of reproductive development and then, after passing a plateau, declines till the end of crop life (Lawrie and Wheeler, 1974; Bethlenfalvay *et al.*, 1978; Bethlenfalvay and Phillips, 1977). These variations are thought to be the result of competition for carbon between nodules and seeds (Jeuffroy and Warenbourg, 1991), and differ according to species, and sometimes to cultivars (Cousin, 1997).

Secondly, BNF has been shown to be closely linked to crop photosynthetic activity through experiments using labelled CO_2 (Warembourg, 1983; Kouchi and Nakaji, 1985; Gordon *et al.*, 1985; Voisin *et al.*, 2003), and thus correlated to crop growth rate (Finn and Brun, 1982; Jensen, 1987; Voisin *et al.*, 2002).

Finally, several abiotic factors have been mentioned to explain BNF inhibition, and especially soil nitrate availability. Indeed, the negative effect of nitrate on BNF has been reported by several authors (Mac Duff *et al.*, 1996; Waterer and Vessey, 1993), soil nitrate availability inhibiting both nodule formation and nitrogenase activity (Sprent *et al.*, 1988). BNF is also limited by soil water deficiency (Pena-Cabriales and Castellanos, 1993) but this effect may be reversible (Guérin *et al.*, 1991). Waterlogging may prevent BNF, while limiting O_2 availability for bacteria (Jayasundara *et al.*, 1998). Low temperatures reduce nodule activity (Rennie and Kemp, 1980) and high temperatures affect bacterial lifespan in the soil (Hungria and Vargas, 2000) and nitrogenase activity.

In STICS, symbiotic N_2 fixation by legumes is simulated considering three criteria. The first of these is the presence of nodules, depending on their own phenology and lifespan and also on an inhibiting effect of excessive nitrate in the soil. The second is the capacity of the plant to feed these supplementary symbiotic organs depending on plant growth rate, and the third is the soil-dependent physicochemical conditions allowing optimal nodule activity: soil nitrate level, water deficit, anoxia and temperature (Debaeke *et al.*, 2001; Voisin *et al.*, 2003). The first two criteria define the potential N_2 fixation while the third defines the actual N_2 fixation.

8.7.1 The potential N_2 fixation

The potential N_2 fixation (FIXPOT in kg N ha^{-1} day^{-1}) is calculated as the product of a phenology-dependent coefficient PROPFIXPOT (between 0 and 1) and the maximal fixation capacity of the crop (FIXMAX in kg N ha^{-1} day^{-1}).

eq. 8.41

$$FIXPOT(I) = PROPFIXPOT(I) \cdot FIXMAX(I)$$

The PROPFIXPOT coefficient varies according to growing degree-days (eq. 8.42) calculated as for root growth (see § 5.1). The fixation process begins at the IDNO stage (defined by the thermal duration STLEVDNO$_p$) and stops at the IFNO stage (defined by the thermal duration STDNOFNO$_p$). The potential curve then decreases until the death of nodules, corresponding to the IFVINO stage, during the $STFNOFVINO_p$ thermal duration. The establishment rate of nodules between IDNO and IFNO stages depends on the potential rate $VITNO_p$ and on growing degree-days.

It may be inhibited by high mineral nitrogen levels, under the control of NODN which is nil when soil mineral nitrate concentration exceeds the threshold $CONCNNODSEUIL_p$ (in kg N ha^{-1} mm^{-1} water), and otherwise is equal to 1.0 (Figure 8.18 and eq. 8.42).

eq. 8.42

$$if \quad I \leq IDNO \quad or \quad I > IFVINO, \quad PROPFIXPOT\,(I) = 0.0$$

$$if \quad IDNO \leq I \leq IFNO,$$
$$PROPFIXPOT\,(I) = PROPFIXPOT\,(I-1) + VITNO_p \cdot (TCULT\,(I) - TCMIN_P) \cdot NODN\,(I)$$

$$if \quad IFNO \leq I \leq IFVINO,$$
$$PROPFIXPOT(I) = PROPFIXPOT\,(I-1) - \frac{PROPFIXPOT\,(IFNO)}{STFNOFVINO_P} \cdot (TCULT\,(I) - TCMIN_P)$$

$$PROPFIXPOT\,(I) \leq 1.0$$

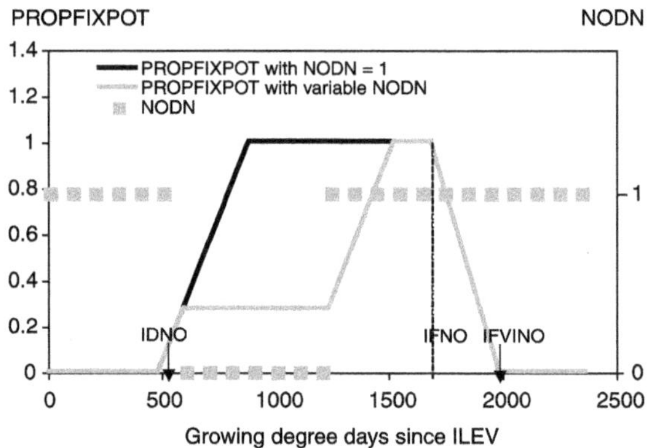

Figure 8.18. Evolution of PROPFIXPOT versus thermal time, for two levels of soil nitrate content: low level (NODN=1), high level (NODN variable). Parameter values: $STLEVDNO_p$=500, $STDNOFNO_p$=1200, $STFNOFVINO_p$=300, $VITNO_p$=0.0025.

The maximal fixation capacity of the crop FIXMAX is calculated from above-ground biomass growth rate (eq. 8.43). The $FIXMAXVEG_p$ parameter defines the N fixed per ton of produced vegetative dry matter and the $FIXMAXGR_p$ parameter defines the amount of N fixed per ton of grain dry matter produced.

eq. 8.43

$$FIXMAX(I) = FIXMAXVEG_P \cdot (DLTAMS\,(I) - DLTAGS\,(I)) + FIXMAXGR_P \cdot DLTAGS(I)$$

DLTAMS is the daily biomass accumulation and DLTAGS is the daily grain filling.

8.7.2. The actual N$_2$ fixation

To calculate the actual N$_2$ fixation (FIXREEL, in kg N ha^{-1} day^{-1}), the potential N$_2$ fixation FIXPOT is multiplied by indices (varying between 0 and 1) corresponding to constraints due to anoxia (FXA), temperature (FXT), water content (FXW) and soil mineral nitrogen (FXN) (eq. 8.44).

eq. 8.44

$$FIXREEL\ (I) = FIXPOT\ (I) \cdot FXT\ (I) \cdot FXA\ (I) \cdot \min\ (FXW\ (I),\ FXN\ (I))$$

Limitation by temperature (FXT) uses the soil temperature in the nodulation zone and is a trapezoidal function defined by four cardinal temperatures (TEMPNOD1$_p$ to TEMPNOD4$_p$) as depicted in Figure 8.19.

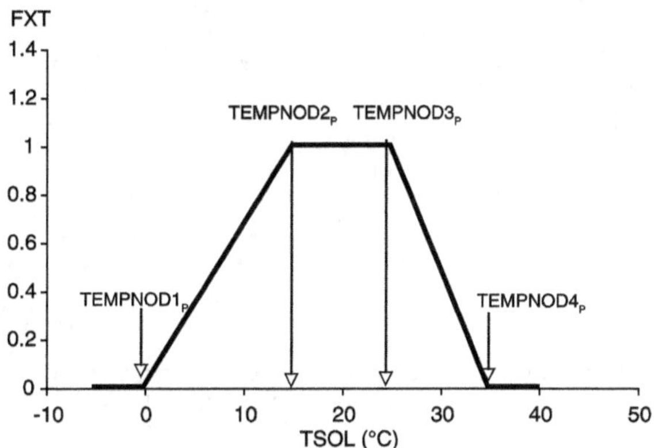

Figure 8.19. Effect of temperature on N$_2$ fixation. Parameter values are TEMPNOD1$_p$= 0°C, TEMPNOD2$_p$ = 15°C, TEMPNOD3$_p$ = 25°C, TEMPNOD4$_p$ = 35°C.

The water stress factor (FXW) is estimated from the proportion of elementary soil layers in the nodulation area whose water content HUR is at least as high as the permanent wilting point HUMIN (eq. 8.45).

eq. 8.45

$$FXW(I) = \frac{1}{PROFNOD_P - PROFSEM_T + 1} \sum_{Z=PROFSEM_T}^{PROFNOD_P} H(I,Z)$$

$$\text{with } \quad H(I,Z)=1 \quad if \quad HUR(I,Z) > HUMIN(Z)$$
$$H(I,Z)=0 \quad if \quad HUR(I,Z) \le HUMIN(Z)$$

Limitation by anoxia (FXA) is calculated according to the same principle, as the proportion of elementary soil layers which are in aerobic conditions using the ANOX variable (see $ 5.2.2.b.).

165

eq. 8.46

$$FXA(I)=1-\frac{1}{PROFNOD_P - PROFSEM_T +1}\sum_{z=PROFSEM_T}^{PROFNOD_P}ANOX(Z,I)$$

Finally the fixation is partially inhibited when the mean amount of mineral nitrogen in the rooting zone (AZORAC/ZRAC, in kg N ha^{-1} cm^{-1} soil) exceeds the threshold AZOZRAC100$_P$, and is fully inhibited when it exceeds the threshold AZOZRAC0$_P$ (Figure 8.20).

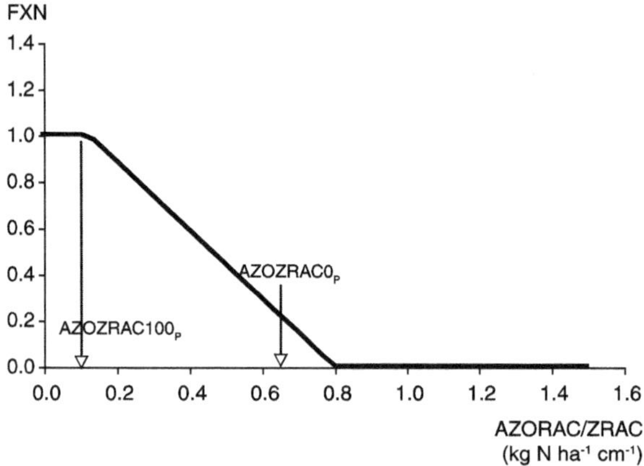

Figure 8.20. Effect of soil mineral N content on N$_2$ fixation. Parameter values are AZOZRAC0$_P$ = 0.80 kg N ha^{-1} cm^{-1} soil and AZOZRAC100$_P$ = 0.125 kg N ha^{-1} cm^{-1} soil.

9 Transfers of heat, water and nitrate

As far as transfer modelling is concerned, two methods are commonly used (Addiscott and Wagenet, 1985): the functional reservoir type model and the flux-gradient model. Most crop models rely successfully on the first of these, whose main limitation is that it does not take explicitly into account the capillary rises of water and nitrate, which can be important in highly conducting soils. In the case of a cultivated soil, this obstacle can be partly overcome if it is assumed that the depth where water and nitrogen are taken up by the plant is a bit deeper than the actual depth of rooting. In order to rigorously simulate the rising flows, it is necessary to work with models that use Darcy's law and the convection-dispersion equation (Addiscott and Wagenet, 1985). Though work on soil transfer functions (deducing the hydrodynamic parameters required for transfer laws from readily available soil data) has progressed (Bruand *et al.*, 2003) the variability of the hydrodynamic parameters in space and between soils is still difficult to assess (Vachaud *et al.*, 1993). Consequently, these mechanistic models are difficult to use and to parameterize. Several studies have shown that the transfer of nitrate can be simulated with a functional as well as with a mechanistic model provided that the dispersivity is weak and that the thickness of elementary layers is small (Vinten and Redman MH, 1990; Van der Ploeg *et al.*, 1995). On the other hand, it is clear that a functional model cannot simulate precisely, and with a small enough time step, the water content of surface layers and their porosity to the air, which can hinder the estimation of soil subsurface phenomena such as plant germination and emergence or nitrogen losses from denitrification. Heat transfers in the soil are very seldom accounted for in crop models, assuming that the soil temperature-dependent processes are sufficiently superficial to respond to air temperature.

9.1 Soil temperature

Temperature variation in soil depends on the surface conditions which determine the daily thermal variation but also thermal inertia related to the environment. This inertia is the cause of the lower daily average temperatures in deep layers compared to those at the surface: this is the annual thermal variation. The temperature at the upper limit for calculating soil temperature is assumed to be TCULT and the daily thermal amplitude (AMPLSURF) at the surface is given by eq. 9.1. Crop temperature calculations are explained in § 6.6.2.

eq. 9.1

$$AMPLSURF(I) = TCULTMAX(I) - TCULTMIN(I)$$

The daily thermal amplitude, AMPLZ, and the soil temperature, TSOL, at depth Z in the soil are calculated using a formalisation suggested by McCann *et al.* (1991). It is a recurrent calculation using the previous day's values.

eq. 9.2

$$AMPLZ(Z,I) = AMPLSURF(I) \cdot exp\left(-Z\sqrt{\frac{7.272 \ 10^{-5}}{2 \ DIFTHERM_G}}\right)$$

$$TSOL(Z,I) = TSOL(Z,I-1) - \frac{AMPLZ(Z,I)}{AMPLSURF(I)}[TCULT(I-1) - TCULTMIN(I)]$$

$$+ 0.1[TCULT(I-1) - TSOL(Z,I-1)] + \frac{AMPLZ(Z,I)}{2}$$

The thermal diffusivity $DIFTHERM_G$ is assumed to be independent of soil water conditions and general throughout the various soil types. A value of 5.37 10^{-3} cm²s⁻¹ is proposed, based on the work by McCann *et al.* (1991).

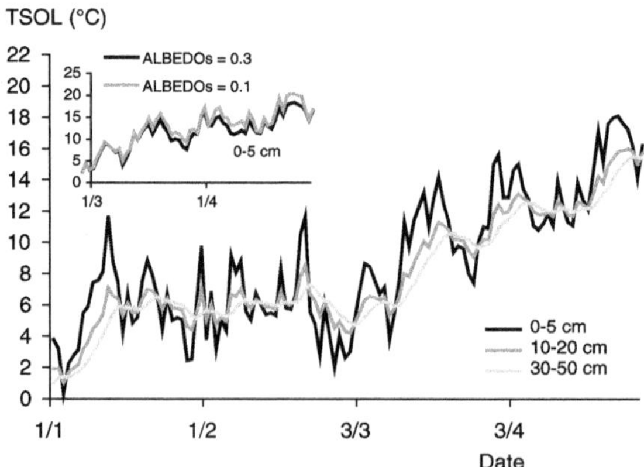

Figure 9.1. Calculation of soil warming in spring (example of a site in northern France) as a function of the considered soil layer and the soil colour represented by its albedo.

9.2 Transfers of water and nitrate in free drained soil

The nitrates circulate with water downwards through the soil. There are also some extreme cases of drought or waterlogging which require the simulation of upward fluxes. The way these transfers are accounted for in STICS relies on the soil compartmental description and on the tipping bucket concept.

9.2.1 Soil compartmentation

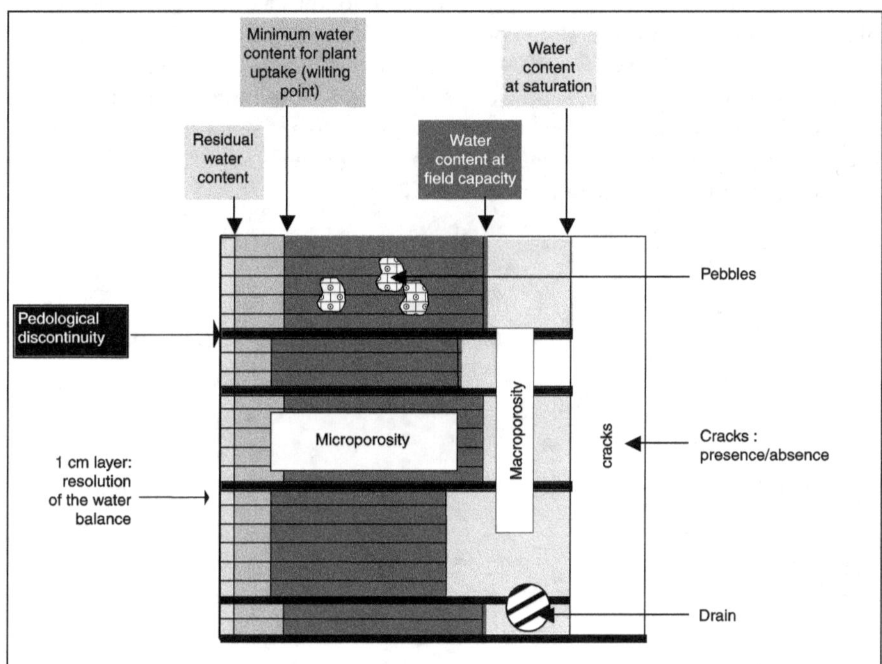

Figure 9.2. Schema of soil pore space components.

As shown in Figure 9.2, the description of the soil can involve up to four compartments: microporosity, macroporosity, cracks (the case of swelling clay soils) and pebbles. However, only the description of microporosity is obligatory, the description of the other compartments being optional.

9.2.2 Soil microporosity: basis for calculating water and nitrogen transfer values

Water transfer in the soil microporosity is calculated per elementary 1 cm layer using a reservoir-type analogy. Water fills the layers by downward flow, assuming that the upper limit of each basic reservoir corresponds to the layer's field capacity. The permanent features of the elementary layers, as well as the initial water contents, are inferred from those of the 5 layers describing the soil: $HMINF_S(H)$ (minimal moisture or wilting point of the layer H), $HCCF_S(H)$ (field capacity moisture of the layer H), $DAF_S(H)$ (bulk density of fine earth for layer H). It is possible to account for pebbles (see § 9.2.3). If the flow is not obstructed, (cf. macroporosity), the excess water above field capacity is drained downward. The soil layers affected by evaporation, i.e. down to a depth of $ZESX_S$, can dry until they reach the residual soil water content. In deeper layers, the water is only extracted by the plant and therefore always remains above the wilting point.

The transfer of nitrates is also described using this reservoir-type analogy, according to the "mixing cells" principle. Any nitrate arriving by convection with water in the elementary layer mixes with the nitrate already present. Excess water then leaves with the new concentration of the mixture. This description produces results which are very similar to the convection-dispersion model, the thickness of layers (EPD_S) being equal to twice the dispersivity (Mary *et al.*, 1999). In the first STICS versions, this thickness was fixed at 1 cm, which often led to too weak a dispersion. A minimum concentration level may exist ($CONCSEUIL_S$), below which mineral nitrogen cannot be leached (eq. 9.3). This can be a simple way to simulate ammonia nitrogen without using the simulation of the ammoniacal phase of mineralisation (see § 8.3).

eq. 9.3

$$if \; \frac{\sum_{k=Z1}^{Z2} NIT(k)}{\sum_{k=Z1}^{Z2} HUR(k)} \leq CONCSEUIL_S \quad then \quad AZLES(Z,I) = 0.$$

$$if \; \frac{\sum_{k=Z1}^{Z2} NIT(k)}{\sum_{k=Z1}^{Z2} HUR(k)} > CONCSEUIL_S$$

$$then \; AZLES(Z,I) = HUSUP(Z-1,I) \cdot \left[\frac{\sum_{k=Z1}^{Z2} NIT(k)}{\sum_{k=Z1}^{Z2} HUR(k)} - CONCSEUIL_S \right]$$

with Z1 and Z2 the limits of the mixing cell , and $Z \in [Z1, Z2]$

The amounts of drained water and leached nitrogen, i.e. leaving via the base of the soil profile (eq. 9.3 for $Z = PROFSOL_S$), are not retrievable by another crop. Upwards nitrate movements occur via plant uptake only (§ 8.6). Capillary rises provided by humid subsoil can be taken into account (Figure 9.3 and § 9.2.4)

9.2.3 Pebbles

In the presence of pebbles, defined by their volumetric percentage ($CAILLOUX_S(H)$) in the layer H, the typical moisture levels and the bulk density of the layers are modified depending on the amount and type of pebbles (see for example

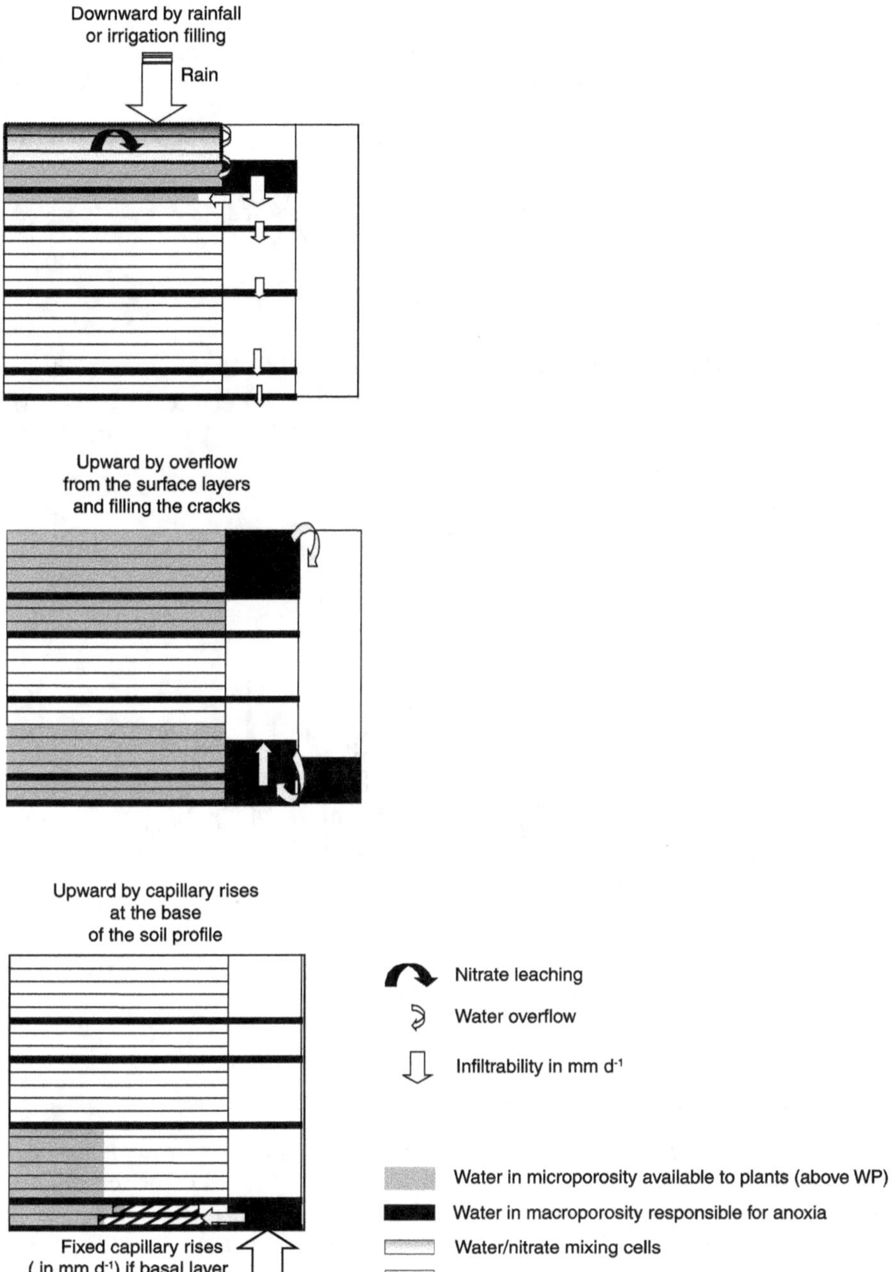

Figure 9.3. Water/nitrate transfers in the soil: mainly downwards but could be upwards if cracks are present or capillary rise occurs.

eq. 9.5 for bulk density), according to Gras, 1994. The type of pebbles is defined by a volumetric mass value ($MASVOLCX_G$) and a field capacity moisture value ($HCCX_G$), assuming that the minimal moisture content of pebbles (HMINCX) is simply calculated with reference to fine earth values (eq. 9.4)

eq. 9.4

$$HMINCX(H) = HUMIN(H) \cdot \frac{HCCX_G(H)}{HUCC(H)}$$

eq. 9.5

$$DA(H) = \frac{1}{100}\left[CAILLOUX_S(H) \cdot MASVOLCX_G\left(TYPECAILLOUX_S(H)\right) + \left(100 - CAILLOUX_S(H)\right) \cdot DAF_S(H)\right]$$

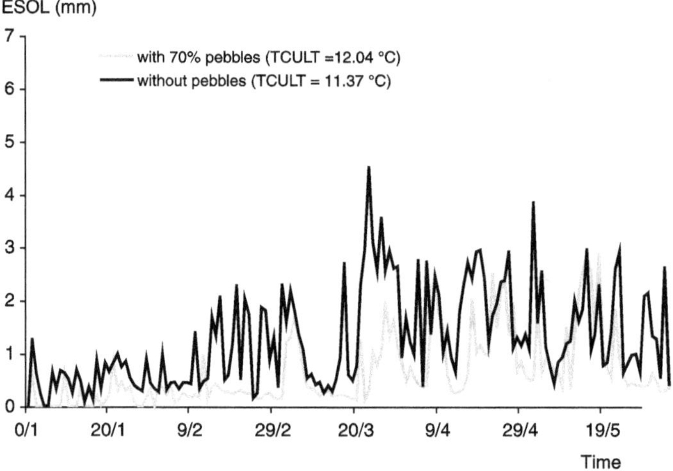

Figure 9.4. Effect of pebbles (70% in volume) on soil evaporation and the consequences on mean spring crop temperature in a vineyard in south-eastern France.

9.2.4 Macroporosity and cracks

Two compartments can be functionally added in the soil: macroporosity and shrinkage cracks, should this occur in the soil ($CODEFENTE_S = 1$). The macroporosity compartment is discretized by layer (but not by the 1 cm layers used for the standard microporosity compartments) whereas the cracks correspond to a single entity. Needless to say, this decomposition is somewhat imaginary and arbitrary; it is only justified in that it makes the modelling more convenient.

At each pedological discontinuity level, a daily infiltrability parameter is defined ($INFIL_S(H)$ in mm day^{-1}). At the soil surface, the daily amount of water penetrating into the soil accounts for soil surface status (see § 6.4) and allows the runoff estimates. Between two discontinuous levels, the "downward" circulation occurs due to

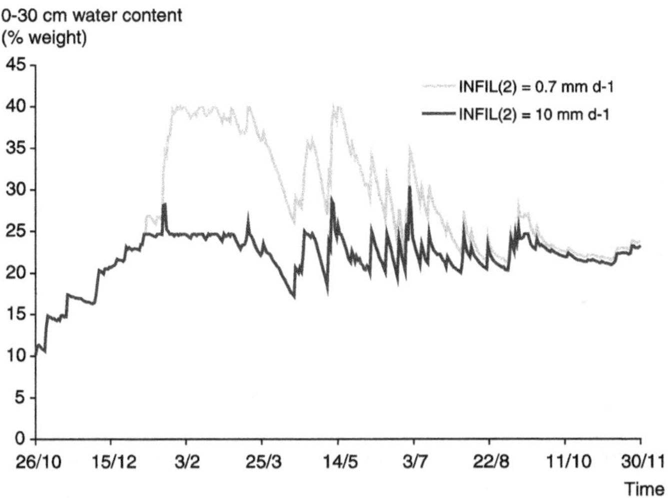

Figure 9.5. Effect of a strong decrease in infiltrability (INFIL) at the base of the second layer located at 30 cm and its consequences on root penetration due to anoxia.

"overflowing" from one layer to the next, as mentioned above (c.f. microporosity). At a discontinuity level, the amount of water which has filtered through is limited by the infiltration parameter $INFIL_S$ (H) which also sustains the macroporosity of the layer. As the infiltrability acts when the microporosity is filled, its value is similar to the saturated hydraulic conductivity, though it cannot be exactly equated to it because of a residual role of sorptivity (Boivin *et al.*, 1987).

The pore space corresponding to the macroporosity of each layer (MACROPOR(H)) is evaluated by one of the following two formulae, depending on the soil swelling properties ($CODEFENTE_S$ =0 or 1):

eq. 9.6

$$if\ CODEFENTE_S = 0\ \ then\ MACROPOR(H) = \left[\left[1 - \frac{DA(H)}{2.66}\right] - \left[\frac{HCC(H)}{100}DA(H)\right]\right] \cdot 10 \cdot EPC(H)$$

$$if\ CODEFENTE_S = 1\ \ then\ MACROPOR(H) = 0.5 \cdot [HCC(H) - HMIN(H)] \cdot DA(H)$$

If the layer's macroporosity has reached saturation, the anoxia index of each layer (ANOX (Z)) is allocated the value of 1 and can restrict root growth. In the case of swelling soils, the fissures, when open, are filled by overflow from the surface layer; water supply by rain interception at the surface is not taken into consideration. The opening of cracks (variable BOUCHON) depends on the combination of two factors in at least one of the layers: empty macroporosity and a root front deeper than the base of the layer.

If the basal soil layer is dry enough (below the $HUMCAPIL_S$ threshold), capillary rise can occur from the subsoil into the soil, at a constant rate ($CAPILJOUR_S$) until the basal layer reaches a moist status (above $HUMCAPIL_S$). As, in the model, these upward transfers take place through the macroporosity (they are considered negative infiltration), they require a zero value of infiltrability at the base of the soil to be active.

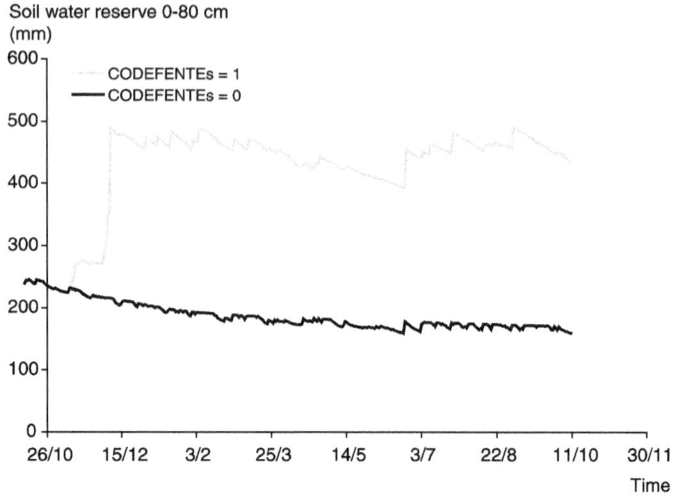

Figure 9.6. Effect of swelling properties (CODEFENTE$_S$=1) in case of a heavy clayey soil of low infiltrability (O.1) on the storage capacity of the soil and its consequences on runoff, transpiration and finally sugar cane production (La Reunion).

9.3 Case of artificially drained soil

The introduction of agricultural drainage into STICS raises two major problems: (1) the time step characteristic of the functioning of a drainage system in the temperate climate of mainland France is about one hour and not one day; (2) the functioning of a drainage system is two- or even three-dimensional and not unidimensional. These two problems require a modification of models usually used in drained soils.

The classic draining system (Figure 9.7) uses the properties of symmetry arising from the presence of lines of drains with spacing (2 LDRAIN$_S$) which is generally constant within a field. Flow is assumed to occur from the space between drains towards the drain following a shape characterized by the parameter BFORMNAPPE$_S$; it occurs within a water table based on an impermeable floor, the depth of which (PROFIMPER$_S$) may be greater than the soil depth considered in STICS.

A simplification of the baseline Hooghoudt equation (1940) is used to simulate the daily water outflow (QDRAIN) at the drain level (eq. 9.7) assuming a single hydraulic conductivity above and below the drains (KSOL$_S$).

eq. 9.7

$$QDRAIN(I) = \frac{KSOL_S \; HMAX(I)^2 + 2KSOL_S \cdot DE(I) \cdot HMAX(I)}{LDRAIN_S^2}$$

This equation relies on the estimation of DE, the equivalent depth of the aquifer below the level of drains, which first requires calculating HMAX (eq. 9.8).

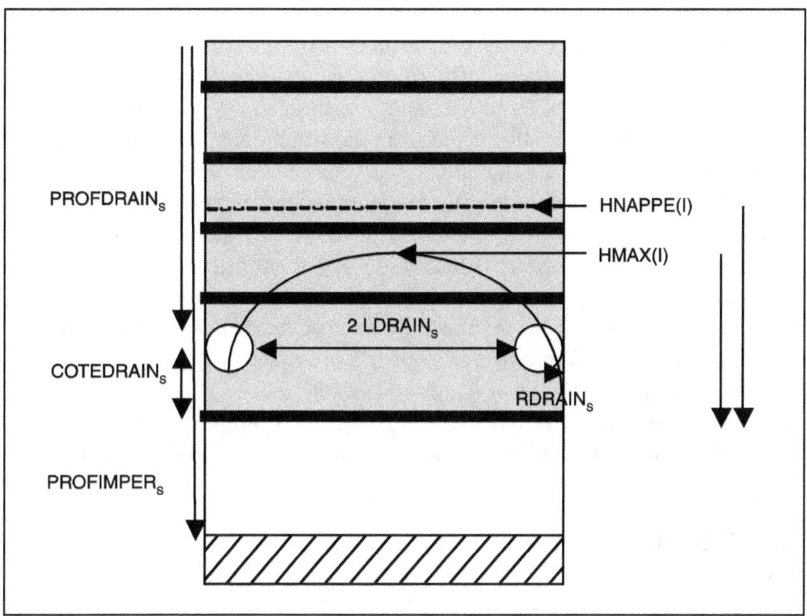

Figure 9.7. Schema of the draining system. In grey is the STICS soil, the water table develops between PROFIMPER$_S$ and the daily level HNAPPE, which is reduced by the presence of the drains to a maximum between-drains level of HMAX. The drains are characterized by their location and spacing in the soil (PROFDRAIN$_S$, LDRAIN$_S$) and their radius (RDRAIN$_S$).

eq. 9.8

$$HMAX(I) = \frac{HNAPPE(I) - COTEDRAIN_S}{BFORMNAPPE_S}$$

$$\text{if} \quad PROFIMPER_S - PROFDRAIN_S \geq \frac{LDRAIN_S}{2} \quad \text{then} \quad DE(I) = DELT(I) \cdot LDRAIN_S$$

$$\text{if} \quad PROFIMPER_S - PROFDRAIN_S < \frac{LDRAIN_S}{2} \quad \text{then}$$

$$DE(I) = DELT(I) \cdot LDRAIN_S \cdot \frac{(PROFIMPER_S - PRFOFDRAIN_S)}{DELT(I) \cdot LDRAIN_S + (PROFIMPER_S - PROFDRAIN_S)}$$

$$\text{and} \quad DELT(I) = \frac{1}{2}\left[\left(\frac{HMAX(I)}{LDRAIN_S}\right)^{0.36} - \frac{HMAX(I)}{LDRAIN_S}\right] \quad \text{and} \quad DELT(I) \leq \frac{\Pi}{4 \cdot \ln\frac{2 \cdot LDRAIN_S}{\Pi \cdot RDRAIN_S}}$$

The Hooghoudt equation is normally valid under a permanent regime, but it was shown (Zimmer, 2001) that for sufficiently large time steps, it provides entirely satisfactory predictions of the flows and water table heights in drainage systems. The operating principle is as follows: when gravity flow begins following saturation of the microporosity in the system, the macroporosity fills and creates a water table, whose level is at the top of the layer whose macroporosity is saturated. If we know the system parameters and the height of the previous table, a quantity of drained water is calculated, to which may be added, if relevant, drainage linked to exchanges with deep layers of the soil. The sum of these two drainage quantities is subtracted from the water contained in the macroporosity, and a new water table height is calculated.

Although it does not appear explicitly in the equations, the porosity of drainage plays an important role in the emptying and filling of soil macroporosity. As a general rule, the simulations are correct only when the value of the soil macroporosity is equal to its drainage porosity.

In order to be able to account for the field heterogeneity due to the drainage system, it is possible to calculate the plant effects of the presence of a water table either at the drain level or at the inter-drain level or for an average level.

Nitrates can be leached through the drains and their amount is calculated assuming that nitrate concentration in the drained water is that of the HNAPPE level.

9.4 Integrated calculations of soil status

9.4.1 Water and nitrogen reserves

By integrating the elementary layer water contents, HUR(Z)+SAT(Z), and the nitrogen contents, NIT(Z)+AMM(Z), over the depth of soil used for taking measurements (PROFMES$_T$) or over the whole depth of soil (PROFSOL$_S$), we obtain the soil water reserve RESMES, and the nitrogen reserve AZOMES.

By integrating the difference between HUR(Z)-HUMIN(Z) over the rooting depth ZRAC, we obtain RESRAC. This same difference integrated over the soil depth, PROFSOL$_S$, and weighted by the difference between HUCC(Z)-HUMIN(Z) gives the soil water status as a proportion of readily available water (RSURRU).

Lastly, the maximal reserve used, RMAXI, corresponds to the integration of the difference between HUCC(Z)-MIN(HUR(Z)) over the rooting depth, where MIN (HUR(Z)) is the lowest water content value in the layer Z encountered during the simulation.

9.4.2 Water and nitrogen balances

The balances are calculated for the three levels represented in the soil: the elementary 1 cm level, the layer and the whole soil from the surface to its basis. Let us take the example of the whole soil (eq. 9.9): the inputs of the water balance are rainfall (TRR), irrigation (AIRG) affected by an efficiency (EFFIRR$_T$) and capillary rises (REMONTEE); while the outputs are soil (ESOL) and plant (EP) evaporations, runoff (RUISSEL) and deep drainage natural (DRAIN) or artificial (QDRAIN). The one day delay between soil and plant evaporation is assumed to account for the soil priority.

eq. 9.9

$$RESMES(I+1) = RESMES(I) + REMONTEE(I) + EFFIRR_T \cdot AIRG_T(I) + TRR(I)$$
$$- ESOL(I+1) - EP(I) - RUISSEL(I) - DRAIN(I) - QDRAIN(I)$$

As far as mineral nitrogen is concerned, the same input-output terms are identified, applied here in a cumulative form from the beginning of the simulation (eq. 9.10). The amounts of nitrogen are those provided by mineralisation of humus (QMINH) or residues (QMINR) in addition to those of the fertilizers (ANIT). In outputs there is the plant uptake (ABSO) and all nitrogen losses by leaching (QLES and QLESD in the artificial drains), as gas (QNVOLENG and QNDENENG for the mineral fertilizers, and QNVOLORG for manure) or by reorganization (QNORGENG)

eq. 9.10

$$AZOMES(I+1) = AZOMES(1) + QMINH(I) + QMINR(I) + \sum_{K=1}^{I} ANIT(K) +$$

$$CONCRR_G \sum_{K=1}^{I} TRR(K) + CONCIRR_T \sum_{K=1}^{I} AIRG(K) - QLES(I) - QLESD(I)$$

$$- \sum_{K=1}^{I} ABSO(K) - QNDENENG(I) - QNORGENG(I) - QNVOLENG(I) - QNVOLORG(I)$$

9.4.3 Predawn plant water potential

At dawn, plant water potential is assumed to be in equilibrium with the soil water. Consequently this measurement is often used as a daily assessment of water stress and a relevant integrated measurement of soil behaviour. In order to be able to compare STICS simulations to this type of measurement, a simple calculation of predawn plant water potential is proposed, based on Brisson *et al.* (1993). Predawn plant potential is calculated as the arithmetic mean over depth of soil water potential, weighted by root density.

The soil potentials (PSISOL) are calculated using the Clapp and Hornberger (1978) formulae, using the points (HUCC, -0.03 MPa) and (HUMIN, -1.5MPa) to calculate the parameters BPSISOL and PSISOLS.

eq. 9.11

$$PSISOL(Z, I) = PSISOLS \left[\frac{HUR(Z, I) + SAT(Z, I)}{WSAT(Z)} \right]^{-BPSISOL}$$

$$with \quad WSAT(Z) = 1 - \frac{DACOUCHE(Z)}{2.66} \quad if \quad CODEFENTE_S = 0$$

$$and \quad WSAT(Z) = \frac{1.5 HUCC(Z) - 0.5 HUMIN(Z)}{10} \quad if \quad CODEFENTE_S = 1$$

The roots participating in predawn potential are the ones located in moist layers (PSISOL above -1.5 MPa)

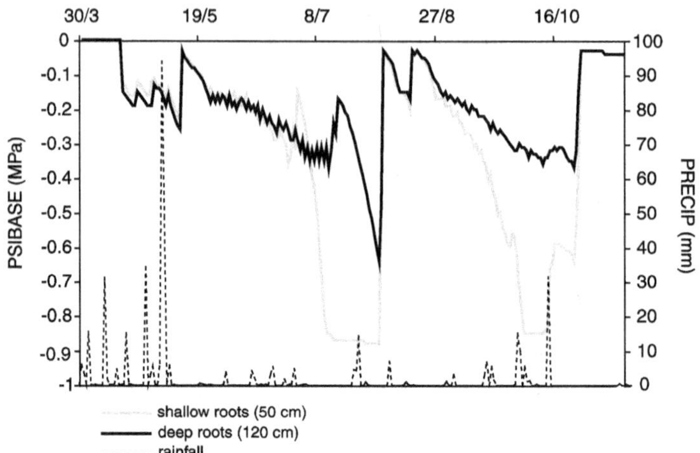

Figure 9.8. Influence of rooting depth on PSIBASE calculation and sensitivity to rainfall. Example of a vineyard in the Rhône valley: the simulated shallow rooted vineyard production is half of that of the deeply rooted vineyard.

$$eq.\ 9.12$$

$$PSIBASE(I) = \frac{\displaystyle\sum_{Z=PROFSEM_T}^{ZRAC(I)} RACPSI(Z,I) \cdot PSISOL(Z,I)}{\displaystyle\sum_{Z=PROFSEM_T}^{ZRAC(I)} RACPSI(I)}$$

$$with \quad RACPSI(Z,I) = LRACZ(Z,I) \quad if \quad PSISOL(Z,I) > -1.5$$

$$and \quad RACPSI(Z,I) = 0 \quad if \quad PSISOL(Z,I) \leq -1.5$$

10 Cropping systems

10.1 The notion of a Unit of SiMulation (USM)

A cropping system is a sequence and/or spatial combination of crops and the corresponding technical operations, applied to a given, uniformly treated agricultural area (Boiffin *et al.*, 2001). "They are identified by involving not only the cash crops themselves, but also periods between crops with bare soil or plant cover". STICS can easily describe the cropping system behaviour, because it integrates the temporal variability of weather, agricultural practices and crops over time. This modelling is implemented at the homogeneous soil unit level. STICS cannot intrinsically investigate the landscape scale, which is the relevant one for economic or environmental diagnosis of agricultural practices. Landscape contains a combination of several cropping systems, called "agro-ecosystem" (Meynard *et al*, 2001). On the assumption that joint soil units can be accounted for independently, the modelling of landscape is possible but requires STICS to be run within a geographical information system, which includes soil mapping of the agro-ecosystem (Nicoullaud *et al.*, 2004; Guérif *et al.*, 2007). Conversely, coupling STICS with a spatially distributed hydrological model allows lateral interactions occurring across the landscape to be described (Durand *et al.*, 2007). The simulation of the course of the cropping system over several cropping seasons needs to simulate succeeding units of simulation, corresponding to both cropped cycles and bare periods.

A Unit of Simulation (USM) is a combination of a given soil/weather situation with a given crop species and a given crop management. A USM gathers all the required information to run a simulation: the daily weather data during the simulation period, the soil characteristics as well as ecophysiological and agronomic characteristics of the crop species and all the techniques applied during crop growth (sowing date and depth, fertilization amount and dates, irrigation amounts and dates, etc.). Another important input is the initial status of the system, i.e. the soil water and nitrogen contents and, if necessary, the plant growth status when the simulation starts. When the simulation should end is also important to be able to correctly chain simulations. In practice, a USM file includes the names of the files to be used for the soil, the plant, the weather and the technical operations.

Depending on the user's objectives, a USM may be created to simulate simple crop cycles, or chained in order to simulate succeeding crops at the same location. Special USMs may be created to simulate intercropping, using the same soil and weather data and the agro-physiological and management properties of the two intercropped species.

10.2 Long term simulations

Numerous biological and physical processes occur over the time course of a given crop succession. Some of them consist of short-term processes such as soil anoxia, denitrification, soil freezing or crop growth. They may have little effect on the soil crop system pattern in the long term. Yet some temporary physical conditions can definitely affect crop behaviour; e.g. drought reducing grass tiller density in Mediterranean conditions (Satger *et al.* 2007). Conversely other conditions and processes only come to light in the long term through their cumulative effects: this is the case of those affecting the Soil Organic Matter (SOM) or the soil structure. The predictive performance of soil carbon and nitrogen turnover by various models: CERES, NCSOIL, SUNDIAL, and STICS has been compared by Gabrielle *et al.*, 2002: "The results highlight a trade-off between the prediction of N mineralization in the short term (day to year) and SOM dynamics in the long term (year to decade)". STICS simulated SOM mineralization rate well when the amounts of incorporated residues were well known. The version of STICS that we will refer to in this section assumes the soil physical parameters to be stable. Making STICS able to simulate the evolution of bulk density or soil permeability is a worthwhile future research project, with some elements already present in the model (Richard *et al.*, 2007 and § 6.5). In order to simulate the soil-crop system behaviour over a long period, we have first to analyze what kind of carry-over effects the model is able to take into account between successive USMs, and second, to define the relevant calendar of each USM in order to get confident initial values. From a practical point of view, there are a lot of ways of splitting the studied period into several successive USMs. But it seems best to link one USM to the next when there is a low level activity in the cropping system and fallow periods in order to minimize the effect of initial conditions. Indeed during the cropping season many processes occur, as crop growth is stimulated by the farmer's practices. Concerning N dynamics we can mention crop uptake, influenced by N fertilisation, as the most influential flow as regards soil mineral N level (Blombäch *et al.*, 2003). If the cropping systems include crop volunteers, green manure or sown catch crops, it may be necessary to define several USMs a year.

STICS is easily able to run monocrop cropping systems by chaining the same agricultural configuration (or USM) over a several years with successive weather files. The other possibility is to account for rotations by chaining various USMs. As far as initial conditions are concerned, in both cases, it is possible either to reset (R: or use the prescribed initial conditions) or run continuously (C: or use the final conditions of the previous simulation as initial condition of the current one) the model over the succeeding USM.

10.2.1 Monocrop *vs* rotations

The monocrop simulation can be performed by the same USM, i.e. the same soil, species and techniques over several years with successive weather files. Using the reset option can be interesting in the case of annual crops for numerical experiments to get rid of cumulative effects. It can also be used to understand the model's sensitivity to cumulative effects, as for instance the amount of winter carbon reserves in forage crops. Of course, using the continuous option is much more realistic, in particular for predicting the behaviour of perennial crops. For instance, the response of various French vineyards to global warming was evaluated by comparing STICS simulations for 1970-2000 to those of 2070-2099 (García de Cortázar Atauri *et al.,* 2006), using this option.

For rotations or successive USMs, the reset option can provide confident diagnosis of the impacts of agricultural practices over a past period when both practices and initial values are well known (Nicoullaud *et al.*, 2004, Beaudoin *et al.*, 2008). This approach takes into account the modelled impacts of cumulative effects of agricultural practices-soil-weather interactions and requires that all state variables are available at the dates chosen for USM initiation: soil water and nitrogen contents, organic matter and bulk density and, if initiation occurs during the cropping period, crop developmental stage and biomass. The availability of such measurements can also be taken as a relevant rule for the calendar chaining of USMs (§ 10.2.3).

In the absence of such measurements or for predictive studies over a long period, it is recommended to simulate rotations with the continuous option (Ducharne *et al*, 2007). In that case the results do not depend on the segmentation of the simulated period into successive USMs but rely greatly on the confidence in the simulated long-term effects of STICS, which can be questionable. For continuous rotations the soil characteristics must be exactly the same for all the USMs and great attention must be paid to the calendar chaining in Julian days. If initial values of the series are missing, it is common to run the model for several years preceding the first year of interest, just to calculate reliable simulated initial values.

The final status of the system used as initial status for the following simulation concerns:
1.   the soil mineral status (water, nitrates, ammonium)
2.   the system thermal status (soil and crop temperature)
3.   the soil organic status in the three pools (humus, biomass and residues for C and N contents and rates of decomposition)
4.   the plant status (LAI, biomass, N content, rooting depth and density, carbon and nitrogen reserves and developmental status including stage and developmental units)

10.2.2 The particular case of crop residues

In the long term, the incorporation or export of crop residues is probably the key process influencing soil organic matter dynamics. It can potentially concern several parts of the plant, from the fine roots to the whole plant from green manure, passing through stems and stubbles (RESSUITE$_T$), and the nature of crop residues (C/N ratio or CSURNRESSUITE) together with their quantity (QRES) are elements of the cropping management accounted for by the model. They can be either prescribed, in the case of the "reset" option or calculated in the case of the "continuous" option. In both cases their incorporation is done by the various soil cultivation operations prescribed in the crop management file, which may occur before the end of the USM simulation (in that case the crop residues are taken into account in the various state variables characterizing the following initial status). An error in the estimation of QRES or, to a lesser extent, of CSURNRESSUITE is likely to be propagated and lead eventually to a false result. The equations detailing the incorporation of crop residues according to the parameter RESSUITE$_T$ and the root simulation options are given in § 6.3.4. Crop residues, such as dead leaves (e.g. for rapeseed), can also be incorporated during crop growth as they are produced.

Crop residues left on the soil surface can act as a mulch after harvest (§ 6.4.1). In that case, the user needs to specify it as a particular technique in the crop management file and prescribe the amount of mulch biomass since it is not automatically implemented in any of the USMs. Yet the mulch will automatically disappear as soon as any form cultivation is done, provided that its depth exceeds 2 cm. In fact, in the model, the physical role of the plant mulch is managed independently from its biological role in residue decomposition.

10.2.3 Examples of long term simulations

The plant carbon reserves (RESPERENNE) are supplied once all the identified organs' demand is satisfied, i.e. leaves, stalks and possibly seeds or fruits (see § 3.5.4). They can also be consumed when photosynthesis is insufficient (see § 3.3.3). The filling or emptying of reserves depends both on the plant parameters (SLAMAX$_p$,

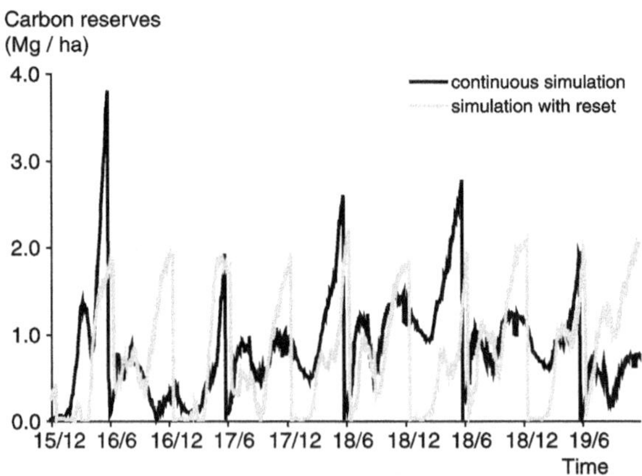

Figure 10.1. Dynamics of the carbon reserve (RESPERENNE) of a temperate grass (Festuca arundinacea), with reset every year or continuous simulation, from 1994 to 2003 (Ruget and Brisson, 2007).

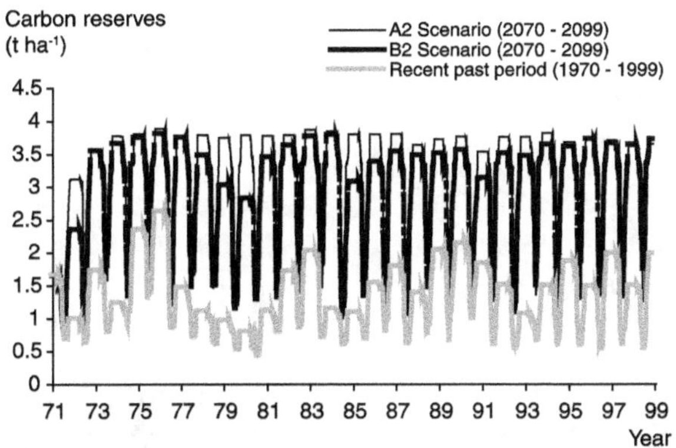

Figure 10.2. Dynamics of the carbon reserve of a vine (*cv* Merlot) at Bordeaux, according to two future climatic scenarios covering the 2070-2099 period (B2-SRES scenario ($[CO_2]$=550 ppm in 2100 and A2-SRES scenario ($[CO_2]$=800 ppm in 2100) as compared to the recent past (1970-1999) (Garcia de Cortazar Atauri, 2006). The future climatic scenarios are calculated by the Arpege GCM model (Gibelin and Déqué, 2003) following the IPCC (2001) recommendations for SRES scenarios and using the method of the "anomalies" for downscaling the large scale GCM outputs.

TIGEFEUILLE$_p$, REMOBRES$_p$) and soil and weather conditions, as it is shown in Figure 10.1. The differential behaviour is particularly important in winter when the meadow's reserves are used for the initiation of spring growth. Cutting causes an abrupt decrease in the carbon reserves in both cases because the remaining organs consist entirely of the whole remaining biomass, leaving little reserves.

The dynamics of the carbohydrate reserves of a vine grown under current conditions in the Bordeaux vineyard (5000 vines per ha, 1.3m high; 183 mm of soil available water) are compared for three climatic scenarios (Garcia de Cortazar Atauri , 2006): the actual recent past (1970-1999) (control) and two possible future scenrios (periods 2070-2099) under two hypotheses of global CO_2 emission (Figure 10.2). The predictions of the vine reserves for the past and future periods differ greatly, both on average as well as in variability. The simulated higher reserves for the future periods is due to stimulating growing conditions, both in terms of air temperature and atmospheric CO_2 concentration, and also to a longer post-harvest period, allowing carbon storage between harvest and leaf fall.

In Figure 10.3 are presented simulated versus measured soil nitrate contents for two contrasting soils within the same field: a shallow sandy stony loam overlying limestone and a deep loamy soil. Year-to-year measured values are more variable on the loamy soil than on the sandy stony loam. Simulations with reset twice a year correctly mimic measurements while the continuous ones agree less well with the measured reality, especially for the loamy soil. The largest discrepancies occur for long fallow periods during dry winters. Hence uncertainties in the model's initial values cause it to generate errors between years when N leaching is low. This study also shows that the model's sensitivity to soil parameters depends greatly on the duration of the simulation.

These data were collected on various soils (36 sampling sites) of a small catchment, named Bruyères, in northern France, and STICS was run on each of them. The mean

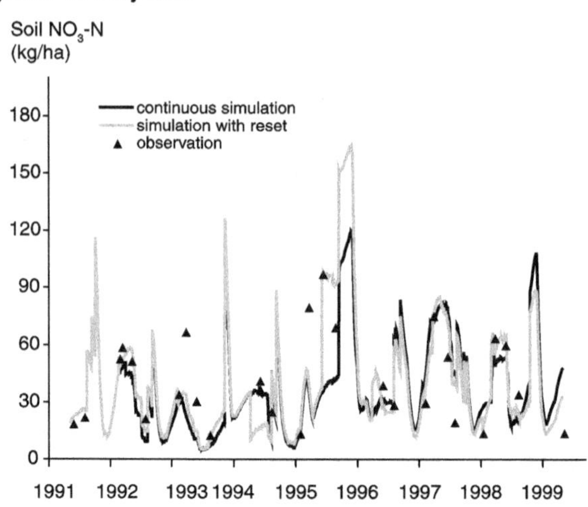

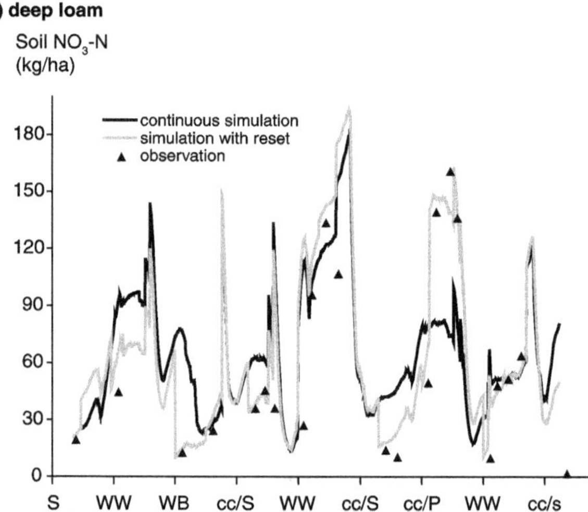

Figure 10.3. Observed and simulated soil nitrate contents during the period 1991-1999 for two contrasting soils: a) shallow sandy loam overlying limestone and b) deep loamy soil from the same field. The succeeding crops include S = Sugarbeet, WW = winter wheat, WB = winter barley, cc= catch crop, P = spring peas (Beaudoin *et al.*, 2008).

prediction by reset simulation (RS) or continuous simulations (RS) of all the variables of interest (yield, N losses) were close (Beaudoin *et al.*, in press). The main difference between RS and CS predictions concerned the residue mineralization despite their predictions, of either biomass or N content, did not greatly differ (Table 10.1). That can reach up to 16 kg × ha⁻¹y⁻¹ for deep loam but only 2 kg × ha⁻¹y⁻¹ for the sandy soil. This difference lies with the automatical addition of the simulated root residues provided by the preceding crop in the amount of crop residues incorporated in CS unlike RS. The N content of root residues is assumed to equal those simulated for the aerial crop

Table 10.1. Bruyères catchment from 1991 to 1999: comparison of STICS predictions of the annual residue biomass, N content and mineralization (kg N ha^{-1} y^{-1}) by reset and continuous simulations. The 36 sampling sites of are of various soil types (Beaudoin *et al.*, 2008).

Mean prediction		Soil type			
		deep loam	shallow sandy loam on limestone	shallow loamy clay on marl and rock	shallow loamy sand on sand
reset simulations					
Residue biomass	t ha^{-1} y^{-1}	8.0	7.3	5.6	5.7
Residue N content	%	0.8	0.9	0.8	1.3
Residue mineralization	kg N ha^{-1} y^{-1}	−5	−9	−5	−8
continuous simulations					
Residue biomass	t ha^{-1} y^{-1}	7.7	6.6	5.2	5.6
Residue N content	%	0.8	0.8	0.8	1.3
Residue mineralization	kg N ha^{-1} y^{-1}	−21	−22	−17	−6

residues. As interaction with the soil type, outputs are quite similar for the sandy soil because the high N content of root residues does not allow significant supplementary N immobilisation.

Evaluating STICS's long-term calculations for bare soil has been possible thanks to the Fagnières lysimeter device in northern France (48°57'N, 4°19'E; Ballif, 1996 cited in Beaudoin, 2006). Water drainage and N leaching have been monitored for 28 years. The evaporation parameter, Q0$_\text{S,}$ was estimated in order to minimize the differences between measured and simulated amounts of drained water (Figure 10.4).The good prediction of

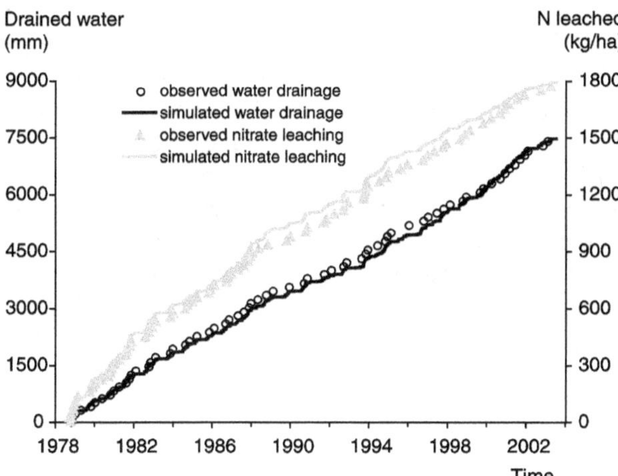

Figure 10.4. Cumulative water drainage and nitrate leaching measured (symbols) and simulated with the STICS model (continuous lines) in the lysimeter experiment of Fagnières, from 1978 to 2003 (Beaudoin *et al.*, 2008).

leaching confirmed the ability of the N mineralization module to correctly simulate soil organic matter dynamics over the long term. The cumulative N mineralized (1770 kg ha^{-1}) represents 32% of the initial organic N content of the biologically active layer (0-27 cm). The rate of N leaching decreases at the end of the period, which is unsurprising due to the depletion of organic nitrogen.

10.3 Intercropping

Intercropping consists of growing several crops (annual or perennial) simultaneously, each crop developing and growing at its own rate as a result of resource partitioning. This practice is traditional in the tropics and is beginning to be used in temperate climates for environmental reasons. Various arrangements of intercrops exist: strip intercrops, alley crops, mixed intercrops or even windbreaks, which exhibit more or less spatial heterogeneity.

Given the complexity of the system, models can be especially helpful for analysing intercropping comprehensively (Caldwell, 1995). The intercrop modelling framework can be summarised using three approaches. The first of these, consistent with de Wit *et al.*'s initial principles (de Wit, 1978 and de Wit *et al.*, 1970) is an extension of sole crop modelling, considering the system to be composed of two species instead of one, simply organised within a kind of elementary pixel supposed to represent the whole field. Actually this is the oldest and more operational approach (Caldwell *et al.*, 1993 and Kiniry *et al.*, 1992), concentrating more on the dynamics of the system than on its spatial heterogeneity. The second approach relies on a description of the intercropping system as a series of discrete crop-based or tree-based points with flow of mass or energy between each. This spatial discretized approach allows big spatial variations to be accounted for, each point generally being simulated under the above-mentioned crop modelling principle, and the field response results from a spatially integrated calculation (Huth *et al.*, 2002). The last possible approach derives from architecture modelling, putting emphasis on a realistic description of the 3D structure of the complex two-species canopy, which leads to fine-scale descriptions of processes (Sonohat *et al.*, 2002) at the organ level. In that approach it is more difficult to account for the system dynamics because of the complexity of organ dynamics in interaction with the whole plant behaviour. The adaptation of the STICS crop model was based on the first approach (Brisson *et al.*, 2004), aiming at producing an operational tool to help managing intercrops, while trying to overcome the problems of unwarranted over-simplification.

The adaptation of STICS's conceptual basis and formalizations to intercropping relies first on a simplified definition of the complex agronomic system of intercropping, and secondly on the adaptation of the modules calculating resource capture (light, water and nitrogen).

10.3.1 Representation of the intercropping system

The intercropping system being complex, some simplifying hypotheses are adopted. The soil-plant-atmosphere system is divided into three sub-systems at the canopy level (Figure 10.5): the dominant canopy (D) and the understorey canopy (U) are divided into

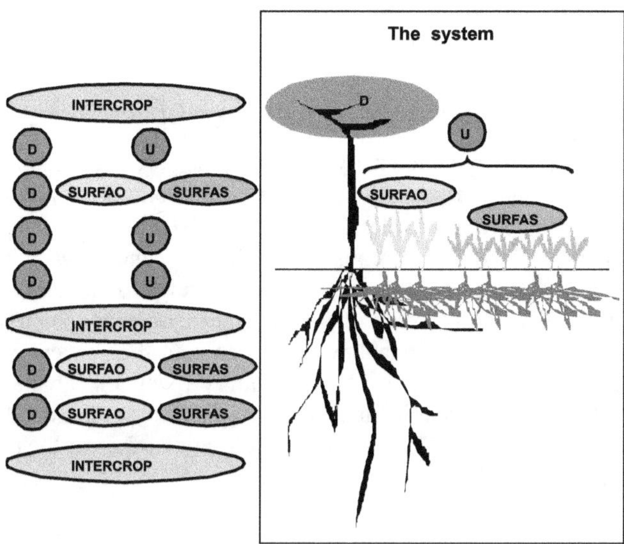

Figure 10.5. Simplified representation of the model with, on the left, the modules (grouped according to the way they are called within the code), and on the right the system with its three sub-systems (D: dominant crop; U: understorey crop divided into a shaded part: SURFAO and a sunlit part: SURFAS) and in the centre the number of calls of each module devoted to a particular part of the system. * corresponds to the modules modified for the adaptation to intercropping.

two parts: a shaded part (SURFAO) and a sunlit part (SURFAS), each of them being defined by a light microclimate. These light microclimates, estimated from a radiation balance (see § 6.6), drive the different behaviours of the sub-systems in terms of growth (dry matter accumulation, LAI) and water and nitrogen budgets (transpiration, nitrogen uptake, stress index). The estimation of the water requirements for both crops relies on light partitioning coupled to a resistive scheme (Figure 7.6) and is applied on a daily time step. The phasic development is considered the same for both parts of the understorey crop. Also the soil environment is assumed to be the same for both crops: that is to say that the horizontal differentiation within the soil profile is neglected in favour of the vertical one. It is assumed that the interactions between the two root systems result from the influence of the soil on each crop root profile through its penetrability and water dynamics.

The application of this theory within the STICS code is done by multiple calls to the elementary subroutines and re-calculation of the state variables as a function of the considered sub-system. Specific modules or options were added to the preceding sole crop version in order to take account of the ecophysiological features of these complex systems. These adaptations are now available in the present version for sole crop simulations as well as for intercrops. They concern radiation interception, energy budget driving water requirements and microclimate, and dynamics of the root system as influenced by soil status. Those modules and options were described in chapters 3, 7, and 5 respectively. The shoot growth was slightly modified to account for the understorey shaded crop growing under limiting radiation. We explain here the value of these formalisms in the case of intercropping system simulation.

10.3.2 The radiation intercepted by the two crops

The objective is to estimate, on a daily time step, the part of the radiation intercepted by the dominant crop and the part transmitted to both components of the understorey crop: the shaded (ROMBRE) and the sunlit (RSOLEIL). To solve this problem, the most complex method for radiation transfers within the canopy was chosen in STICS (see details in § 3.2.2). While for sole crops the basic level of calculation is the soil, in the case of intercropping, it is the top of the understorey canopy. On a daily time step, the shaded part of the understorey canopy corresponds to the vertical projection of the dominant foliage at the soil surface. The elementary pixel for calculation consists of the LARGEUR/2 part of the dominant crop (see § 3.2.2.b and.Figure 3.9), the shaded surface of the understorey crop (SURFAO) and the sunlit surface of the understorey crop (SURFAS) (Figure 10.6).

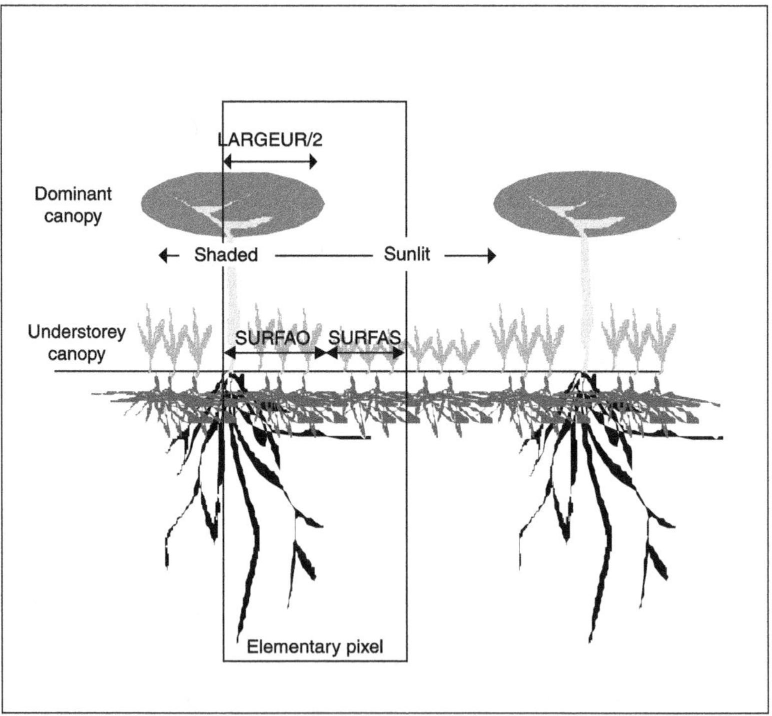

Figure 10.6. Simplified representation of an elementary pixel of the system (LARGEUR/2 represents the half-width of the dominant crop part, SURFAO represents the shaded surface of the understorey crop part, and SURFAS represents the sunlit surface of the understorey crop part of this elementary pixel).

10.3.2a Radiation intercepted by both crops

The radiation intercepted by the dominant crop and its complementary part transmitted to the understorey canopy must be calculated using the radiation transfer formalisms, using the series of equations and crop geometry given in § 3.2.2. Those equations lead to the simple calculations of ROMBRE and RSOLEIL, assuming a discretization of the inter-row distance in 20 points.

Hence for 20 points spread equally along the inter-row, XSH points are on the shaded part of the understorey crop, and ROMBRE (eq. 10.2) is the average value for those XSH points of the transmitted radiation (which is the sum of the radiation not intercepted by the dominant crop (RDROIT), and the transmitted radiation through the dominant crop RTRANSMIS), while RSOLEIL is the complementary value for the 20-XSH points (eq. 10.1: see eq. 3.18 and eq. 3.19 for RDROIT and RTRANSMIS calculations).

eq. 10.1

$$ROMBRE(I) = \frac{1}{XSH} \sum_{X=1}^{XSH} (RDROIT(I,X) + RTRANSMIS(I,X))$$

and

$$RSOLEIL(I) = \frac{1}{20 - XSH} \sum_{X=1}^{20-XSH} (RDROIT(I,X) + RTRANSMIS(I,X))$$

Then the proportion of income radiation intercepted by the dominant ($FAPAR_D$) and the understorey crop ($FAPAR_U$) can be simply derived from eq. 10.1 coupled to eq. 3.16 (Beer law analog applied to the understorey crop): eq. 10.2.

eq. 10.2

$$FAPAR_D(I) = [1 - ROMBRE(I) \cdot SURFAO(I) + RSOLEIL(I) \cdot SURFAS(I)]$$

and

$$FAPAR_U(I) = \frac{[RAINT_U(I,AO) \cdot SURFAO(I) + RAINT_U(I,AS) \cdot SURFAS(I)]}{PARSURRG_C \cdot TRG(I)}$$

The FAPAR of both crops depends greatly on their respective heights, which not only depend on the plant characteristics but also on the growth conditions as demonstrated in Figure 10.7.

10.3.2b Crop geometry

When both canopies (dominant and understorey) are vertically mixed, the sole upper part of the dominant crop, located above the understorey crop, is accounted for in the radiative transfer calculations. Thereby, an efficient shape is defined for the dominant crop; in the case of the "upside-down" triangle, the efficient shape is trapezoidal but it is assumed to be rectangular to simplify the geometrical calculations.

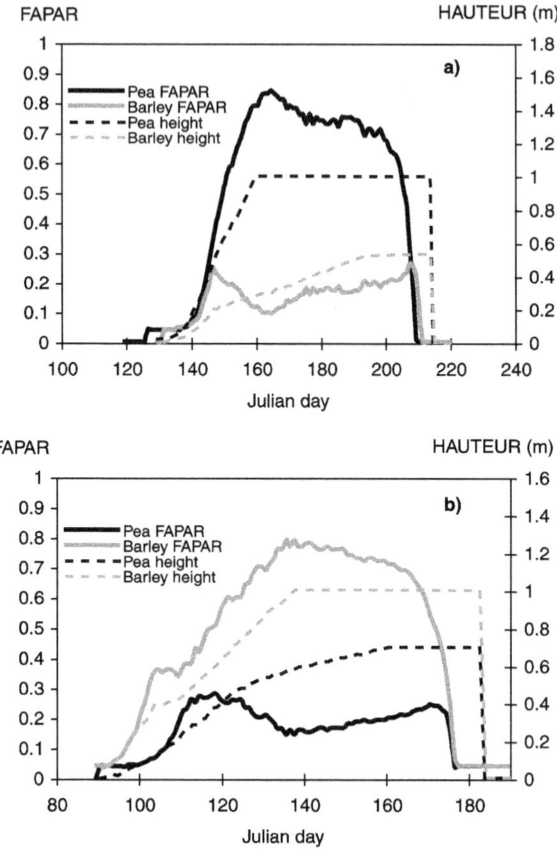

Figure 10.7. Comparison of pea-barley intercrops in Denmark (a) and France (b) in terms of respective crop heights (HAUTEUR) and proportion of intercepted radiation (FAPAR).

In order to allow inversion of dominancy of both crops, the intercrop plant status (dominant or understorey) is a function of the respective plant heights, which can change several times throughout the growing cycle as a function of growth rates of each crop.

10.3.3 Energy budget and microclimate

In the model, the energy budget is used to estimate the crop water requirement through the "resistive approach" option (see § 7.2). This approach is particularly relevant in the case of intercrops, because it allows for microclimatic effects on water requirements: convection beneath the dominant canopy and decrease in the vapour pressure deficit due to transpiration from the understorey plants. Then actual soil evaporation and plant transpiration are calculated independently by means of a soil water balance (see 7.1 and 7.3). These fluxes are then re-introduced into the energy budget to calculate crop temperature, which is a driving variable for growth and development of the plant (see chapters 2 to 5). The required adaptations for intercrops concern the first stage.

10.3.3a Theoretical basis

Following the relative position of the dominant and the understorey crops, the energy budget calculations rely on slightly different resistance networks (Figure 10.8). This simplification aims at limiting to two the number of sites playing the role of water vapour

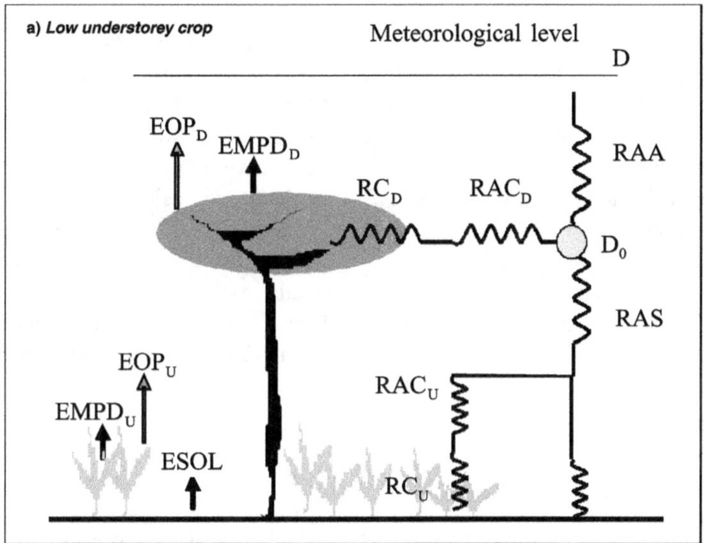

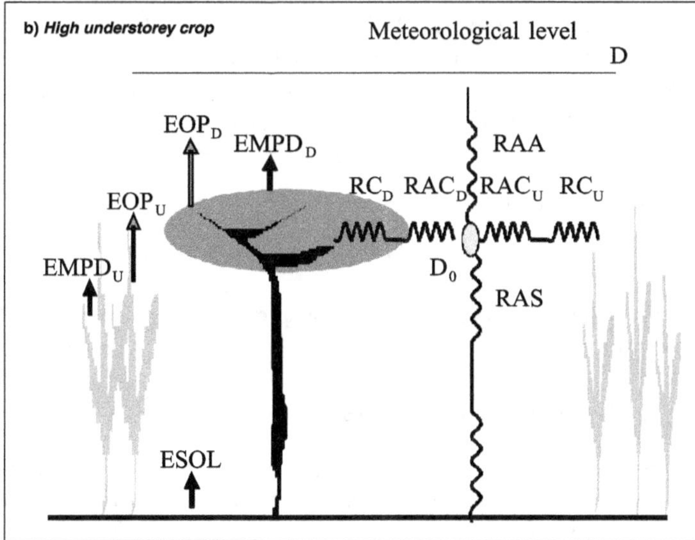

Figure 10.8. The two possible schemes of resistance networks used to estimate water requirements for intercrops (right-hand side of the schemes) and the fluxes (left-hand side of the schemes). (a) the understorey crop is near to the ground, (b) the understorey crop is nearly as high as the dominant crop.

sources. The resistance scheme for a low understorey crop (Figure 10.8.a) is an extrapolation of the original model by Shuttleworth and Wallace (1985) and the resistance scheme for a high understorey crop (Figure 10.8.b) is an extrapolation of the model proposed by Wallace (1995) for intercrops. Those two schemes are applied at a daily time step relying on Monteith's theory (Monteith, 1965) and its consequences (Allen, 1994) and on a previous study (Brisson *et al.*, 1998b), in which full details of definitions and formulations are given.

The calculations involve five evaporative fluxes: soil evaporation (ESOL), maximal plant transpiration for dominant crop (EOP_D), maximal plant transpiration for understorey crop (EOP_U), direct evaporation of the water intercepted by the "dominant" leaves ($EMPD_D$) or by the "understorey" leaves ($EMPD_U$), three net radiation budgets: RNETS, $RNETP_D$ and $RNETP_U$ for the soil, the dominant crop and the understorey crop respectively and three types of resistance (eddy diffusion resistances: RAS and RAA, bulk boundary layer resistances of both crops: RAC_D and RAC_U and surface resistances: soil resistance is accounted for in the soil evaporation calculation, RC_D, and RC_U). Each flux is calculated using a formula such as the ones given in eq. 7.12 and eq. 7.13 (Brisson *et al.*, 2004). The combining of the three subsystems (soil and both crops) into two requires varying the bulk boundary layer resistance applied to the lower level: either (RAC_U + RAS) and RAC_U for the low and the high configurations respectively.

10.3.3b Available energy and its distribution.

In order to evaluate the distribution of available energy between the soil and both crops, we base our method on the hypothesis that we know the proportion of global radiation intercepted by the crops ($FAPAR_D$ and $FAPAR_U$: eq. 10.2), whose values were calculated in the radiative transfers module. In the case of intercrops, the net radiation corresponding to plants (RNETP1 in eq. 7.14) consists of the net radiation of the dominant and understorey crops, $RNETP_D$ and $RNET_U$ respectively (eq. 10.3).

eq. 10.3

$$RNETP1(I) = RNETP_D(I) + RNETP_U(I)$$
$$= 0.83 \cdot FAPAR_D(I) \cdot RNET(I) + 0.83 \cdot FAPAR_U(I) \cdot RNET(I)$$

But the energy actually available for crop transpiration must also account for possible direct water evaporation from the leaves.

10.3.3c Water persistence on foliage.

The simulation of rainfall interception is not usually included in crop models, while it is an important process in forestry models (Bussière, 1995). A common idea is that evaporation of intercepted water compensates exactly for the decrease in evaporative demand, especially for herbaceous canopies (McMillan and Burgy, 1960). As far as intercrops are concerned, the processes are more complex and the above-mentioned compensation is not so likely, depending on rain events, evaporative demand and intercrop structure. Including these processes in an intercrop model seems to be worthwhile to correctly predict water use by canopies, especially in humid tropical climates (high evaporative demand combined with frequent rainfall). Through the simulation of stemflow and direct water evaporation from leaf surfaces, the objective is rather to correctly evaluate the amount of water that will reach the soil than to partition water between the two crops. Indeed, once in the soil the water is assumed to be evenly available for both root systems, neglecting horizontal variability of soil water content.

To account for these processes, water persistence and direct evaporation from the dominant ($EMPD_D$) and understorey ($EMPD_U$) crop foliage as well as the stemflow along the dominant stems are simulated as described in § 6.2.2 and 7.2. Then another value of net radiation is derived (RNETP2) using eq. 7.14

10.3.3d Specific considerations in the calculation of the eddy diffusion resistance (RAA and RAS).

The particular aspects of the application of formalisms described in § 7.2 to intercropping concern the roughness for crop and soil (Z0 and Z0S) and displacement height (DH), which are evaluated as follows:

For the low understorey crop (Figure 10.8.a):

eq. 10.4

$$DH(I) = 0.66 \cdot HAUTEUR_D\,(I)$$
$$Z0(I) = 0.10 \cdot HAUTEUR_D\,(I)$$
$$Z0S(I) = 0.10 \cdot HAUTEUR_U\,(I)$$

For the high understorey crop (Figure 10.8.b):

eq. 10.5

$$D(I) = 0.66\left(HAUTEUR_U\,(I) + (HAUTEUR_D\,(I) - HAUTEUR_U\,(I))/2\right)$$
$$Z0(I) = 0.10\left(HAUTEUR_U\,(I) + (HAUTEUR_D\,(I) - HAUTEUR_U\,(I))/2\right)$$
$$Z0S(I) = Z0SOLNU_S$$

where $HAUTEUR_D$ and $HAUTEUR_U$ are the heights of the dominant and the understorey crops respectively. The threshold height for the "low" understorey crop is arbitrarily fixed at 0.2 m. The reference height taken from meteorological data is 2m. If the plant canopy height exceeds this threshold, a wind speed value is recalculated at a reference height of over 2 m by applying a logarithmic profile. The other meteorological values are not recalculated.

10.3.3e Surface resistances

The resistances of the boundary layers are calculated for dominant and understorey crops as functions of the leaf area index of each crop, as described for a sole crop in eq. 7.19. Concerning the canopy resistances, (eq. 7.20) the saturation deficit is the same for both crops, corresponding to the D0 level (Figure 10.8), while the incident radiations differ for each crop.

10.3.4 Leaf growth of the understorey crop

In the case of sole crops, a strong correlation between intercepted radiation and temperature implicitly links the LAI and the biomass accumulation processes, which makes the separate calculation of LAI and biomass accumulation realistic. In the case of an understorey crop, this correlation no longer exists because of the shade of the dominant crop. It is therefore important to limit leaf expansion when not enough structural biomass is available to expand leaves at the rate predicted by temperature and also to account for light quality effects. This is done by means of:

– a trophic limitation on leaf expansion, using the notion of the maximum leaf expansion allowed per unit of biomass accumulated in the plant, and described in § 3.1.1.b.

– the calculation of an equivalent plant density for the understorey crop (DENSITEUeq), which accounts for the presence of the dominant crop. If $DENSITE_D$ and $DENSITE_U$ are the planting densities of the dominant and the understorey crops respectively and $BDENSD_P$ and $BDENSU_P$ are the threshold densities for inter-plant competition, the equivalent density is calculated as in eq. 10.7:

eq. 10.6

$$DENSITEUeq\ (I) = DENSITE_U\ (I) + DENSITE_D\ (I) \frac{BDENSU_P}{BDENSD_P}$$

This empirical relationship allows the inter-plant competition to be increased (DELTAIdens function, as described in § 3.1.1.a and eq. 3.4) compared to the mono-crop situation (Figure 10.9).

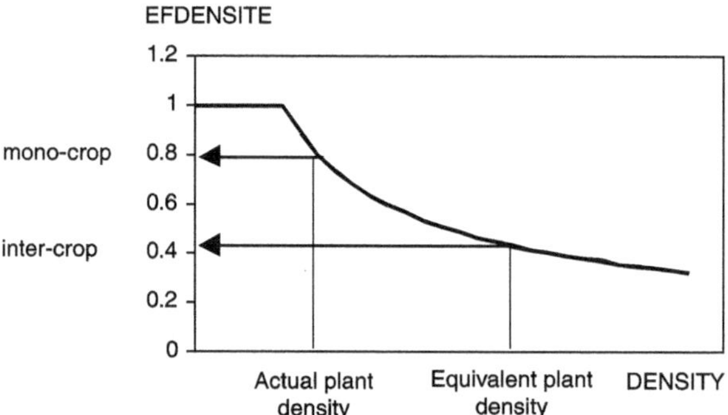

Figure 10.9. Illustration of the calculation of the equivalent plant density for the understorey crop.

10.3.5 Root profiles

The soil volume occupied by each crop is different in sole crops and intercrops. Adiku *et al.* (2001) showed that root systems of component plants in a mixture may intermingle considerably in well-watered situations whereas a tendency for the root systems to cluster within their 'own' zones may be observed under water-stressed conditions. This behaviour does not fit the notion of a standard root profile (see § 5.2.1). In our model, we did not consider allelopathy but we assumed that for intercrops the influence of the crop root systems on each other results from the influence of the soil status on the root distribution. Consequently, the "true density" option (see § 5.2.2) has to be chosen to calculate the root distribution profile in the case of intercrops.

Figure 10.10 illustrates the simulation of root profile dynamics in a case study of pea (Pisum sativum) and barley (Hordeum vulgare) sole crops and a pea-barley intercrop. This example illustrates the plasticity of the "true density" approach through:

– a limitation of root growth in the case of pea intercrop (compared to pea sole crop), with the decrease of temperature induced by the shade effect of the shrub,

– the lower capacity of pea inter-crop to colonize the soil profile compared to barley intercrop, which underlines its low competitiveness.

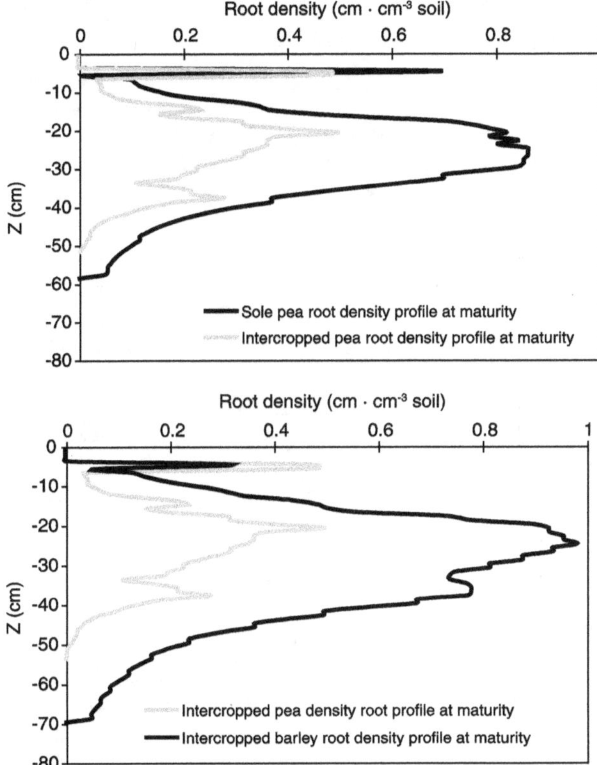

Figure 10.10. Simulation of root density profiles at maturity in the case study of Pea-Barley intercrop in 2004, from a) sole and intercropped peas, and from b) intercropped pea and barley.

10.3.6 Simulation examples

Figure 10.10 illustrates the simulation of root profile dynamics in a case study of pea (Pisum sativum) and barley (Hordeum vulgare) sole crops and a pea-barley intercrop. This example illustrates the plasticity of the "true density" approach through:

– a limitation of root growth in the case of pea inter-crop (compared to pea sole crop), with the decrease of temperature induced by the shade effect of the shrub,

– the lower capactiy of pea intercrop to colonize the soil profile compared to barley intercrop, which underlines its low competitiveness.

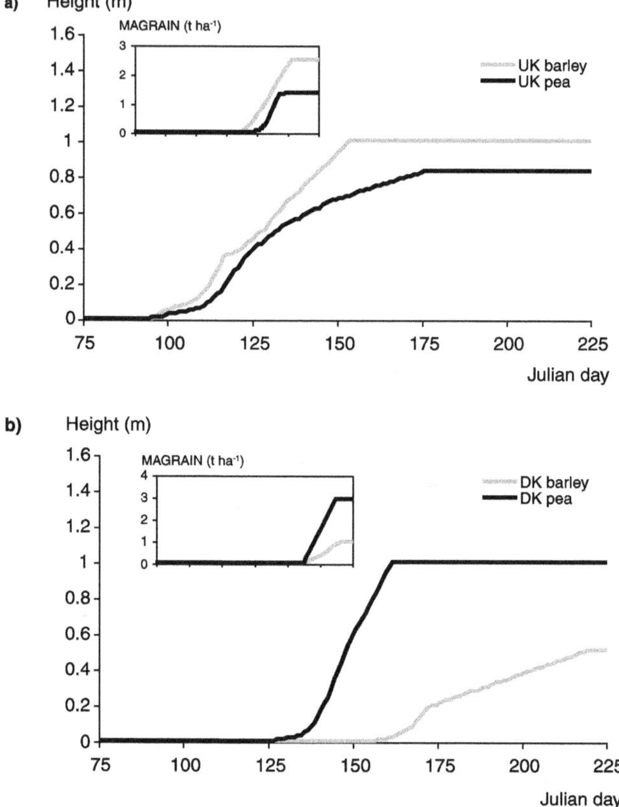

Figure 10.11. Simulation of crop height (HAUTEUR) and yield (MAGRAIN) in the case study of pea-barley intercrop in United Kingdom (a) and Denmark (b) sites in 1999 (Launay *et al.*, 2008).

With the soil and climatic conditions of UK and Denmark, the model simulates pea and barley emergence on the same date in UK whereas peas emerge before barley in Denmark, leading to barley dominance in UK (Figure 10.11.a), and peas appears as a better competitor for light than barley in Denmark (Figure 10.11.b). This result leads to a bigger pea yield in the Danish intercrop, while pea and barley yields remain nearly the same in UK (Launay *et al.*, 2008).

11 Involvement of the user in the model operation

11.1 Driving options

11.1.1 Regular weather driving variables

Like most dynamic crop models, STICS is driven by the weather on a daily time step. The minimal set of weather variables required to run the model comprises minimum and maximum temperature (°C), global radiation ($MJm^{-2}d^{-1}$) and rainfall (mm d^{-1}). There are four possible ways of estimating evapotranspiration (TETP), requiring different numbers of additional variables.

The least demanding option is the Priestley-Taylor (1972) calculation, followed by that of imposing a pre-calculated value. The two last options require two additional primary weather variables, namely wind speed (ms^{-1}) and vapour pressure (mbars). One of them is the calculation of the Penman formula and the other does not rely on the notion of evapotranspiration but directly calculates water requirements at the plant level through a resistive approach (§ 7.2.2). For the three first options, let us point out the close dependence between the value of reference evapotranspiration and the $KMAX_p$ value (because $KMAX_p$ is experimentally calculated with a given reference evapotranspiration), so that a change in this option should theoretically lead to a change in $KMAX_p$. In a comparative work cited by Smith *et al.* (1996), many reference evapotranspiration calculations were compared to lysimeter measurements from 11 locations. The Penman formula exhibits a 0.60-0.70 mm error whatever the environment while Priestley-Taylor formula appears far better in humid environments (0.68 mm) than in arid ones (1.89 mm).

11.1.1a Calculation of Priestley-Taylor reference evapotranspiration

This calculation (eq. 11.1) is recommended in the absence of wind speed and humidity measurements but it takes poor account of convective factors. It relies on a site-dependent coefficient $ALPHAPT_C$, whose value for many soil surface conditions is 1.26, and an empirical calculation for the net radiation ($RNET_{PT}$) and a constant value of the latent heat of vaporization (2.5 MJ kg^{-1}). The other variables have already been defined (see eq. 6.31)

$$\text{eq. 11.1}$$

$$TETP(I) = ALPHAPT_C \cdot \frac{DELTAT(I)}{2.5 \cdot (DELTAT(I) + GAMMA)} RNET_{PT}(I)$$

$$with \quad RNET_{PT}(I) = 0.8 \cdot 0.72 \cdot TRG(I) - 0.9504$$

11.1.1b Calculation of Penman evapotranspiration

The formulae eq. 11.2 is from Penman (1948) fully described in Brochet and Gerbier (1968)

$$\text{eq. 11.2}$$

$$TETP(I) = \frac{DELTAT(I)}{L(I) \cdot (DELTAT(I) + GAMMA)} RNET_{PE}(I)$$

$$+ \frac{GAMMA}{DELTAT(I) + GAMMA} \cdot 0.26 \cdot (1 + 0.54 \cdot TVENT(I)) \cdot DSAT(I)$$

where $RNET_{PE}$ is estimated by combining eq. 6.22 and eq. 6.27 (Brunt formula), using a value of 0.20 for the albedo.

11.1.2 Driving the model by weather data for high altitude climates

The model is driven by standard weather variables (radiation, minimum and maximum temperatures, rainfall, reference evapotranspiration and possibly wind speed and humidity) on a daily time step. These meteorological data are obtained from a weather station and entered in an input file. The difference in altitude between the weather station and the simulation site can be taken into account but only in terms of recalculation of temperatures, the other weather readings remaining unchanged.

As a general rule, temperatures in mountain regions show a gradual fall with altitude, and a difference in temperature between the south-facing and the north-facing slopes. In addition, account must be taken of the temperature inversion phenomenon which affects minimum temperatures. Different studies (Antonioletti, 1986; Antonioletti and Seguin, 1988; Douguedroit, 1986) have been made on temperatures in mountain regions, and the values used are thus taken from these studies. Differences in incident radiation also occur between south and north-facing slopes but they are not accounted for by the model.

11.1.2a Parameterization of the various phenomena

$ALTISIMUL_C$ and $ALTISTATION_C$ are the altitudes of the simulation site and the weather station, respectively, with the assumption that $ALTISIMUL_C > ALTISTATION_C$.

To account for the gradual fall in temperature with increasing altitude (adiabatic gradient), we have used the values provided by Douguedroit (1986) who proposed a reduction of 0.55°C (+/-0.08°C) per 100 m at night and a reduction of 0.61°C ($\pm$ 0.03°C) during the day. These mean values were assigned to the parameters $GRADTX_G$ and $GRADTN_G$, which affect the calculation of maximum and minimum temperature respectively.

For the differences between south and north-facing slopes, the problem is more complex, and studies are lacking. According to Antonioletti and Seguin (1988), the difference between south and north-facing slopes is mainly found to affect maximum temperatures. In the case of Mont Ventoux, these exhibit an almost constant difference of about 1.4°C. The parameter $OMBRAGETX_G$ represents this constant difference, and is removed to the maximum temperature when the simulation site is on the north-facing slope (parameter $CODEADRET_C = 2$ otherwise $=1$).

The phenomenon of temperature inversion is due to the circulation of air masses during clear weather at night, which causes a flux of cold air into valleys. This leads to a rise in temperature as the altitude increases (approximately 1.3°C per 100 m up to an altitude of between 400 and 900 m) (Antonioletti, 1986). This has been included in the model through the parameter $GRADTNINV_G$ up to the threshold altitude $ALTINVERSION_G$. The notion of "clear weather" has been taken into account from calculation of the complement of cloud cover FRACINSOL (eq. 6.26.), which has to reach at least the value of the parameter $CIELCLAIR_G$.

11.1.2b Calculation of the maximal temperature

The maximal temperature at $ALTISIMUL_C$ (TMAX) depends on the reference maximal temperature measured by the weather station (TMAXS) and the adiabatic gradient ($GRADTX_G < 0$) corrected by $OMBRATX_G$ (<0) in case of north-facing slope (eq. 11.3)

eq. 11.3

$$TMAX(I) = TMAXS(I) + GRADTX_G \cdot (ALTISIMUL_C - ALTISTATION_C)$$
$$/100 + (CODADRET_C - 1)\, OMBRAGETX_G$$

11.1.2c Calculation of the minimal temperature

The relative position of the inversion altitude requires defining two functions, corresponding to the adiabatic (ADIA) and to the inversion (INV) gradients (eq. 11.4).

eq. 11.4

$$ADIA(TN, ALT1, ALT2) = TN + GRADTN_G \cdot (ALT2 - ALT1)/100$$
$$INV(TN, ALT1, ALT2) = TN + GRADTNINV_G \cdot (ALT2 - ALT1)/100$$

where ALT1 and ALT2 are two increasing altitudes (ALT2>ALT1), $GRADTN_G$ <0 and $GRADTNINV_G$ >0. Then 2 cases must be considered to calculate the resulting minimal temperature of the simulated site (TMIN) as a function of the reference measured one (TMINS)

1. $ALTINVERSION_G > ALTISTATION_C, ALTISIMUL_C$
   $$TMIN(I) = INV(TMINS(I), ALTISIMUL_C, ALTISTATION_C)$$
2. $ALTISIMUL_C > ALTINVERSION_G > ALTISTATION_C$
   $$TMIN(I) = ADIA[INV(TMINS(I), ALTINVERSION_G, ALTISTATION_C),$$
   $$ALTISIMUL_C, ALTINVERSION_G]$$

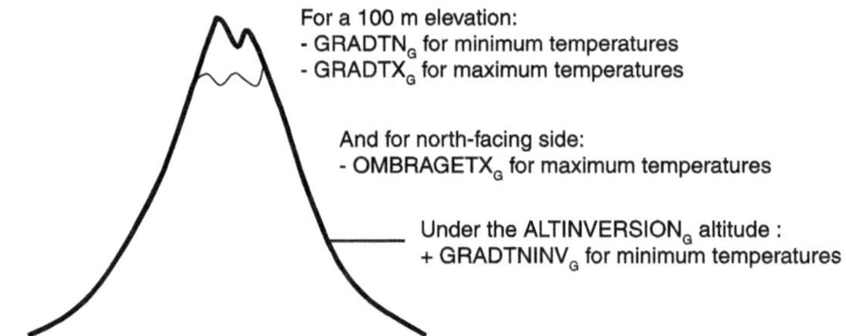

Figure 11.1. Temperature variations at high altitudes.

11.1.3 Driving the model by observed stages.

The model can be driven using the observed phenological stages. In this case, the dates of the different stages in the techniques file are given the observed values. The model can equally be asked to calculate or use the observation dates for any of the following vegetative stages ILEV, IAMF, ILAX, or harvested organ stages IDRP, IMAT, IDEBDES and IREC (see § 2, Table 2.1). The flowering stage IFLO is always calculated by the model from a given $STFLODRP_V$ parameter corresponding to the thermal duration between flowering and the onset of filling of the harvested organs IDRP. There is no point in driving this flowering stage date IFLO with observed data, because it doesn't trigger any calculation in the model. Should there be any disagreement between the calculated dates and the observed dates for consecutive stages, then the model will cease to run.

11.1.4 Driving the model by the LAI.

In this case, another model is used, called STICS-feuille, which uses the observed LAI data as inputs (Ripoche *et al.*, 2001). This way of driving it can be very useful when developing the model. By imposing the LAI, water and nitrogen requirement levels suitable to cope with stress are also imposed.

As it is usually difficult to obtain daily observed LAI data, a tool was developed to interpolate LAI measurements using a statistical relationship representing the time course of LAI (eq. 11.5). This function is fitted to calculate daily LAI from measurements.

eq. 11.5

$$LAI(I) = K_{LAI}\left(\frac{1}{1+exp\left(-B_{LAI}\left(ST_{LAI}(I)-TI_{LAI}\right)\right)} - exp\left(A_{LAI}\left(ST_{LAI}(I)-TF_{LAI}\right)\right)\right)$$

where ST_{LAI} is the growing degree-days since emergence, K_{LAI} is the maximal LAI produced (which is higher than the maximal measured LAI because of the effect of senescence), TI_{LAI} and TF_{LAI} are the growing degree-days for the point of inflexion of the growth curve and complete senescence respectively, and A_{LAI} and B_{LAI} describe the shapes of the growth and senescence curves. (Figure 11.2). The first component of eq. 11.5 stands for the green leaf appearance and the second for leaf senescence.

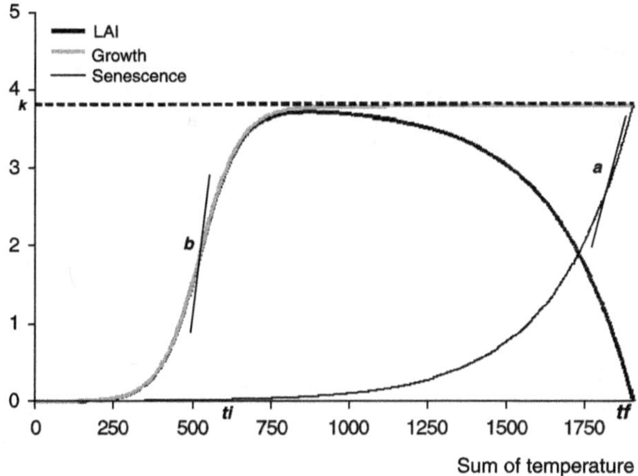

Figure 11.2. Empirical relationship representing the time course of LAI, and its two components, green leaf appearance and leaf senescence.

In the following example of a wheat crop (Figure 11.3), daily prescribed LAI data were calculated by fitting the 5 parameters of eq. 11.5 to LAI measurements. Then those daily LAIs were used to drive the model, improving the simulation of above-ground biomass (MASEC).

11.2 Simulation options

11.2.1 Water or nitrogen stress activation or deactivation

Model users are allowed to deactivate water and/or nitrogen stress effects in order to simulate crop growth and development without water or nitrogen stress (Figure 11.4). These options allow "potential" yield to be predicted for example, or to organize these stresses into a hierarchy (by deactivating them separately). Only water and nitrogen stress effects (Table 3.1 and Table 3.2) may be deactivated, while trophic, temperature, water-logging or stress effects linked to the soil structure remain. However, even if water or

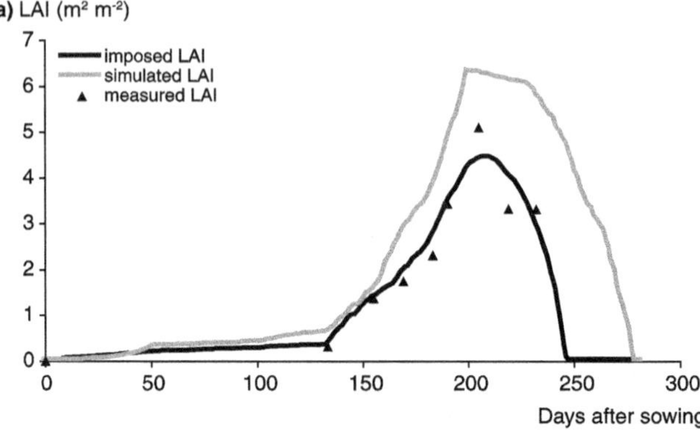

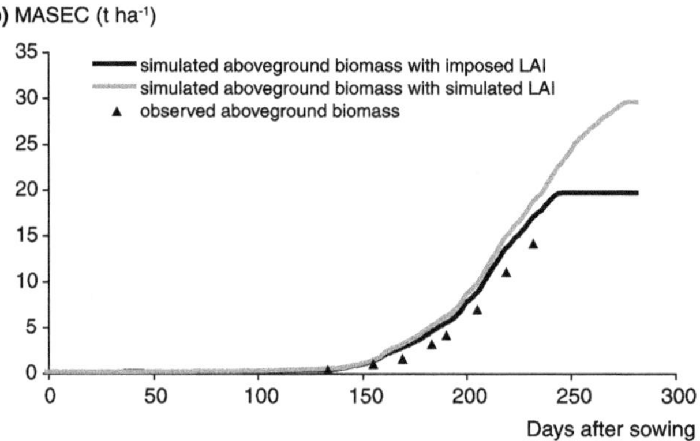

Figure 11.3. Improvement of the above-ground biomass values (MASEC) when the model is run by prescribed LAI as compared to free simulation of LAI. a) LAI and b) above-ground biomasses.

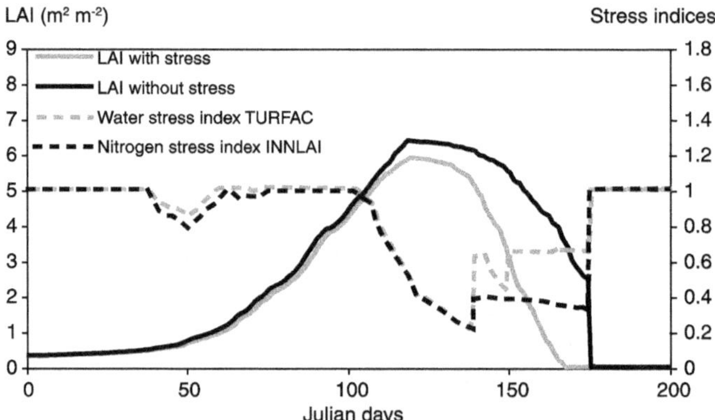

Figure 11.4. Comparison of stressed and unstressed LAI dynamics and evolution of water and nitrogen stress indices over the cropping season (calculated when influencing crop growth).

nitrogen stress is deactivated and does not influence crop growth and development, it is still calculated by the model and is available to the user through the model outputs. But in that case, these stress index values may not be considered as "actual" values because they depend on the crop growth which is calculated as a "potential" one.

11.2.2 Smoothing of initial profiles

In order to avoid discontinuities of water and nitrogen content between soil layers, a smoothing option may be activated. This option smoothes initial profiles of water and nitrogen contents (Figure 11.5), thanks to a spline function.

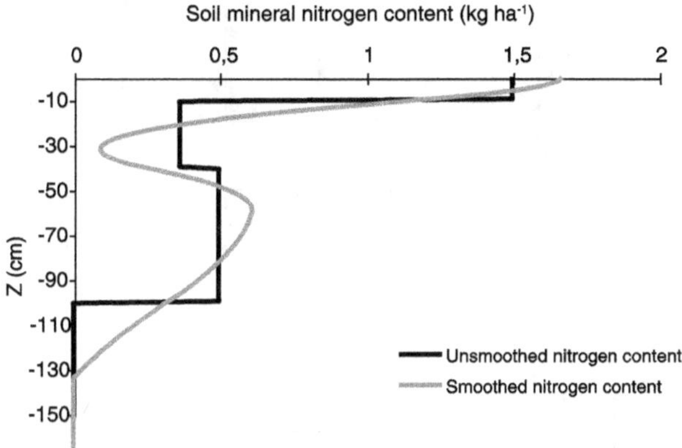

Figure 11.5. Nitrogen content partitioning with soil depth with or without activating the smoothing option.

11.3 Formalisation options

The generality of the model is allowed first through the availability of formalisation options, some of which are exclusive. Moreover some formalisation choices are linked to each other, e.g. the option allowing the calculation of the amount of foliage simply as ground cover precludes the use of the mechanistic option of calculating water balance with a resistive approach. These choices drive the efficient parameter number.

These options address plant and variety ecophysiology (Table 11.1), soil biophysics (Table 11.2) or cropping management (Table 11.3). If the first set of options is pre-determined by species specialists, the last two need input from the user as descriptions of the local cropping system conditions.

Table 11.1.a List of the formalisation options which address plant and variety ecophysiology for development and shoot growth.

Chapter	Paragraph	Sub-paragraph	Options	Sub-options	Sub-sub-options	Reference	Links between options
Development	Emergence and starting of crop development and growth		Annual plant	Emergence	Germination (Yes/No)	§2.2.1	
				Emergence	Subsoil plantlet growth (Yes/No)	§2.2.1	
				Onset of crop development after planting		§2.2.2	
				Frost damage on plantlet density (Yes/No)		eq. 2.6	
			Perennial plant			§2.2.3	
	Aboveground development		Action of photoperiod (Yes/No)			§2.3.3	
			Cold requirement (Yes/No)	Vernalisation		§2.3.4	
				Dormancy	Bidabe formalism	eq. 2.13	
				Dormancy	Richardson formalism	eq. 2.14	
			Action of stress (Yes/No)			§2.3.5	
Shoot growth	Leaf dynamics		Cover rate			§3.1.4	*"Radiation transfers"*, *"Trophic stress"*, *"Resistive"* and *"Energy balance"* approaches are forbidden
			Leaf surface	Trophic stress (Yes: indeterminate / No: determinate crop)		§3.1.1	
				Frost damage on leaf growth	During the juvenile phase (Yes/No)	§3.4.4	
					During the vegetative and reproductive phasis (Yes/No)	§3.4.4	
	Radiation interception		Beer's law			§3.2.1	
			Radiation tranfers			§3.2.2	has to be chosen with *"Resistive"* and *"Energy balance"* approaches
	Shoot biomass growth		Effect of atmospheric CO_2 concentration (Yes/No)			§3.3.2	
			Remobilisation of reserve from the preceeding cycle (Yes: perennial plant / No: annual plant)			§3.3.3	Linked to the option *"perennial/annual"* plant
	Stress indices	Nitrogen deficiency	Nitrogen nutrition index			§3.4.2	
			Instantaneous nitrogen nutrition index			eq. 3.33	
	Partitioning of biomass in organs	Dimensioning of organs	Monocotyledon plant			§3.5.2	
			Dicotyledon plant			§3.5.2	
		Reserves	Use of reserves (Yes: perennial or indeterminate plants / No: determinate annual plants)			§3.5.4	Linked to the options *"perennial/annual"* and *"determinate/indeterminate"* plant

Table 11.1.b Formalisation options for plant and variety ecophysiology for yield formation, root growth, water balance, management and crop environnement, and nitrogen transformation processes.

Chapter	Paragraph	Sub-paragraph	Options	Sub-options	Sub-sub-options	Reference	Links between options
Yield formation			Determinate growing plant			$4.1	
			Indeterminate growing plant	Imposed number of inflorescences		$4.2	
				Trophic dependant number of inflorescences		$4.2.1	
			Nitrogen stress on harvested organ number (Yes/No)			$4.1 & $4.2	
			Frost damage on harvested organs (Yes/No)			eq. 4.3 & eq. 4.8	
			Thermal stress on harvested organs (Yes/No)			eq. 4.5 & eq. 4.10	
Root growth	Root front growth		Crop temperature dependent			eq. 5.2	
			Soil temperature dependent			eq. 5.3	
	Growth in root density		"Standard" profile			$5.2.1	
			"True" density	Self-governing production of root length		eq. 5.7	
				Trophic-linked production of root length	just acting as a threshold	eq. 5.7	
					driving the root length production	eq. 5.11	
				Mineral nitrogen effect on root distribution in the profile (Yes/No)		$5.2.2 & fig.5.6	
Water balance	Plant water requirements		Crop coefficient approach			$7.2.1	
			Resistive approach	Effect of atmospheric CO_2 concentration (Yes/No)		$7.2.2	has to be chosen with *"Radiation transfers"* and *"Energy balance approach"*
Nitrogen transformations	Nitrogen uptake by plants		Classical dilution curve			$8.2	
			Dilution curve for very little dense crop			$8.2	
	Nitrogen fixation		Leguminous (Yes/No)	Potential fixation being imposed		$8.3	
				Potential fixation depending on aboveground growth		$8.3	

Table 11.2 Formalisation options for soil biophysics.

Chapter	Paragraph	Sub-paragraph	Options	Sub-options	Reference	Links between options
Management and crop environment	Net nitrogen supply	Denitrification	Mechanistic model of denitrification from NEMIS		§6.3.2.a & §8.5 & eq.8.24	
			Constant proportion of the mineral fertiliser that can be denitrified		§6.3.2.a & eq.6.4	
Water functionning	Soil evaporation	Potential evaporation	Beer's law equivalent approach		§7.1.1 & eq. 7.1	
			Energy balance approach		§7.1.1 & eq. 7.2	has to be chosen with *"Radiation transfers"* and *"Resistive approach"*
Nitrogen functionning	Mineralisation		Nitrification (Yes/No)		§8.3	
Transfers of heat, water and nitrates	Tranfers of heat, water and nitrates in undrained soil	Stones	Pebble presence (Yes/No)	Pebble types (8 given types)	§9.2.3 & tab.11.10	
		Macroporosity and cracks	Macroporosity presence (Yes/No)	Cracks presence (Yes/No)	§9.2.4	
		Capillary rise	Capillary rise (Yes/No)		§9.2	
	Case of drained soils		Drainage system presence (Yes/No)		§9.3	

Table 11.3.a Formalisation options for crop management and crop environment through effects on plants.

Sub–Chapter	Paragraph	Sub-paragraph	Options	Sub-options	Reference	Links between options
Effects on plants	Planting design & Simulation of the decision to sow		Sole crop / Intercrop		§6.1.1 & §10.3	
			Crop establishment	Sowing grains	§6.1.1	
				Planting plantlets	§6.1.1	
				Sowing delay depending on the soil physical conditions (Yes/No)	§6.1.2	
			No establishment (perennial plants)		§6.1.1	
			Geometrical pattern of the crop (Yes/No)		§6.1.1	Has to be chosen whith "Radiation transfers" option
	Yield regulation		Foliage regulation	Topping	§6.1.3.a	
				Leaf removal	§6.1.3.a	
			Fruit removal		§6.1.3.b	
			No yield regulation		§6.1.3.b	
	Harvest	Harvest policy	Cutting		§6.1.4.a	
			Picking		§6.1.4.a	
			One cutting or picking		§6.1.4	
			Several cuttings or pickings		§6.1.4	
			Buffer date of harvest		§6.1.4.d	
		Particular case of forage crops	Automatic calculation of cutting dates		§6.1.4.b	
			Imposed cutting dates	Calendar dates	§6.1.4.b	
				Physiological dates	§6.1.4.b	
		Simulation of the decision to harvest	Physiological maturity		§6.1.4.d	
			Maximum or minimum water content in harvested organs		§6.1.4.d	
			Minimum sugar content in harvested organs		§6.1.4.d	
			Minimum nitrogen content in harvested organs		§6.1.4.d	
			Minimum oil content in harvested organs		§6.1.4.d	
			Harvest delay depending on the soil physical conditions (Yes/No)		§6.1.4.d	
	Pruning		Pruning (Yes/No)		§6.1.5	

Table 11.3.b Formalisation options for crop management and crop environment through soil water and net nitrogen supplies.

Sub-Chapter	Paragraph	Sub-paragraph	Options	Sub-options	Reference	Links between options
Soil water supply	Irrigation	Irrigation amount	Imposed quantities and dates	Calendar dates	$6.2.1	
				Physiological dates	$6.2.1	
			Calculated quantities and dates		$6.2.1	
		Irrigation system	Above foliage irrigation system		$6.2.1	
			Under foliage irrigation system		$6.2.1	
			Underground irrigation system		$6.2.1	
	Interception of water by foliage		Interception of water by foliage (Yes/No)		$6.2.2	
Net nitrogen supply	N inputs from mineral fertilisers	Nitrogen use efficiency	Type of mineral fertiliser (7 different given types + 1 "blank" type)		$6.3.2	
			Imposed fertiliser efficiency		$6.3.2.a	Only available with the fertiliser type number 8
			Calculated fertiliser efficiency		$6.3.2.a	Only available with the first 7 types of fertilisers
			Fertiliser localisation at the soil surface		$6.3.2	
			Fertiliser localisation in the soil		$6.3.2	
		Fertilisation calendar	Imposed quantities and dates	Calendar dates	$6.3.2.b	
				Physiological dates	$6.3.2.b	
			Calculated quantities and dates	Rainfall test	$6.3.2.b	
				Test on water availability in the upper soil layer	$6.3.2.b	
	N inputs from organic residues		Type of organic residue (7 different given types + 1 "blank" type)		$6.3.3	
	Crop residues for the following crop		Type of crop residues (roots / straw and fine roots / stubble and fine roots / whole crop)		$6.3.4	

Table 11.3.c Formalisation options for crop management and crop environment through physical soil surface conditions, soil structure modification and microclimate.

Sub-Chapter	Paragraph	Sub-paragraph	Options	Sub-options	Reference	Links between options
Physical soil surface conditions		Presence of a plant or plastic cover	Vegetal mulch	Corn residue mulch	$6.4.1 & eq.6.15 & eq.6.16	Water directly evaporated from the mulch and soil evaporation under mulch are calculated differently according to the option chosen to calculate the soil potential evaporation (see $7.2.1 & $7.2.2)
				Sugar cane residue mulch	$6.4.1 & eq.6.15 & eq.6.16	
				Vine-shoot residue mulch	$6.4.1 & eq.6.15 & eq.6.16	
			Plastic mulch		$6.4.1 & eq.6.16	
			No mulch		$6.4	
Soil structure modification	Compaction under the influence of sowing and harvesting machines		Compaction (Yes/No)		$6.5.2	
	Fragmentation under the effects of soil tillage implements		Fragmentation (Yes/No)		$6.5.3	
Microclimate	Calculation of net radiation	Long wave radiation	Brunt's formula		$6.6.1.b & eq.6.27	
			Brutsaert's formula		$6.6.1.b & eq.6.28	
	Calculation of crop temperature		Empirical approach		$6.6.2.a & eq.6.29	
			Energy balance		$6.6.2.b & eq.6.30	
	Estimation of microclimate under shelter		Crop grown under a shelter (Yes/No)		$6.6.4 & eq.6.33 to eq.6.38	

11.4 Parameterization

Running the model without the proper parameter set makes it inoperative or leads to incorrect results, and yet this part is rarely documented in scientific literature because the parameterization is not regarded as novel. Expressing phenomena as equations is considered a much nobler task and is the object of many scientific papers. The actual parameter values are mostly only available in the technical documentation of crop models, or worse, in the code. Unlike statisticians for whom a model comprises equations and parameters, so that changing parameters brings about a change of the model, for crop scientists the model does not include the parameter values.

However the robustness of the model, as well as its ability to be extrapolated, is closely linked to the parameter values, mainly their spatial and temporal validity as well as their validity throughout various cropping systems. But spatial and temporal validity must not be mistaken for biophysical meaning. The closer the parameters are to the processes, the better their biophysical meaning. We can assume that such parameters can be arrived at independently from the model through experiments, especially those carried out in controlled environments. These parameters are valuable even though they can be strictly soil- or plant-dependent. On the other hand parameters that encompass many processes are difficult to measure by experiments and must be evaluated with the model by mathematical optimization techniques (Makowski *et al.*, 2006). Although not always the case, those parameters may be ones which confer robustness to the model, e.g. the maximal radiation use efficiency or the harvest index.

The number of parameters for a model is often a subject of discussion. A widely accepted idea is that the more the parameters, the more complicated the model. It is true that the number of parameters more or less reflects the number of processes simulated, which can be regarded as a source of complication. At the same time, parameters may be easy to access, so we prefer the notion of cost of availability of the parameters rather than their number. Readily available parameters are, for example, parameters of a biophysical nature connected to largely shared databases, such as soil parameters deduced from soil databases using soil transfer functions. It is very important to know the sensitivity of the model to the considered parameter in order to determine the required precision for its value (see for STICS Ruget *et al.*, 2002). Note that each sensitivity analysis is applicable only to the soil and weather conditions explored.

In STICS we have adopted the commonly-used definitions for parameters and variables, i.e. parameters can be considered as constant throughout the simulation while variables vary over time. In STICS, some parameters were converted to variables as new processes were added.

The following paragraphs focus on the parameters characterizing the three parts of the cropping system, i.e. the plant, the soil and the cropping techniques, and attempt to recommend methodologies to assign them values.

11.4.1 Plant parameterization

11.4.1a Methodology

First is the choice of formalisations. Some formalisations are prescribed from agro-physiological knowledge of the plant, while others are chosen by the user as a function of his point of view concerning the output variables of interest. The most important processes should be as mechanistically simulated as possible, the available information on the plant or its cropping conditions providing default values.

The second element of the recommended methodology is the sensitivity analysis, which allows the parameters to be ranked according to their influence on the variable of interest and to quantify the magnitude of this influence. It also reveals the agricultural conditions which maximise this influence, such as nutrient availability, weather conditions etc. Some methods of applying sensitivity analysis to crop models are described in Monod *et al.* (2006).

Finally the inventory of available experimental data or published parameters for the considered plant or similar species should determine the means of specifying parameter values (Table 11.4). Some methods for estimating parameters are detailed in Makowski *et al.* (2006).

Table 11.4. Summary of the various means available for assigning parameter values according to the data available. 1 means: "parameter estimation" is used in its mathematical meaning, referring to statistical methods for finding the parameter giving the best fit between observed and simulated output variables.

Parameter	High sensitivity	Low sensitivity
Available in literature for the plant of interest or for an analogous species Available as a parameter of another model	Estimation[1] recommended if the validity domain of available information does not match the pursued objective.	Use of available parameter values from literature or modelling documentation
Measured or calculated directly from available experimental data	Use of measurements compulsory	Use of measurements optional
Unavailable through the above-mentioned means	Estimation[1] compulsory using dedicated experimental data	Analogous species parameters (always possible with STICS)

11.4.1b Set of plant parameters for some species

In order to give some examples of plant parameterizations, we propose in tab. 11.5 the plant parameter values for five different crops. Two perennial crops, forage (herbaceous) and grapevines (*Vitis vinifera* L., ligneous), and three annual crops, spring pea (*Pisum sativum* L.), sugar beet (*Beta vulgaris* L.) and winter wheat (*Triticum aestivum* L.) are taken as examples. In this table, some parameter values are not given if the equations for which they are needed have not yet been formulated for the species concerned. Others are given for different varieties when they are variety-dependent.

Table 11.5. Plant parameter values for forage, spring pea, sugarbeet, two varieties of vine and two varieties of winter wheat (grey boxes correspond to the parameters that are not activated).

Chapter	Parameter	Unit	Value						
			Forage	Pea	Sugar-beet	Vineyard		Winter wheat	
						Grenache	Chardonnay	Talent	Shango
Development	TDMIN$_P$	°C	0	0	2	10		0	
	TDMAX$_P$	°C	30	25	25	37		28	
	STLEVAMF$_V$	degree·day	116	230	500	25		245	235
	STAMFLAX$_V$	degree·day	3500	418	5000	1123	927	260	340
	STLEVDRP$_V$	degree·day	1000	900	450	363	305	692	837
	STFLODRP$_V$	degree·day		216	0	50	51	0	0
	STDRPMAT$_V$	degree·day		490				700	700
	COEFLEVAMF$_P$	ND						1	
	COEFAMFLAX$_P$	ND						1	
	COEFLEVDRP$_P$	ND						1	
	COEFDRPMAT$_P$	ND						1	
	COEFFLODRP$_P$	ND						1	
	PHOBASE$_P$	h						6.3	
	PHOSAT$_P$	h						20	
	SENSIPHOT$_V$	ND						0	0
	STRESSDEV$_P$	ND	0.2			0.2		0.2	
	JVCMINI$_P$	day						7	
	JULVERNAL$_P$	julian day						274	
	TFROID$_P$	°C						6.5	
	AMPFROID$_P$	°C						10	
	Q10$_P$	ND				2.17			
	IDEBDORM$_P$	julian day				213			
	STDORDEBOUR$_P$	degree day				9174	6577		

Chapter	Parameter	Unit	Forage	Pea	Sugar-beet	Vineyard: Grenache	Vineyard: Chardonnay	Winter wheat: Talent	Winter wheat: Shango
	JVC_V	day				102.2	101.2	50	45
	$TGMIN_P$	°C		0	3			0	
	$STPLTGER_P$	degree day		60	45			50	
	HUMECGRAINE	MPa		−1.6	−1.6			−1.6	
	$NBJGERLIM_P$	days		15	15			15	
	PROPJGERMIN	$day.day^{-1}$		1	1			1	
	$BELONG_P$	degree day		0.0115	0.0115			0.012	
	$CELONG_P$	ND		4.57	4.57			3.2	
	$ELMAX_P$	cm		7.21	7.21			8.0	
	$NLEVLIM1_P$	day		10	10			10	
	$NLEVLIM2_P$	day		50	50			50	
	$VIGUEURBAT_P$	ND						1.0	
	LAIPLANTULE	m^2 leaf m^{-2} soil							
	NBFEUILPLANT	nb leaves $plant^{-1}$							
	MASECPLANTULE	$t\ ha^{-1}$							
	ZRACPLANTULE	m							
Shoot growth	$PHYLLOTHERME_P$	degree·day	120	120	120	25		120	
	$ADENS_V$	ND	−0.6	−0.45	−0.47	0.0	0.0	−0.6	−0.6
	$BDENS_P$	$plants·m^{-2}$	7.0	10.0	7.0	1.0		7.0	
	$LAICOMP_P$	$m^2·m^{-2}$	0.30	0.0	0.75	0.0		0.30	
	$HAUTBASE_P$	m	0.02	0.0	0.0	0.6	0.3	0.0	
	$HAUTMAX_P$	m	0.3	1.0	0.6	2.5		1.20	
	$VLAIMAX_P$	ND	2.2	2.2	2.2	2.2		2.2	
	$PENTLAIMAX_P$	ND	5.5	5.5	5.5	5.0		5.5	
	$UDLAIMAX_P$	ND	3.0	3.0	3.0	3.0		3.0	

Chapter	Parameter	Unit	Value						
			Forage	Pea	Sugar-beet	Vineyard		Winter wheat	
						Grenache	Chardonnay	Talent	Shango
Shoot growth	RATIODURVIEI$_P$	ND	1.0	0.8	1.0	1.0		0.8	
	TCMIN$_P$	°C	0.0	0.0	2.0	10.0		0.0	
	TCMAX$_P$	°C	40.0	30.0	30.0	37.0		40.0	
	RATIOSEN$_P$	ND	0.8	0.25	0.87	0.8		0.8	
	ABSCISSION$_P$	ND	0.0	0.0	0.28	1.0		0.0	
	PARAZOFMORTE$_P$	ND	13.0	13.0	13.0	13.0		13.0	
	INNTURGMIN$_P$	ND	−0.2	0.3	0.3	0.3		−0.2	
	DLAIMAXBRUT$_P$	m^2 leaf plant^{-1} degree day^{-1}	$3.8 \cdot 10^{-4}$	$3.5 \cdot 10^{-4}$	$4.4 \cdot 10^{-3}$	0.015		$4.4 \cdot 10^{-4}$	
	DURVIEF$_V$	ND	170	160	129	400	400	200	200
	DURVIESUPMAX$_P$	ND	0.4	0.4	0.0	0.1		0.4	
	INNSEN$_P$	ND	0.35	0.35	0.3	0.87		0.35	
	RAPSENTURG$_P$	ND	0.5	0.5	0.5	0.05		0.5	
	EXTIN$_P$	ND	0.55		0.58			0.5	
	KTROU$_P$	ND		1.0		1.7			
	FORME$_P$	code		1		2			
	RAPFORME$_P$	ND		8.0		1.5			
	ADFOL$_P$	m^{-1}		1.0		3.16			
	DFOLBAS$_P$	m^2 leaf·m^{-3}		5.0		1.5			
	DFOLHAUT$_P$	m^2 leaf·m^{-3}		5.0		11.5			
	TEMIN$_P$	°C	0.0	0.0	2.0	10.0		0.0	
	TEMAX$_P$	°C	40.0	30.0	30.0	37.0		40.0	
	TEOPT$_P$	°C	24.0	15.0	15.0	25.0		12.0	
	TEOPTBIS$_P$	°C	24.0	20.0	26.0	25.0		17.0	
	ALPHACO2$_P$	ND	1.2	1.2	1.2	1.2		1.2	

<table>
<tr><th rowspan="3">Chapter</th><th rowspan="3">Parameter</th><th rowspan="3">Unit</th><th colspan="7">Value</th></tr>
<tr><th rowspan="2">Forage</th><th rowspan="2">Pea</th><th rowspan="2">Sugar-beet</th><th colspan="2">Vineyard</th><th colspan="2">Winter wheat</th></tr>
<tr><th>Grenache</th><th>Chardonnay</th><th>Talent</th><th>Shango</th></tr>
<tr><td rowspan="21">Shoot growth</td><td>EFCROIJUV$_P$</td><td>g·MJ^{-1}</td><td>2.0</td><td>1.2</td><td>4.23</td><td colspan="2">1.2</td><td colspan="2">2.2</td></tr>
<tr><td>EFCROIVEG$_P$</td><td>g·MJ^{-1}</td><td>2.5</td><td>2.7</td><td>3.0</td><td colspan="2">1.04</td><td colspan="2">4.25</td></tr>
<tr><td>EFCROIREPRO$_P$</td><td>g·MJ^{-1}</td><td>2.2</td><td>3.3</td><td>5.7</td><td colspan="2">2.25</td><td colspan="2">4.25</td></tr>
<tr><td>REMOBRES$_P$</td><td>ND</td><td>0.2</td><td>0.2</td><td>1.0</td><td colspan="2">0.073</td><td colspan="2">0.2</td></tr>
<tr><td>COEFMSHAUT$_P$</td><td>ND</td><td>25.0</td><td></td><td></td><td colspan="2"></td><td colspan="2"></td></tr>
<tr><td>SLAMAX$_P$</td><td>cm^2·g^{-1}</td><td>100</td><td>350</td><td>200</td><td colspan="2">235</td><td colspan="2">350</td></tr>
<tr><td>SLAMIN$_P$</td><td>cm^2·g^{-1}</td><td>80</td><td>180</td><td>30</td><td colspan="2">100</td><td colspan="2">180</td></tr>
<tr><td>TIGEFEUILLE$_P$</td><td>ND</td><td>0.1</td><td>0.0</td><td>1.41</td><td colspan="2">0.9</td><td colspan="2">0.5</td></tr>
<tr><td>ENVFRUIT$_P$</td><td>ND</td><td>0.0</td><td>0.1</td><td>0.0</td><td colspan="2">0.0</td><td colspan="2">0.10</td></tr>
<tr><td>SEA$_P$</td><td>cm^2·g^{-1}</td><td>100.0</td><td>0.0</td><td>0.0</td><td colspan="2">0.0</td><td colspan="2">0.0</td></tr>
<tr><td>TLETALE$_P$</td><td>°C</td><td>−25.0</td><td>−25.0</td><td>−25.0</td><td colspan="2">−20.0</td><td colspan="2">−25.0</td></tr>
<tr><td>TDEBGEL$_P$</td><td>°C</td><td>−5.0</td><td>−4.0</td><td>−4.0</td><td colspan="2">−1.5</td><td colspan="2">−4.0</td></tr>
<tr><td>NBFGELLEV$_P$</td><td>nb leaves plant^{-1}</td><td>2</td><td>2</td><td></td><td></td><td></td><td colspan="2">2</td></tr>
<tr><td>TGELLEV10$_P$</td><td>°C</td><td>−6.0</td><td>−4.0</td><td></td><td></td><td></td><td colspan="2">−4.0</td></tr>
<tr><td>TGELLEV90$_P$</td><td>°C</td><td>−20.0</td><td>−20.0</td><td></td><td></td><td></td><td colspan="2">−20.0</td></tr>
<tr><td>TGELJUV10$_P$</td><td>°C</td><td>−10.0</td><td>−10.0</td><td></td><td></td><td>−2.0</td><td colspan="2">−10.0</td></tr>
<tr><td>TGELJUV90$_P$</td><td>°C</td><td>−20.0</td><td>−20.0</td><td></td><td></td><td>−5.0</td><td colspan="2">−20.0</td></tr>
<tr><td>TGELVEG10$_P$</td><td>°C</td><td>−4.5</td><td>−4.5</td><td></td><td></td><td>−2.0</td><td colspan="2">−4.5</td></tr>
<tr><td>TGELVEG90$_P$</td><td>°C</td><td>−10.0</td><td>−10.0</td><td></td><td></td><td>−5.0</td><td colspan="2">−10.0</td></tr>
<tr><td>TGELFLO10$_P$</td><td>°C</td><td>−4.5</td><td>−4.5</td><td></td><td></td><td>−2.0</td><td colspan="2">−4.5</td></tr>
<tr><td>TGELFLO90$_P$</td><td>°C</td><td>−6.5</td><td>−6.5</td><td></td><td></td><td>−5.0</td><td colspan="2">−6.5</td></tr>
<tr><td rowspan="4">Yield
Formation</td><td>NBJGRAIN$_P$</td><td>days</td><td></td><td>15</td><td></td><td colspan="2"></td><td colspan="2">30</td></tr>
<tr><td>CGRAIN$_P$</td><td>nb grains·g^{-1} day</td><td></td><td>0.031</td><td></td><td colspan="2"></td><td colspan="2">0.0357</td></tr>
<tr><td>NBGRMIN$_P$</td><td>grains·m^{-2}</td><td></td><td>447</td><td></td><td colspan="2"></td><td colspan="2">6000</td></tr>
<tr><td>NBGRMAX$_V$</td><td>nb grains</td><td></td><td>$3.5 \cdot 10^3$</td><td></td><td colspan="2"></td><td>$3 \cdot 10^5$</td><td>$2.25 \cdot 10^5$</td></tr>
</table>

<table>
<thead>
<tr><th rowspan="3">Chapter</th><th rowspan="3">Parameter</th><th rowspan="3">Unit</th><th colspan="7">Value</th></tr>
<tr><th rowspan="2">Forage</th><th rowspan="2">Pea</th><th rowspan="2">Sugar-beet</th><th colspan="2">Vineyard</th><th colspan="2">Winter wheat</th></tr>
<tr><th>Grenache</th><th>Chardonnay</th><th>Talent</th><th>Shango</th></tr>
</thead>
<tbody>
<tr><td rowspan="22">Yield formation</td><td>VITIRCARB$_P$</td><td>g grain·g plant^{-1} day^{-1}</td><td></td><td>0.022</td><td></td><td colspan="2"></td><td colspan="2">0.011</td></tr>
<tr><td>IRMAX$_P$</td><td>ND</td><td></td><td>0.65</td><td></td><td colspan="2"></td><td colspan="2">0.55</td></tr>
<tr><td>PGRAINMAXI$_V$</td><td>g</td><td></td><td>0.35</td><td>1519</td><td>1.96</td><td>2.0</td><td>0.0388</td><td>0.0478</td></tr>
<tr><td>NBOITE$_P$</td><td>ND</td><td></td><td></td><td>50</td><td colspan="2">10</td><td colspan="2"></td></tr>
<tr><td>ALLOCAMX$_P$</td><td>ND</td><td></td><td></td><td>0.9</td><td colspan="2">1</td><td colspan="2"></td></tr>
<tr><td>AFPF$_P$</td><td>ND</td><td></td><td></td><td>0.5</td><td colspan="2">0.55</td><td colspan="2"></td></tr>
<tr><td>BFPF$_P$</td><td>ND</td><td></td><td></td><td>1.0</td><td colspan="2">18.0</td><td colspan="2"></td></tr>
<tr><td>CFPF$_P$</td><td>ND</td><td></td><td></td><td>0.0</td><td colspan="2">15.0</td><td colspan="2"></td></tr>
<tr><td>DFPF$_P$</td><td>ND</td><td></td><td></td><td>0.2</td><td colspan="2">0.2</td><td colspan="2"></td></tr>
<tr><td>AFRUITSP$_V$</td><td>nb fruits degree.day^{-1}</td><td></td><td></td><td>0.004</td><td>1.15</td><td>2.12</td><td colspan="2"></td></tr>
<tr><td>DUREEFRUIT$_V$</td><td>degree·day</td><td></td><td></td><td>2000</td><td>1472</td><td>1280</td><td colspan="2"></td></tr>
<tr><td>STDRPNOU$_P$</td><td>degree.day</td><td></td><td></td><td>250</td><td>91</td><td>90</td><td colspan="2"></td></tr>
<tr><td>SPFRMIN$_P$</td><td>ND</td><td></td><td></td><td>$1 \cdot 10^{-5}$</td><td colspan="2">0.75</td><td colspan="2"></td></tr>
<tr><td>SPFRMAX$_P$</td><td>ND</td><td></td><td></td><td>$2 \cdot 10^{-5}$</td><td colspan="2">1.0</td><td colspan="2"></td></tr>
<tr><td>SPLAIMIN$_P$</td><td>ND</td><td></td><td></td><td>$5.7 \cdot 10^{-3}$</td><td colspan="2">0.63</td><td colspan="2"></td></tr>
<tr><td>SPLAIMAX$_P$</td><td>ND</td><td></td><td></td><td>1.0</td><td colspan="2">1.0</td><td colspan="2"></td></tr>
<tr><td>NBINFLO$_P$</td><td>nb·plant^{-1}</td><td></td><td></td><td>1</td><td>15</td><td>22</td><td colspan="2"></td></tr>
<tr><td>INFLOMAX$_P$</td><td>nb·plant^{-1}</td><td></td><td></td><td>5</td><td colspan="2">15</td><td colspan="2"></td></tr>
<tr><td>PENTINFLORES$_P$</td><td>ND</td><td></td><td></td><td>0.8</td><td colspan="2">4.5</td><td colspan="2"></td></tr>
<tr><td>TMINREMP$_P$</td><td>°C</td><td></td><td>10.0</td><td></td><td colspan="2">0.0</td><td colspan="2">0.0</td></tr>
<tr><td>TMAXREMP$_P$</td><td>°C</td><td></td><td>40.0</td><td></td><td colspan="2">37.0</td><td colspan="2">38.0</td></tr>
<tr><td>VITPROPSUCRE$_P$</td><td>g sugar·g DM^{-1}·day^{-1}</td><td></td><td>0.0</td><td>8×10^{-4}</td><td colspan="2">0.0029</td><td colspan="2">0.0</td></tr>
</tbody>
</table>

Chapter	Parameter	Unit	Forage	Pea	Sugar-beet	Vineyard: Grenache	Vineyard: Chardonnay	Winter wheat: Talent	Winter wheat: Shango
	VITPROPHUILE$_P$	g oil·g DM^{-1}·day^{-1}		0.0	0.0	0.0		0.0	
	VITIRAZO$_P$	g grain·g plant^{-1}·day^{-1}		0.022	0.0	0.004		1.76·10^{-2}	
Root growth	CROIRAC$_V$	cm degree.day^{-1}	0.06	0.12	0.14			0.12	0.1381
	SENSANOX$_P$	SD	0.0	0.0	0.0	0.0		1.0	
	STOPRAC$_P$	stage	SEN	LAX	SEN	LAX		SEN	
	SENSRSEC$_P$	SD	0.0	0.4	0.0	0.0		0.0	
	CONTRDAMAX$_P$	SD	0.3	0.34	0.3	0.3		0.3	
	ZLABOUR$_P$	cm	25.0		102				
	ZPENTE$_P$	Cm	25.5		119				
	ZPRLIM$_P$	Cm	40.0		150				
	DRACLONG$_P$	cm root·plant^{-1}·degree.day^{-1}				40		80.0	
	STDEBSENRAC$_P$	degree·day				2000		1000	
	LVFRONT$_P$	cm root.cm^{-3} soil				5·10–3		5·10^{-2}	
	LONGSPERAC$_P$	cm·g^{-1}		3300		1021		18182	
	KREPRAC$_P$	SD		0.003					
	REPRACMAX$_P$	SD		1.0					
	REPRACMIN$_P$	SD		0.03					
	MINEFNRA								
	MINAZORAC	kg N·ha^{-1}·mm^{-1}							
	MAXAZORAC	kg N·ha^{-1}·mm^{-1}							
Water balance	PSISTO$_P$	bar	15.0	10.0	5.0	15.0		15.0	
	PSITURG$_P$	bar	4.0	2.0	2.0	6.0		4	

<table>
<tr><th rowspan="3">Chapter</th><th rowspan="3">Parameter</th><th rowspan="3">Unit</th><th colspan="7">Value</th></tr>
<tr><th rowspan="2">Forage</th><th rowspan="2">Pea</th><th rowspan="2">Sugar-beet</th><th colspan="2">Vineyard</th><th colspan="2">Winter wheat</th></tr>
<tr><th>Grenache</th><th>Chardonnay</th><th>Talent</th><th>Shango</th></tr>
<tr><td></td><td>H2OFEUILVERTE$_p$</td><td>g water·g^{-1} FW</td><td>0.9</td><td>0.9</td><td>0.9</td><td colspan="2">0.75</td><td colspan="2">0.9</td></tr>
<tr><td></td><td>H2OFEUILJAUNE$_p$</td><td>g water·g^{-1} FW</td><td>0.15</td><td>0.15</td><td>0.15</td><td colspan="2">0.5</td><td colspan="2">0.15</td></tr>
<tr><td></td><td>H2OTIGESTRUC$_p$</td><td>g water·g^{-1} FW</td><td>0.6</td><td>0.60</td><td>0.6</td><td colspan="2">0.7</td><td colspan="2">0.60</td></tr>
<tr><td></td><td>H2ORESERVE$_p$</td><td>g water·g^{-1} FW</td><td>0.7</td><td>0.70</td><td>0.8</td><td colspan="2">0.7</td><td colspan="2">0.70</td></tr>
<tr><td></td><td>H2OFRVERT$_p$</td><td>g water·g^{-1} FW</td><td>0.4</td><td>0.55</td><td>0.8</td><td colspan="2">0.925</td><td colspan="2">0.4</td></tr>
<tr><td></td><td>STDRPDES$_p$</td><td>degree·day</td><td></td><td>700</td><td>700</td><td>0</td><td>0</td><td colspan="2">700</td></tr>
<tr><td></td><td>DESHYDBASE$_p$</td><td>g water·g FW^{-1}·°C^{-1}</td><td></td><td>0.008</td><td>0.0</td><td>1.58·10^{-3}</td><td>1.32·10^{-3}</td><td colspan="2">0.008</td></tr>
<tr><td></td><td>TEMPDESHYD$_p$</td><td>% eau·°C^{-1}</td><td></td><td>0.005</td><td>0.0</td><td colspan="2">1.6·10^{-4}</td><td colspan="2">0.005</td></tr>
<tr><td></td><td>KMAX$_p$</td><td>ND</td><td>1.0</td><td></td><td>1.4</td><td colspan="2"></td><td colspan="2">1.0</td></tr>
<tr><td></td><td>RSMIN$_p$</td><td>s·m^{-1}</td><td></td><td>100</td><td></td><td colspan="2">250</td><td colspan="2"></td></tr>
<tr><td></td><td>MOUILLABIL$_p$</td><td>mm·LAI^{-1}</td><td></td><td></td><td></td><td colspan="2"></td><td colspan="2"></td></tr>
<tr><td></td><td>STEMFLOWMAX$_p$</td><td>ND</td><td></td><td></td><td></td><td colspan="2"></td><td colspan="2"></td></tr>
<tr><td></td><td>KSTEMFLOW$_p$</td><td>ND</td><td></td><td></td><td></td><td colspan="2"></td><td colspan="2"></td></tr>
<tr><td rowspan="10">Nitrogen transfor–mations</td><td>VMAX1$_p$</td><td>µmole·cm^{-1}·h^{-1}</td><td>1.8·10^{-3}</td><td>0.012</td><td>1.8·10^{-3}</td><td colspan="2">1.8·10^{-3}</td><td colspan="2">1.8·10^{-3}</td></tr>
<tr><td>KMABS1$_p$</td><td>µmole. cm root^{-1}</td><td>50</td><td>50</td><td>50</td><td colspan="2">50</td><td colspan="2">50</td></tr>
<tr><td>VMAX2$_p$</td><td>µmole·cm^{-1}·h^{-1}</td><td>5·10^{-2}</td><td>0.12</td><td>5·10^{-2}</td><td colspan="2">5.8·10^{-3}</td><td colspan="2">5·10^{-2}</td></tr>
<tr><td>KMABS2$_p$</td><td>µmole.cm root^{-1}</td><td>25000</td><td>20000</td><td>25000</td><td colspan="2">25000</td><td colspan="2">25000</td></tr>
<tr><td>ADIL$_p$</td><td>% DM</td><td>4.8</td><td>5.08</td><td>5.21</td><td colspan="2">3.3</td><td colspan="2">5.35</td></tr>
<tr><td>BDIL$_p$</td><td>ND</td><td>0.32</td><td>0.32</td><td>0.56</td><td colspan="2">0.44</td><td colspan="2">0.44</td></tr>
<tr><td>ADILMAX$_p$</td><td>% DM</td><td>7.8</td><td>8.0</td><td>7.21</td><td colspan="2"></td><td colspan="2">8.5</td></tr>
<tr><td>BDILMAX$_p$</td><td>ND</td><td>0.32</td><td>0.32</td><td>0.56</td><td colspan="2"></td><td colspan="2">0.44</td></tr>
<tr><td>MASECNMAX$_p$</td><td>t·ha^{-1}</td><td>1.0</td><td>1.0</td><td>1.0</td><td colspan="2">1.6</td><td colspan="2">1.54</td></tr>
<tr><td>NMETA$_p$</td><td>% DM</td><td></td><td></td><td></td><td colspan="2">6.0</td><td colspan="2"></td></tr>
</table>

Chapter	Parameter	Unit	Value						
			Forage	Pea	Sugar-beet	Vineyard		Winter wheat	
						Grenache	Chardonnay	Talent	Shango
	$NRES_P$	% DM				1.6			
	$INNIMIN_P$	ND				−0.49	−0.49	−0.5	
	$INNMIN_P$	ND	0.3	0.3	0.3	0.3		0.3	
	$INNGRAIN1_P$	ND						1.0	
	$INNGRAIN2_P$	ND						1.2	
	$STLEVDNO_P$	degree·day		0.0					
	$STDNOFNO_P$	degree·day		2000					
	$STFNOFVINO_P$	degree·day		0.0					
Nitrogen	$VITNO_P$	degree·day^{-1}		1.0					
functionning	$PROFNOD_P$	cm		30.0					
	$CONCNNODSEUIL_P$	kg ha^{-1} mm^{-1}		6.0					
	$AZOZRAC0_P$	kg ha^{-1} mm^{-1}		2.75					
	$AZOZRAC100_P$	kg ha^{-1} mm^{-1}		0.04					
	$HUNOD_P$	mm·cm^{-1} soil		0.2					
	$TEMPNOD1_P$	°C		0.0					
	$TEMPNOD2_P$	°C		15.0					
	$TEMPNOD3_P$	°C		25.0					
	$TEMPNOD4_P$	°C		35.0					
	$FIXMAX_P$	kg·ha^{-1}·day^{-1}		8.9					
	$FIXMAXVEG_P$	kg·(t DM)$^{-1}$		30.0					
	$FIXMAXGR_P$	kg·(t DM)$^{-1}$		0.0					

Table 11.6. List of soil parameters with help to assign them. PDT is soil transfer rules or functions. [1]signifies the optional character of the parameter. The sensitivity levels are just suggestions and depend on the purpose of the simulation.

Parameter	Recommended assigning method	Default value	Links between parameters	Sensitivity level
ARGI$_s$	Soil analysis	Fig. 11.7		**
NORG$_s$	Soil analysis	Vineyard soils = 0.06 Arable crop soils = 0.14 Pasture soils = 0.20	PROFHUM$_s$	***
CALC$_s$	Soil analysis	Non calcareous = 1 Limestone = 10 Chalk = 60		**
PROFHUM$_s$	Depth of soil tilling	30 cm	NORG$_s$	**
[1]CONCSEUIL$_s$	Estimation by fitting to the mineral nitrogen content profile	1.01 in temperate soil 0.20 in tropical soil	EPD$_s$ (Z)	*
[1]pH$_s$	Soil analysis	7		*
ALBEDO$_s$	Reflectance measurements	Table 11.9		*
Q0$_s$	Estimation by fitting to bare soil water reserve measurements	Fig. 11.7	DAF$_s$(1)	***
EPDs(Z)	Estimation by fitting to soil nitrate contents during infiltration periods	10 cm	CONCSEUIL$_s$	**
Z0SOLNUs	Measurements by a roughness meter Estimated as the 1/10 of the average asperity height.	10^{-3} m for sowed soil 10^{-2} m for a ploughed soil	DAF$_s$ (1)	***
CFESs	Estimation by fitting to water content profiles during evaporation periods	1		*
ZESXs		30 cm	DAF$_s$ (1)	**
PLUIEBATs, MULCHBATs	Sensitivity to crusting estimated by fitting to emergence in terms of date and density	50mm, 3 cm : insensitive 3 mm, 0.5 cm : high sensitivity 10 mm, 1.5 cm : low sensitivity	Q0$_s$	**

Parameter	Recommended assigning method	Default value	Links between parameters	Sensitivity level
RUISOLNU$_s$		Table 11.8		*
OBSTARAC$_s$	Either mechanical constraint or chemical toxicity for roots	200 cm		*
EPC$_s$(Z)	Soil description	30 cm		
HCCF$_s$(Z)	Soil analysis or in situ measurements of water content in winter	Table 11.7		***
HMINF$_s$ (Z)	Taken as wilting point : soil analysis or PDT		**
DAF$_s$ (Z)	In situ measurements or PDT		EPC$_s$(Z)	***
^{1}CAILLOUX$_s$ (Z)	In situ estimation	0.0		***
^{1}TYPECAILLOUX$_s$ (Z)		Table 11.10		**
^{1}INFIL$_s$ (Z)	Estimation by fitting to the water content profile during rainy events	Table 11.7	DAF$_s$ (Z)	***
HUMCAPIL$_s$	Threshold soil water content under which caoillary rises occur	0.0		*
CAPILJOUR$_s$	Capillary rises	0.0		**
PROFDENIT$_s$	Thickness of the denitrifying layer	20		*
VPOTDENIT$_s$	Total denitrification potential rate	16		**
PROFIMPER$_s$	Depth of the impermeable floor	150		***
LDRAIN$_s$	Between drain ½ spacing	150		*
KSOL$_s$	Hydraulic conductivity in the soil above and below the drains	1		**
PROFDRAIN$_s$	Drain depth	80		**

Commencement of growth for forage and vines is usually simulated after the winter rest (dormancy and budding having being parameterized for vines, Garcia de Cortazar, 2006), when perennial reserves are remobilised. The vine root system is considered to be already completely established (Garcia de Cortazar, 2006), whereas that of forage crops is partially established and will continue to grow during the cropping period. Forage crop parameterization was done for a grass mixture with an ecophysiology similar to tall fescue (*Festuca arundinacea* Schreb.) and cocksfoot (*Dactylis glomerata* L.) (Ruget *et al.*, 2006). Sugar beet is regarded as an annual crop by the model because of the way in which it is grown and despite the fact that it completes its vegetative cycle in two years (Launay and Brisson, 2004). Parameters controlling the photoperiod slowing effect ($PHOBASE_p$, $PHOSAT_p$ and $SENSIPHOT_p$) and vernalisation requirement ($JVCMINI_p$, $JULVERNAL_p$, $TFROID_p$ and $AMPFROID_p$) are activated for winter wheat only (Brisson *et al.*, 2002a).

Shoot growth, and especially leaf production are unrestricted throughout the cropping period for forage and sugar beet, which is simulated by a high $STLAMFLAX_p$ parameter value (Graux, 2006 or Launay and Brisson[2], 2004). Considering the row-planting arrangement of vines and the need to simulate intercropping with peas, those two crops were parameterized in order to use the radiation transfer formalisation (see $3.2.2) and the associated resistive approach, involving the estimation of $KTROU_p$, $FORME_p$, $RAPFORME_p$, $ADFOL_p$, $DFOLBAS_p$, $DFOLHAUT_p$ and $RSMIN_p$ parameters (Table 11.5).

Considering yield formation, forage, spring pea and winter wheat are simulated as determinate crops, whereas sugar beet and vines are simulated as indeterminate (see chapter 4). In the case of forage, the parameterization was not targeted on grain production but on the above-ground biomass prediction since this is the harvested part of the crop (Ruget *et al.*, 2006). For sugar beet, we assumed that only one tuber (storage and harvested root) was set by each plant ($NBINFLO_p=1$), and the trophic stress effects on tuber setting were cancelled by means of very low $SPFRMIN_p$ and $SPFRMAX_p$ parameter values; storage root growth was assumed to be linear over the growth cycle ($BFPF_p=1$) (Launay and Brisson[2], 2004).

Finally, root length growth was simulated as trophic-linked for spring peas, as shown in trials comparing sole and intercropped peas (Corre-Hellou *et al.*, 2007). Symbiotic N uptake formalisation was also parameterized for this leguminous plant (Corre-Hellou *et al.*, 2007). The nitrogen stress index relying on the daily accumulation of nitrogen rather than on the plant nitrogen concentration, named INNI (see eq.3.33 and $8.2), was chosen to avoid the notable inertia of the INN dynamics in the case of vines and winter wheat (Garcia de Cortazar, 2006 and Mary and Guerif 2005).

11.4.2 Soil parameterization

Table 11.6 summarises the various soil parameters and recommends some methods to assign them. The hydrodynamic parameters need to be discretized by layers, whose maximal number is 5.

When soil information is missing, some soil parameters, considered as permanent characteristics, can be accessed by soil transfer functions or rules (Bruand *et al.*, 2003, Wösten *et al.*, 1999), i.e. their values can be inferred from readily available soil data such as texture, particle size and organic matter content. It has been much used for field capacity and wilting point and to a lesser extent for bulk density. A review of the literature on the soil transfer function was carried out by Bastet *et al.* (1998) and many of them are available in the SOILPAR program by Acutis and Donatelli (2003: http://www.isci.it/tools). For approximate values for non-tilled layers, the pioneer work by Jamagne *et al.* (1977) can be used (tabl.). Databases of hydraulic soil properties, such as HYPRES (Wösten *et al.*, 1999) at the European scale, constitute another source of data to assign some soil parameters as well as databases of agricultural soil analyses, such as BDAT developed in France (http://www.gissol.fr/programme/bdat/bdat.php).

Yet some parameters closely associated with soil structure are difficult to assign with only database information and textural characteristics. This is the case for field capacity and bulk density values, especially for sub-surface layers whose hydrodynamic characteristics depend heavily on soil structure, much more than on soil texture (Bruand *et al.*, 2003).

In order to enable STICS users to parametrize their soil, at least roughly for test runs, we have constructed soil transfer tables based on well-known literature. They mostly use textural information (see Figure 11.6) so that they are likely to change with soil structure and organic matter content.

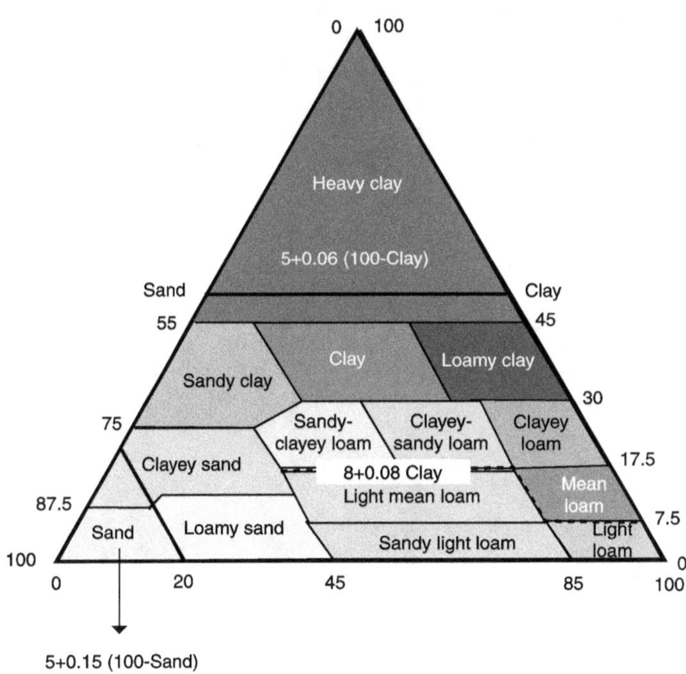

Figure 11.6. Textural triangle and classification by Jamagne *et al.* (1977). Soil transfer functions to assess the $Q0_s$ parameter as a function of clay or sand content.

Table 11.7. Pedotransfer table to estimate hydrodynamic parameters as functions of textural classes (Figure 11.6) based on Jamagne *et al.* (1977) and the ICRISAT technical manual to run Ritchie's model (1985) available at http://www.icrisat.org/gt-aes/oneds/DataNeeds.htm (*: for heavy clay soils air porosity is estimated taking into account the cracks: eq. 9.6).

	Textural class	$HCCF_s$ (% dry soil)	$HMINF_s$ (% dry soil)	DAF_s (g cm^{-3})	$INFIL_s$ for various layer thickness											
					5 cm	10 cm	15 cm	20 cm	25 cm	30 cm	40 cm	50 cm	60 cm	80 cm	100 cm	120 cm
topsoil	coarse	13	5	1.3	50.00	50.00	50.00	50.00	50.00	45.93	34.48	27.59	22.99	17.24	13.79	13.69
	medium	22	12	1.3	22.69	12.66	8.54	6.42	5.13	4.28	4.18	4.18	4.18	4.18	4.18	4.18
	medium fine	25	10	1.3	8.17	4.29	2.87	2.28	2.28	2.28	2.28	2.28	2.28	2.28	2.28	2.28
	fine	31	21	1.3	4.22	2.12	1.41	1.06	0.85	0.71	0.58	0.58	0.58	0.58	0.58	0.58
	very fine	38	26	1.3	1.81	0.91	0.60	0.45	0.36	0.30	0.27	0.27	0.27	0.27	0.27	0.27
subsoil	coarse	8	3	1.45	50.00	50.00	46.97	36.09	29.10	24.31	18.25	14.60	13.56	13.56	13.56	13.56
	medium	19	10	1.45	24.43	13.63	9.20	6.91	5.53	4.61	3.45	3.33	3.33	3.33	3.33	3.33
	medium fine	23	11	1.45	6.20	3.26	2.18	1.63	1.48	1.48	1.48	1.48	1.48	1.48	1.48	1.48
	fine	28	20	1.5	1.04	0.52	0.43	0.43	0.43	0.43	0.43	0.43	0.43	0.43	0.43	0.43
	very fine	32	24	1.45	0.27	0.27	0.27	0.27	0.27	0.27	0.27	0.27	0.27	0.27	0.27	0.27

In Table 11.7 the permeability classes proposed by Ritchie (1985) are arbitrarily associated with textural classes: they correspond to a percentage of the amount of water stored in the macroporosity that infiltrates from one day to the next. The calculations show that the effect of layer thickness on $INFIL_S$ disappears with decreasing permeability.

The values of $RUISOLNU_S$ (Table 11.8) derived from the USDA Runoff Curve Number method are rather low because they represent only Hortonian (surface) runoff, which only depends on obstacles created by plants and on the water velocity on a slope field. The other component of runoff, i.e. resistance to infiltration, as well as the presence of a plant mulch, are taken into account by the model (see § 6.4).

Table 11.8. Values of the parameter $RUISOLNU_S$ as the proportion of Hortonian runoff to incoming precipitation minus the $PMINRUIS_G$ threshold, based on the USDA CN approach described in Chapman and Lake (2003) and at http://www.icrisat.org/gt-aes/oneds/DataNeeds.htm

Soil cover	Slope classes			
	0-2%	2-5%	5-10%	>10%
smooth soil	0.05	0.07	0.10	0.13
ploughed soil	0.00	0.03	0.06	0.08
row crop in direction of slope	0.05	0.07	0.10	0.13
row crop perpendicular to slope	0.00	0.03	0.06	0.09
homogeneous crop	0.00	0.03	0.07	0.10

The albedo parameter applies to a dry soil, the effect of water content being simulated (eq. 6.24). There are two criteria to assign this parameter, either texture or colour, the latter being read from a Munsell chart (Table 11.9).

Table 11.9. Values of the dry soil albedo ($ALBEDO_S$) using either textural classes or colours, based on Richard and Cellier (1998), Jacquemoud and Baret (1992) and http://www.icrisat.org/gt-aes/oneds/DataNeeds.htm

Soil type	$ALBEDO_S$
TEXTURE	
Limestone	0.31
Loamy sand	0.25
Clayey loam	0.18 – 0.22
Mean loam	0.22 – 0.23
Crusted mean loam	0.28
Clay	0.28
COLOUR	
Brown soil	0.27
Red soil	0.29
Black soil	0.17
Grey soil	0.29
Yellow soil	0.35

Pebbles are characterized by their water retention ability. Some laboratory experiments were done to assess them for pebbles frequently met in French agricultural fields (Table 11.10).

Table 11.10. Water retention characteristics of various pebbles, used as classes of stone types (Gras, 1995; Nicoullaud *et al.*, 1995; Beaudoin, 2006).

Stone type	Volumetric mass g cm^{-3}	Field capacity % in mass
Non-porous limestone	2.20	7
Porous limestone	1.80	16
Lutetian semi-porous limestone	2.00	11
Lutetian stones	2.30	7
Morainic gravel	2.60	3
Silex, sandstone or unaltered granite	2.65	0
Altered granite	2.30	10
Rendzinic porous calcareous	1.20	28

11.4.3 Crop management parameterization

While management data are probably the easiest input to provide, the links between practices and the proper state variables in the model can require the implementation of transfer rules. For example the interactions between the fertilizers and the soil-crop system depend very much on the type of fertilizer, of course either organic or mineral, but also within each of these types their proper biochemical and physical behaviour. As listed in Table 11.11, the practices accounted for in STICS concern bare soil and cropping

Table 11.11. List of techniques accounted for by the model and the corresponding parameters. [1] technique that can be either prescribed or partly calculated using some decision rules (see table 11.15). [2] Several cultivation operations can be planned to bury or mix in residues (of different types) in the soil. These operations include new residues or serve simply to modify the structural and moisture conditions of residues previously added.

technique	Compulsory Parameters if the technique is applied	Optional parameters if the technique is applied	Typology	
			codification	dependant parameters
[2] soil tillage and residue incorporation	JULTRAV$_T$ PROFRES$_T$ PROFTRAV$_T$ CODERES$_T$ DACHISEL$_T$ DALABOUR$_T$ DASEM$_T$ DAREC$_T$ RUGOCHISEL$_T$ RUGOLABOUR$_T$		CODERES$_T$ (cf tabl. 11.12)	QRES$_T$ CSURN$_T$ CRESPC$_T$ NMINRES$_T$ EAURES$_T$ all mineralization parameters (cf tabl.11.12)
[1] sowing	DENSITE$_T$ VARIETE$_T$	if annual : IPLT$_T$, PROFSEM$_T$ if row crop : INTERRANG$_T$ ORIENTRANG$_T$ if sowing date is calculated : NBJMAXAPRESSEMIS$_T$, NBJSEUILTEMPREF$_T$, HUMSEUILTASSSEM$_T$, PROFHUMSEM$_T$	VARIETE$_T$	all varietal parameters
[1] irrigation	EFFIRR$_T$	If prescribed by dates : JULAPI$_T$, DOSEI$_T$ If prescribed by phasic stages : UPVTTAPI$_T$, DOSEI$_T$ if calculated : RATIOL$_T$, DOSIMX$_T$ If in the soil : LOCIRRIG$_T$		
[1] fertilisation	DOSEN$_T$ ENGRAIS$_T$	If fert-irrigation : CONCIRR$_T$ If calendar application : JULAPN$_T$ If phasic application : UPVTTAPN$_T$ if in the soil : LOCFERTI$_T$	ENGRAIS$_T$ (cf tabl. 11.13)	ENGAMM$_T$ ORGENG$_T$ DENENG$_T$ VOLENG$_T$

technique	Compulsory Parameters if the technique is applied	Optional parameters if the technique is applied	Typology	
			codification	dependent parameters
[1]harvesting	$RESSUITE_T$	if non physiological : $IRECBUTOIR_T$ If water content dependent : $H2OGRAINMIN_T$ or $H2OGRAINMAX_T$ If sugar dependent : $SUCREREC_T$ If nitrogen dependent : $CNGRAINREC_T$ If oil dependent : $HUILREC_T$ If several pickings : $CADENCEREC_T$ if compaction dependent : $NBJMAXAPRESRECOLTE_T$, $NBJSEUILTEMPREF_T$, $HUMSEUILTASSREC_T$, $PROFHUMREC_T$		
[1]forage cutting	$MSCOUPEMINI_T$	If prescribed cuts : $HAUTCOUPE_T$, $LAIRESIDUEL_T$, $MSRESIDUEL_T$, $ANITCOUPE_T$ If calendar prescription : $JULFAUCHE_T$ if phasic prescription : $TEMPFAUCHE_T$ If calculated : STADEFAUCHET, $HAUTCOUPEDEFAUT_T$		
mulching		If plastic : $ALBEDOMULCH_T$ If plant : $JULAPPMULCH_T$, $QMULCH0_T$, $COUVERTMULCH_T$, $TYPEMULCH_T$	$TYPEMULCH_T$ (cf tabl. 11.14)	$DECOMPOSMULCH_T$ $QMULCHRUIS0_T$ $MOUILLABILMULCH_T$ $KCOUVMLCH_T$ $ALBEDOMULCH_T$
trellising crops	$HAUTMAXTEC_T$ $LARGTEC_T$			
[1]tactical shape control	$LARGROGNE_T$ $HAUTROGNE_T$ $BIOROGNEM_T$	if prescribed : $JULROGNE_T$ If calculated : $MARGEROGNE_T$		
[1]leaf removal	$CODHAUTEFF_T$	if prescribed : $JULEFFEUIL_T$ $LAIEFFEUIL_T$ if calculated $LAIDEBEFF_T$ $EFFEUIL_T$		
fruit removal	$JULECLAIR_T$ $NBFROTE_T$			
pruning	$JULTAILLE_T$			
shelter	$TRANSPLASTIC_T$	if shelter punctually opened (3 times maximum) $SURFOUVRE_T$ $JULOUVRE_T$		

periods for industrial crops as well as fruit crops and vegetables. There is no information about the sanitary status of the crop.

Table 11.12. List of available mineral fertilizers and corresponding parameters.

Code	1	2	3	4	5	6	7	8
Type	ammonium Nitrate	UAN solution	urea	anhydrous ammonia	ammonium sulfate	ammonium phosphate	calcium nitrate	Fixed efficiency[1]
ENGAMM$_T$	0.50	0.75	1.00	1.00	1.00	1.00	0.00	0.50
DENENG$_T$	0.11	0.13	0.10	0.10	0.10	0.10	0.20	0.05
VOLENG$_T$	0.12	0.30	0.35	0.35	0.25	0.25	0.00	0.05
ORGENG$_T$	30.0	33.8	37.7	37.7	37.7	37.7	25.0	0.20

[1] With this option the DENENG, VOLENG and ORGENG values represent the proportion of fertilizer which is denitrified, volatilized and immobilized in soil, respectively.

Table 11.13. List of organic residues and corresponding default parameters. The CODERES$_T$ number refer to mineralization dynamics (§ 6.3.3).

	Residue code	Average rate	Carbon content	C/N ratio	Mineral N content	Water content	Reference (pers. com.)
		t FM ha^{-1}	% DM		% FM	% FM	
	CODERES$_T$	QRES$_T$	CRESPC$_T$	CSURNRES$_T$	NMINRES$_T$	EAURES$_T$	
RESIDUES OF MATURE CROPS							
Cereals (straw)	1	9	42	90	0	7	
sugarbeet (leaves and crowns)	1	40	42	22	0	90	J.M. Machet
grain maize (stalks)	1	12	43	60	0	25	
soybean (straw and roots)	1	5	44	75	0	10	
proteaginous pea (foliage and roots)	1	4	42	28	0	10	B. Nicolardot
rapeseed (roots, pods and straw)	1	6	44	45	0	10	E. Justes
RESIDUES OF CATCH CROPS							
wheat, rye (cereals)	2	8	42	15	0	80	J.M. Machet
mustard (cruciferous)	2	10	42	15	0	70	
phacelia (cruciferous)	2	15	42	20	0	80	
radish, oil seed (cruciferous)	2	10	42	16	0	80	E. Justes
ryegrass (grass)	2	18	40	25	0	80	
MANURE							
bovine manure	3	45	32	20	0	75	
ovine manure	3	45	45	20	0	75	T. Morvan
poultry manure	3		22			45	

COMPOSTS							
rubbish compost	4	10	25	19	0.08	44	
green waste compost	4	10	26	18	0.04	30	S. Houot
compost of sewage plant	4	10	37	19	0.04	50	
SEWAGE SLUDGE							
non processed sludge	5	60	30	8	0.12	90	
limed sludge	5	25	25	10	0.13	70	V. Parnaudeau
physico-chemical sludge	5	20	30	15	0.05	75	
CONCENTRATED VINASSE	6	3	40	8	0	50	J.M. Machet
GROUND HORN	7		40	3.8	0	10	B. Nicolardot
LIQUID MANURE							
porcine liquid manure	8	50	35	15	0.35	91	T. Morvan
bovine liquid manure	8	50	25	18	0.10	94	
FEATHER FLOUR	9	0.5	37	4	0	10	B. Nicolardot

Table 11.14. Some plant mulches and corresponding parameters.

	$decomposmulch_T$	$qmulchruis0_T$	$mouillabilmulch_T$	$kcouvmlch_T$	$albedomulch_T$
maize stalk	0.0070	1.0	0.4	0.367	0.10
sugar cane	0.0070	1.0	0.4	0.367	0.50
vine stems	0.0070	1.0	0.0	0.050	0.08

Many of the techniques mentioned offer some possibility of calculation using simple decision rules (Table 11.15).

Table 11.15. Decision rules to help to implement practices.

technique	Possible decision rules
sowing	**date** as a function of soil water status and temperature
irrigation	calendar **dates** or **phenological stages** and **amounts** as a function of water stress
fertilisation	calendar **dates** or **phenological stages** and **amounts** as a function of nitrogen stress and soil surface water status
harvesting	**date** as a function of plant physiology and soil water status
forage cutting	calendar **dates** or **phenological stages** with a minimum level of biomass
tactical shape control	**dates** and **amounts** as a function of the purposed shape
leaf removal	**dates** and **amounts** as a function of the leaf quantity to remove

References

ABBAD H., EL JAAFARI S.A., BORT J., ARAUS J.L., 2004. Comparative relationship of the flag leaf and the ear photosynthesis with the biomass and grain yield of durum wheat under a range of water conditions and different genotypes. *Agronomie*, 24, 19-28.

ACUTIS M., DONATELLI M., 2003. *SOIL*PAR 2.00: software to estimate soil hydrological parameters and functions. *Eur. J. Agron.* 18, 373-377.

ADDISCOTT T.M., WAGENET R.J., 1985. Concepts of solute leaching in soils: a review of modeling approaches. *Journal of Soil Science* 36, 411-424

ADIKU S.G.K., OZIER-LAFONTAINE H., BAJAZET T., 2001. Patterns of root growth and water uptake of a maize-cowpea mixture grown under greenhouse conditions. *Plant and Soil* 235, 85-94.

ALLEN R.G., 1994. An update for the calculation of the reference evapotranspiration. *ICID Bull.* 43, 35-91.

ALLEN R.G., PEREIRA L.S., RAES D., SMITH M. 1998. Crop evapotranspiration. Guidelines for computing crop water requirements. *FAO Irrigation and Drainage Paper* 56. FAO, Rome.

ALLEN W.A., RICHARDSON A.J., 1968. Interaction of light with a plant canopy. *J. Opt. Soc. Am.* 58, 1023-1028.

ALM D.M., STOLLER E.W., WAX L.M., 1993. An Index Model for Predicting Seed Germination and Emergence Rates. *Weed Technology* 7, 560-569.

AMIR J., SINCLAIR T.R., 1991a. A Model of the Temperature and Solar-Radiation Effects on Spring Wheat Growth and Yield. *Field Crop. Res.* 28, 47-58.

AMIR J., SINCLAIR T.R., 1991b. A Model of Water Limitation on Spring Wheat Growth and Yield. *Field Crop. Res.* 28, 59-69.

ANGUS J.F., MACKENZIE D.H., MORTON R., SCHAFER C.A., 1981. Phasic development in field crops. II. Thermal and photoperiodic responses of spring wheat. *Field Crop. Res.* 4, 269-283.

ANTONIOLETTI R., 1986. Contribution à l'étude du mont Ventoux. *In Note de l'INRA*, 147 rue de l'Université, Paris, France. 43 p.

ANTONIOLETTI R., SEGUIN. B., 1988. Quelques éléments sur le climat du mont Ventoux, *Bulletin Clim. et Agroclim. de Vaucluse*, 26-34.

ARAUS J.L., BROWN HR., FEBRERO A., BORT J., SERRET M.D., 1993. Ear photosynthesis, carbon isotope discrimination and the contribution of respiratory CO_2 to differences in grain mass in durum wheat. *Plant Cell and Environment* 16: 383-392.

AURA E., 1996. Modelling non-uniform soil water uptake by a single plant root. *Plant and Soil* 186, 237-243.

BAIER W., 1969. Concepts of soil moisture availability and their effect on soil moisture estimates from a meteorological budget. *Agricultural Meteorology*, 6: 165-178.

BAKER D.N., 1980. Simulation for research and crop management. *In World Soybean Research Conference II*, Corbin F.T. ed., 533-546.

BALDOCCHI D.D., LUXMOORE R.J., HATFIELD J.L.D., 1991. Discerning the forest from trees: an essay on scaling canopy stomatal conductance. *Agricul. For. Meteorol.* 54, 197-226.

BALESDENT J., RECOUS S., 1997. Les temps de résidence du carbone et le potentiel de stockage de carbone dans quelques sols cultivés français. *Can. J. Soil Sci.* 77: 187-193.

BARBER S.A., SILBERBUSH M., 1984. Plant root morphology and nutrient uptake. *ASA Special publ.* 49, 65-87.

BARBOTTIN A., LECOMTE C., BOUCHARD C., JEUFFROY M.H., 2005. Nitrogen remobilisation during grain filling in wheat: genotypic and environmental effects. *Crop Science* 45, 1141-1150.

BARET F., 1986. Contribution au suivi radiométrique des cultures de céréales. Thèse Univ. Orsay, 182 p.

BARET F., ANDRIEU B., FOLMER J.C., HANOCQ J.F., SARROUY C., 1993. Gap fraction measurement from hemispherical infrared photography an its use to evaluate PAR interception efficiency. *In Crop structure and light microclimate*, Valert-Grancher C., Bonhomme R., Sinoquet H. (eds). INRA Editions, Versailles, France, 359-372.

BARET F., OLIOSO A., LUCIANI J.L., 1992. Root Biomass Fraction as a Function of Growth Degree Days in Wheat, *Plant and Soil* 140, 137-144.

BASTET G., BRUAND A., QUÉTIN P., COUSIN I., 1998. Estimation des propriétés de rétention en eau des sols à l'aide de fonctions de pédotransfert (FPT): une analyse bibliographique. *Étude et gestion des Sols* 5: 7-28.

BEAUDOIN N., LAUNAY M., SAUBOUA E. , PONSARDIN G., MARY B., 2008. Evaluation of the soil crop model STICS over 8 years against the "on farm" database of Bruyères catchment. *European Journal of Agronomy*, 29, 1, 46-57.

BEAUDOIN N., 2006. Caractérisation expérimentale et modélisation des effets des pratiques culturales sur la pollution nitrique d'un aquifère en zone de grande culture. Thèse INA-PG, Paris, France, 192 p.

BENFEY P.N., MITCHELL-OLDS T. (2008) From genotype to phenotype: Systems Biology meets natural variation. *Science* 320 495-497.

BERTIN N., GARY C. 1993. Evaluation of TOMGRO, a Dynamic Model of Growth and Development of Tomato (*Lycopersicon esculentum* Mill) at Various Levels of Assimilate Supply and Demand. *Agronomie* 13, 395-405.

BETHLENFALVAY G.J., ABU-SHAKRA S.S., FISHBECK K., PHILLIPS D.A., 1978. The effect of source-sink manipulations on nitrogen fixation in peas. *Physiol. Plant.*, 43, 31-34.

BETHLENFALVAY G.J., PHILLIPS D.A., 1977. Ontogenic interactions between photosynthesis and symbiotic nitrogen fixation in legumes. *Plant Physiol.*, 60, 419-421.

BIDABE B., 1965. L'action des températures sur l'évolution des bourgeons de l'entrée en dormance à la floraison, Congrès Pomologique, Editeur, 51-56.

BINDI M., SINCLAIR T.R., J. HARRISON, 1999. Analysis of seed growth by linear increase in harvest index. *Crop Science* 39, 486-493.

BINGHAM I.J., 1995. A comparison of the dynamics of root growth and biomass partitioning in wild oat (*Avena fatua* L.) and spring wheat. *Weed research* 35, 57-66.

BLOMBÄCK K., ECKERSTEN H., LEWAN E., ARONSSON H., 2003. Simulations of soil carbon and nitrogen dynamics during seven years in a catch crop experiment. *Agricultural Systems* 76, 95-114.

BLUM A., 1996. Crop responses to drought and the interpretation of adaptation. *Plant Growth Regulation* 20, 135-148.

BOESTEN J.J.T., STROOSNIJDER L., 1986. Simple model for daily evaporation from fallow tilled soil under spring conditions in a temperate climate. *Netherland Journal of Agricultural Science* 34, 75-90.

BOIFFIN J., MALÉZIEUX E., PICARD D., 2001. Cropping systems for the future. *In CAB International*, Crop Science (eds). Nösberger, H.H. Geiger and P.C. Struik.

BOIVIN P., TOUMA J., ZANTE P., 1987. Mesure de l'infiltrabilité du sol par la méthode du double anneau.1. Résultats expérimentaux. *Cahier ORSTOM, série Pédologie* 24, 17-25.

BONACHELA S., 1996. Root growth of triticale and barley grown for grain or for forage-plus-grain in a Mediterranean climate. *Plant and Soil* 183, 239-51.

BONHOMME R., DERIEUX M., EDMEADES G.O., 1994. Flowering of diverse maize cultivars in relation to temperature and photoperiod in multilocation field trials. *Crop Science* 34, 156-164.

BONHOMME R., RUGET F., DERIEUX M., VINCOURT P., 1982. Relations entre production de matière sèche aérienne et énergie interceptée chez différents génotypes de maïs. *C.R. Acad. Sci. Paris,* France, 294, III, 393-398.

BOOTE K.J., JONES J.W., 1987. Equations to define canopy photosynthesis from quantum efficiency maximum leaf rate light extinction LAI and photon flux density. *In Progress in photosynthesis research, Proceedings of the VIIth international congress on photosynthesis,* Providence Rhode, Island USA August 10-15 1986, Biggins J. (ed.), 415-418.

BOOTE K.J., PICKERING N.P., 1994. Modeling photosynthesis of row crop canopies. *Hort. Science* 29, 1423-1434.

BOOTE K.J., JONES J.W., HOOGENBOOM G. 1998. Simulation of crop growth: CROPGRO Model. *In Agricultural Systems Modeling and Simulation.* Peart R.M. and Curry R.B. (eds.). Marcel Dekker, Inc, New York, 651-692.

BOUAZIZ A., BRUCKLER L., 1989. Modeling wheat seedling growth and emergence: Seedling growth affected by soil water potential, *Soil Science Society America Journal* 53, 1832-1838.

BOULARD T., WANG S., 2000. Greenhouse crop transpiration model from external climate conditions. *Agricultural and Forest Meteorology,* 100, 25-34.

BRADFORD K.J., 1990. A water relations analysis of seed germination rates. *Plant Physiology* 94, 840-849.

BRADFORD K.J., 2002. Applications of hydrothermal time to quantifying and modelling seed germination and dormancy. *Weed Science* 50: 248-260.

BRADFORD K.J., HSIAO T.C., 1982. Physiological responses to moderate water stress. *In Physiological Plant Ecology II*. Springler Verlag (ed.). Berlin New Series 12B, 263-324.

BRISSON N., RUGET F., GATE P., LORGEOU J., NICOULLAUD B., TAYOT X., PLENET D., JEUFFROY M.H., BOUTHIER A., RIPOCHE D., MARY B., JUSTES E., 2002a. STICS: a generic model for the simulation of crops and their water and nitrogen balances. II. Model validation for wheat and corn. *Agronomie*, 22, 69-93.

BRISSON N., 1998. An analytical solution for the estimation of the critical soil water fraction for the water balance under growing crops. *Hydrology and Earth System Science* 2, 221-231.

BRISSON N., BONHOMME R., AMEGLIO T., GAUTIER H., OLIOSO A., DROUET J.L. (1997a). Modèle de culture : simulations sous contraintes à partir d'un potentiel de production. *In Tome 2 du couvert végétal à la petite région agricole*, Cruiziat P., Lagouarde J.P., (eds), Ecole chercheurs INRA en bioclimatologie, Le Croisic, 1996/04/03-07, Service de formation permanente, INRA.

BRISSON N., BUSSIÈRE F., OZIER-LAFONTAINE H., SINOQUET H., TOURNEBIZE R., 2004. Adaptation of the crop model STICS to intercropping. Theoretical basis and parameterisation. *Agronomie* 24: 409-421.

BRISSON N., DELÉCOLLE R., 1991. Développement et modèles de simulation de culture. *Agronomie* 12, 253-263.

BRISSON N., ITIER B., L'HOTEL J.C., LORENDEAU J.Y., 1998b. Parameterisation of the Shuttleworth-wallace model to estimate daily maximum transpiration for use in crop models. *Ecological Modelling* 107, 159-169.

BRISSON N., MARY B., RIPOCHE D., JEUFFROY M.H., RUGET F., GATE P., DEVIENNE-BARRET F., ANTONIOLETTI R., DURR C., NICOULLAUD B., RICHARD G., BEAUDOIN N., RECOUS S., TAYOT X., PLENET D., CELLIER P., MACHET J.M., MEYNARD J.M., DELÉCOLLE R., 1998a. STICS: a generic model for the simulation of crops and their water and nitrogen balance. I. Theory and parametrization applied to wheat and corn. *Agronomie* 18, 311-346.

BRISSON N., OLIOSO A., CLASTRE P., 1993. Daily Transpiration of Field Soybeans as Related to Hydraulic Conductance Root Distribution Soil Potential and Midday Leaf Potential. *Plant and Soil* 154, 227-237.

BRISSON N., PERRIER A., 1991. A semi-empirical model of bare soil evaporation for crop simulation models. *Water Resources Research* 27, 719-727.

BRISSON N., REBIÈRE B., ZIMMER D., RENAULT P., 2002b. Response of the root system of a winter wheat crop to waterlogging. *Plant and Soil*, 243, 43-55.

BRISSON N., SEGUIN B., BERTUZZI P., 1992b. Agrometeorological soil water balance for crop simulation models. *Agricultural and Forest Meteorology* 59, 267-287.

BRISSON N., WERY J., BOOTE K.J., 2005. An overview of the fundamental concepts of crop models by a comparative approach. *Ze Book* (in press).

BRISSON N., DOREL M., OZIER-LAFONTAINE H. 1998d. Effects of soil management and water regime on the banana growth between planting and flowering Simulation using the STICS model. *In Proc. Int. Symp. Banan in Subtropics*. Galan Sauco V. (ed). *Acta Hort.* 490, 229-238.

BRISSON N., GARY C., JUSTES E., ROCHE R., MARY B., RIPOCHE D., ZIMMER D., SIERRA J., BERTUZZI P., BURGER P., BUSSIÈRE F., CABIDOCHE Y.M., CELLIER P., DEBAEKE P., GAUDILLÈRE J.P., MARAUX F., SEGUIN B., SINOQUET H., 2003. An overview of the crop model STICS. *Eur. J. Agron.* 18, 309-332.

BRISSON N., KING D., NICOULLAUD B., RUGET F., RIPOCHE D., DARTHOUT R. 1992a. A crop model for land suitability evaluation: a case study of the maize crop in France. *Eur. J. Agron.* 1, 163-175.

BROCHET P., GERBIER N., 1968. L'évapotranspiration. *Monographies de la Météorologie Nationale* 65, 67 p.

BRONSON K.F., TOUCHTON J.T., HAUCK R.D., KELLY K.R., 1991. Nitrogen-15 recovery in winter wheat as affected by application timing and dicyandiamide. *Soil Sci. Soc. Am. J.* 55, 130-135.

BRUAND A., PEREZ FERNANDEZ P., DUVAL O., 2003. Use of class pedotransfer functions based on texture and bulk density of clods to generate water retention curves. *Soil Use and Management*, 19, 232-242.

BRUNT D., 1932. Notes on radiation in the atmosphere. *Quart. J. R. Meteorol. Soc.* 58, 389-418.

BRUTSAERT W.H., 1982. Evaporation into the atmosphere. *In Theory, history and application.* D. Reidel Publishing Company, Boston, Massachusets, 299 p.

BURCH, G.J., SMITH, R.C.G., MASON, W.K., 1978. Agronomic and physiological responses of soybean and sorghum crops to water deficits. II crop evaporation, soil water depletion and root distribution. *Aust. J. Plant Physiol.*, 5: 169-177.

BURGER P., 2001. Analyse de la variabilité de la teneur en protéines de la graine de soja: approche par voie d'enquête et étude expérimentale de conduites de culture visant de hautes teneurs en protéines. Thèse INA-PG, France, 171 p.

BUSSIÈRE F., 1995. Rainfall interception by plant canopies, consequences for water partitioning in intercropping systems. *In Ecophysiology of Tropical Intercropping.* Sinoquet H. and Cruz P. (eds.). INRA Editions, Versailles, France, 162-174 p.

CALDWELL R.M., 1995. Simulation models for intercropping systems. *In Ecophysiology of Tropical Intercropping.* Sinoquet H. and Cruz P. (eds.). INRA Editions, Versailles, France, 353-368 p.

CALDWELL R.M., HANSEN J.W., 1993. Simulation of multiple cropping systems with CropSys. *In Systems Aproaches for Agricultural Development.* Penning de Vries F., Teng P., Metselaar K. (eds.). Kluwer Academic Publishers, Dordrecht, The Netherlands, pp. 397-412.

CASALS M.L., 1996. Introduction des mécanismes de résistance à la sécheresse du Blé dur au fonctionnement phénologique et trophique de la plante dans un modèle dynamique de croissance. Thèse INA-PG, Paris, France, 130 p.

CELLIER P., RUGET F., CHARTIER M., BONHOMME R., 1993. Estimating the Temperature of a Maize Apex During Early Growth Stages. *Agricultural and Forest Meteorology* 63, 35-54.

CHANTIGNY M.H., ROCHETTE P., ANGERS D.A., MASSÉ D., CÔTÉ D., 2004. Ammonia volatilization and selected soil characteristics following application of anaerobically digested pig slurry. *Soil Sci. Soc. Am. J.* 68, 306-312.

CHAPMAN D.C., LAKE D.W., 2003. Computing runoff. *In New York standards specifications for erosion and sediment control.* New York State Soil & Water Conservation Committee, Section 4, 16 p.

CHAPMAN S.C., HAMMER G.L., PALTA J.A., 1993. Predicting leaf area development of sunflower. *Field Crop. Res.* 34, 101-112.

CHAUSSOD R., NICOLARDOT N., CATROUX G., CHRÉTIEN J., 1986. Relations entre les caractéristiques physico-chimiques et microbiologiques de quelques sols cultivés. *Science du Sol* 2, 213-226.

CHENU K., CHAPMAN S.C., HAMMER G.L., MCLEAN G., BEN HAJ SALAH H., TARDIEU F., 2008. Short-term responses of leaf growth rate to water deficit scale up to whole-plant and crop levels: an integrated modelling approach in maize. *Plant Cell Environment* **31**, 378-391

CLAPP R.B., HORNBERGER G.M., 1978. Empirical equations for some soil hydraulic properties. *Water Resources Research* 14, 601-604.

COLEMAN K., JENKINSON D.S., 1996. RothC-26.3 – A model for the turnover of carbon in soil. *In Evaluation of Soil Organic Matter Models*. Using Existing Long-Term Datasets, Powlson DS, P Smith & JU Smith (eds.). Springer-Verlag Berlin, 237-246.

CORDERY I., GRAHAM A.G., 1989. Forecasting wheat yields using a water budgeting model. *Australian Journal of Agricultural Research* 40, 715-728.

CORRE-HELLOU G., BRISSON N., LAUNAY M. FUSTEC J. CROZAT Y., 2007. Effect of root depth penetration on soilnitrogen competitive interactions and dry matter production in pea-barley intercrops given different soil nitrogen supplies. *Field Crops Research*. 103, 76-85.

CORRE-HELLOU, G, BRISSON N, LAUNAY M, FUSTEC J, CROZAT Y (2007) Effect of root depth penetration on soil nitrogen competitive interactions and dry matter production in pea-barley intercrops given different soil nitrogen supplies. *Field Crop Research* 103, 76-85

COUSIN R., 1997. Peas (*Pisum sativum* L.). *Field Crop Res.,* 53, 111-130.

COWAN I.R., 1968. The interception and absorption of radiation in plant stands. *J. Appl. Ecol.,* 5, 367-379.

DALE R.F., COELHO D.T., GALLO K.P., 1980. Prediction of daily green leaf area index for corn. *Agron. J.,* 72, 999-1005.

DE COCKBORNE A.M., JAUZEIN M., STENGEL P., GUENNELON R., 1988. Variation du coefficient de diffusion de NO_3^- dans les sols: influence de la teneur en eau et de la porosité. *Agronomie* 8: 905-914.

DE TOURDONNET S., 1999. Control of quality and N pollution in greenhouse lettuce production: a modelling study. *Acta Hort.* 507, 263-270.

DE VILLÈLE O., 1974. Besoin en eau des cultures sous serre. Essai de conduite de l'arrosage en fonction de l'ensoleillement. *Acta Horticulturae* 35, 23-139.

DE WIT C.T., 1978. Simulation of assimilation respiration and transpiration of crops. *In Simulation Monographs*. Pudoc, Wageningen, The Netherlands. 141 p.

DE WIT C.T., BROUWER R., PENNING DE VRIES F.W.T., 1970. The simulation of photosynthetic systems. *In Prediction and measurement of photosynthetic productivity*. Setlik I. (ed.). Proceeding IBP/PP Technical Meeting Trebon, 1969. Pudoc, Wageningen, The Netherlands, 47-50.

DEBAEKE P., BURGER P., BRISSON N., 2001. A simple model to simulate N2 symbiotic fixation of annual grain legumes: application to soybean. In *Proceedings of the second international symposium on modelling cropping system*, Firenze, Italy, 17-18.

DEFOSSEZ P., RICHARD G., BOIZARD H., 2003. Modeling change in soil compaction due to agricultural traffic as function of soil water content. *Geoderma* 116, 89-106.

DELPHIN J.E., 1986. Évaluation du pouvoir minéralisateur de sols agricoles en fonction de leurs caractéristiques physico-chimiques. *Agronomie* 6, 453-458.

DENMEAD O.T., 1973. Relative significance of soil and plant evaporation in estimating evapotranspiration. *In Plant response to climatic factors Proceedings Uppsala (Sweden) Symposium 1970*. UNESCO Paris, France, 505-511.

DENMEAD O.T., SHAW R.H., 1962. Availability of soil water to plants as affected by soil moisture content and meteorological conditions. *Agron. J.*, 45, 385-390.

DESCLAUX D., ROUMET P., 1996. Impact of drought stress on the phenology of two soybean (*Glucine max L.* Merr) cultivars. *Field Crops Research* 46, 61-70.

DEVIENNE-BARRET F., JUSTES E., MACHET J.M., MARY B., 2000. Integrated control of nitrate uptake by crop growth rate and soil nitrate availability under field conditions. *Annals of Botany* 86, 995-1005.

DOORENBOS J., KASSAM A.H., 1979. Yield response to water. *In FAO Irrig. Drain*. Paper N° 33, Food and Agric. Organization, United Nations, Rome, Italy, 235 p.

DORSAINVIL F., 2002. Évaluation par modélisation de l'impact environnemental des cultures intermédiaires sur les bilans d'eau et d'azote. Thèse, INA-PG, Paris, France, pp. 183.

DOUGUEDROIT A., 1986. Les topoclimats thermiques de moyenne montagne. *In Agrométéorologie des régions de moyenne montagne*. Les colloques de l'INRA (ed), Toulouse, France, 198-213.

DROUET J.L., PAGES L. 2003. GRAAL: a model of GRowth, Architecture and carbon ALlocation during the vegetative phase of the whole maize plant. Model description and parameterisation. *Ecological* 165, vol. 2-3, 147-173.

DUCHARNE A., BAUBION C., BEAUDOIN N., BENOIT M., BILLEN G., BRISSON N., GARNIER J., KIEKEN H., LEBONVALLET S., LEDOUX E., MARY B., MIGNOLET C., POUX X., SAUBOUA E., SCHOTT C., THÉRY S., VIENNOT P., 2007. Long term prospective of the Seine river system: confronting climatic and direct anthropogenic changes. *Science of the Total Environment*, 375, 292-311.

DUNCAN W.G., 1971. Leaf angles, leaf area, and canopy photosynthesis. *Crop Science* 11, 482-485.

DURAND P., TORTRAT F., VIAUD V., SAADI Z., 2006. Modélisation de l'effet des pratiques agricoles et de l'aménagement du paysage sur les flux d'eau et de matière dans les bassins versants. *In Qualité de l'eau en milieu rural, Savoirs et pratiques dans les bassins versants*, INRA Update, Mérot P. (ed.), 193-209.

DURAND R., 1967. Action de la température et du rayonnement sur la croissance. *Ann. Physiol. Veg.* 9, 5-27.

DURR C., AUBERTOT J.N., RICHARD G., DUBRULLE P., DUVAL Y., BOIFFIN J., 2001. SIMPLE: a model for simulation of plant emergence predicting the effects of soil tillage and sowing operations. *Soil Sci. Soc. Am. J.* 65, 414-423.

DURU M., DUCROCQ H., TIRILLY V., 1995. Modelling growth of cocksfoot (*Dactylis glomerata L.*) and tall fescue (*Festuca arundinacea* Schreb.) at the end of spring in relation to herbage nitrogen status. *Journal of Plant Nutrition* 18, 2033-2047.

DUVAL, Y., BOIFFIN, J., 1990. Influence of soil crusting on the emergence of sugar beet seedlings. *In Proceedings of the First Congress of the European Society of Agronomy*, 1990/12/05-07, Paris, France. Scaife A. (ed.), 311-312.

DWYER L.M., STEWART D.W., 1986. Leaf area development in field-grown maize. *Agron. J.,* 78, 334-343.

EAGLEMAN J.R., 1971. An experimentally derived model for actual evapotranspiration. *Agric. Meteorol.,* 8, 385-394.

FARQUHAR G.D., VON CAEMMERER S., BERRY J.A. 1980. A biochemical model of photosynthetic CO2 assimilation in leaves of C3 species. *Planta* 149, 78-90.

FEDDES R.A., 1987. Crop factors in relation to making reference crop evapotranspiration. Evaporation and Weather. *In Proc and Inform, TNO Comm. On Hydrol. Res.* Hooghart J.C. (ed). The Hague, The Netherlands, 39, 33-45.

FINN G.A., BRUN W.A., 1982. Effect of atmospheric CO_2 enrichment on growth, non-structural carbohydrate content and root nodule activity in soybean. *Plant Physiol.,* 69, 327-331.

FISHER D.K., ELLIOTT R.L., 1996. Modeling range land/pasture evapotranspiration using the Shuttleworth-Wallace approach. *In Proceedings of the International Conference on Evapotranspiration and irrigation scheduling.* Camp C.R., Sadler E.J. and R.E. Yoder (eds), 664-672.

FONTAINE S., BAROT S., BARRÉ S., BDIOUI N., MARY B., RUMPEL C., 2007. Stability of organic carbon in deep soil layers controlled by fresh carbon supply. *Nature,* in press.

FRIEND A.D., 1991. Use of a Model of Photosynthesis and Leaf Microenvironment to Predict Optimal Stomatal Conductance and Leaf Nitrogen Partitioning. *Plant Cell and Environment* 14, 895-905

GABRIELLE B., MARY B., ROCHE R., SMITH P., GOSSE G., 2002. Simulation of carbon and nitrogen dynamics in arable soils: a comparison of approaches. *Eur. J. Agron* 18, 107-120.

GARCIA DE CORTAZAR ATAURI I., 2006. Adaptation du modèle STICS à la vigne (*Vitis Vinifera* L.). Utilisation dans le cadre d'une étude d'impact du changement climatique à l'échelle de la France. Thèse ENSAM, 292 p.

GARDNER W.R., 1991. Modeling Water Uptake by Roots. *Irrigation Science* 12, 109-114.

GARRIDO F., HÉNAULT C., GAILLARD H., PÉREZ S., GERMON J.C., 2002. N_2O and NO emissions by agricultural soils with low hydraulic potentials. *Soil Biology & Biochemistry* 34, 559-575.

GARY C., JONES J.W., LONGUENESSE J.J., 1993. Modelling daily changes in specific leaf area of tomato: the contribution of the leaf assimilate pool. *Acta Horticulturae* 328, 205-210.

GARY C., DAUDET F.A., CRUIZIAT P., BRISSON N., OZIER-LAFONTAINE H., BREDA N., 1996. Le concept de facteur limitant. In de la plante au couvert végétal. Tome 1, Ecole chercheurs INRA en bioclimatologie, 1995/04/03-07, Le Croisic. Cruiziat P. and Lagouarde J.P. (eds), 121-128.

GÉNERMONT S., CELLIER P., 1997. A mechanistic model for estimating ammoniacal volatilization from slurry applied to bare soil. *Agric. For. Meteorol.* 88, 145-167.

GEWITZ A., PAGE E.R., 1974. An empirical mathematical model to describe plant root systems. *Journal of Applied Ecology* 11, 773-781.

GIACOMINI S., RECOUS S., MARY B., AITA C., 2007. Simulating the effects of N availability, straw particle size and location in soil on C and N mineralization. *Plant and Soil,* in press.

GIAUFFRET C., DERIEUX M., 1991. Variabilité génétique de la croissance de la jeune racine en fonction de la température. *In Physiologie et production du maïs, 13-15 nov. 1990, Pau, France. Picard D. (ed), 69-74.*

GIBELIN A.L., DÉQUÉ M., 2003. Anthropogenic climate change over the Mediterranean region simulated by a global variable resolution model. *Climate Dynamics* 20, 327-339.

GIRARD M.L., 1997. Modélisation de l'accumulation de biomasse et d'azote dans les grains de blé tendre d'hiver (*Triticum aestivum* L.), simulation de leur teneur en protéines à la récolte. Thèse INA-PG, 97 p.

GORDON A.J., RYLE G.J.A., MITCHELL D.F., POWEL C.E., 1985. The flux of 14-C labelled photosynthate through soybean root nodules during N_2 fixation. *J. Exp. Bot.*, 36, 756-769.

GOUDRIAAN J., 1986. Boxcartrain methods for modelling of ageing, development, delays and dispersion. *In The dynamics of physiologically structured populations*. Metz J.A.J. and Diekman O. (eds), Springer-Verlag, Berlin, 543-473.

GRAS R. M., 1994. Sols caillouteux et production végétale, *Collection mieux comprendre,* INRA éditions, Paris, France, 174 p.

GRAUX A.I. 2005. Etude des mécanismes de remobilisation d'assimilats chez la betterave sucrière en réponse aux stress abiotiques: expérimentation et modélisation. Mémoire de DEA, ENSAIA. INRA Avignon, France, 44 p.

GREBET P., 1993. Functions, subroutines and a program for easy computation of commonly required data related to the sun. *In "Crop structure and light microclimate"*, Varlet-Grancher C., Bonhomme R., Sinoquet H. (eds.), 501-511.

GREENWOOD D.J., GASTAL F., LEMAIRE G., DRAYCOTT A., MILLARD P., NEETESON J.J., 1991. Growth rate and N % of field grown crops: Theory and experiments. *Annals of Botany,* 67, 181-190.

GUÉRIF M., HOULÈS V., MARY B., BEAUDOIN N., MACHET J.M., MOULIN S., NICOULLAUD B., 2007. Modulation intra-parcellaire de la fertilisation azotée du blé fondée sur le modèle de culture STICS. intérêt de la démarche et méthodes de spatialisation. *In Hétérogénéité parcellaire et gestion des cultures : vers une agriculture de précision,* Guérif M., King D. (eds), Editions Quae, collection Update Sciences and Technologies, 225-248.

GUÉRIN V., PLADYS D., TRINCHANT C., RIGAUD J.M., 1991. Proteolysis and nitrogen fixation in faba bean (*Vicia faba*) nodules under water stress. *Physiol. Plant,* 82, 360-366.

GUYOT G., 1997. Climatologie de l'environnement : de la plante aux écosystèmes. Masson (ed.). Paris, 505 p.

HAH H., 2000. Ureolytic microorganisms and soil fertility: A review. *Communications in soil science and plant analysis* 31, 2565-2589.

HALLAIRE M., 1964. Le potentiel efficace de l'eau dans le sol en régime de dessèchement. *In L'eau et la production végétale*. INRA, Paris, France (ed.), 27-62.

HAMMER G., COOPER M., TARDIEU F., WELCH S., WALSH B., EEUWIJK F., CHAPMAN S. & PODLICH D., 2006. Models for navigating biological complexity in breeding improved crop plants *Trends in Plant Sciences* **11**, 587-593.

HAMMER G.L., MUCHOW R.C., 1994. Assessing Climatic Risk to Sorghum Production in Water-Limited Subtropical Environments 1 Development and Testing of a Simulation Model. *Field Crop. Res.* 36, 221-234.

HANSEN S., JENSEN H.E., NIELSEN N.E., SWENDEN H., 1990. DAISY – Soil Plant Atmosphere System Model NP0 Research *in the NAEP report*. The Royal Veterinary and Agricultural University (ed.) Nr A10, 272 p.

HÉNAULT C., BIZOUARD F., LAVILLE P., GABRIELLE B., NICOULLAUD B., GERMON J.C. and CELLIER P., 2005. Predicting in situ soil N_2O emission using NOE algorithm and soil database. *Global Change Biology* 11, 115-127.

HOOGHOUDT S.B., 1940. General consideration of the problem of field drainage by parallel drains, ditches, watercourses and channels. *In Contribution to the knowledge of some physical parameters of the soil.* Publ. N° 7 Bodemkundig Instituut, Groningen, The Netherlands.

HUCL P., 1993. Effects of Temperature and Moisture Stress on the Germination of Diverse Common Bean Genotypes. *Canadian Journal of Plant Science* 73, 697-702.

HUNGRIA M., VARGAS M.A.T., 2000. Environmental factors affecting N_2 fixation in grain legumes in the tropics with emphasis on Brazil. *Field Crop Res.* 65, 151-164.

HUNT L.A., PARARAJASINGHAM S., 1995. CROPSIM-WHEAT: A model describing the growth and development of wheat. *Canadian Journal of Plant Science* 75, 619-632.

HUTH N.I., CARBERRY P.S., POULTON P.L., BRENNAN L.E., KEATING B.A., 2002. A framework for simulating agroforestry options for the low rainfall areas of Australia using APSIM. *Eur. J. Agron.* 18, 171-185.

IDSO S.B., 1991. A general relationship between CO_2-induced increases in net photosynthesis and concomitant reductions in stomatal conductance. *Environmental and Experimental Botany* 31, 381-383.

IDSO S.B., JACKSON R.D., REGINATO R.J., 1978. Extending the "degree-day" concept of plant phenological development to include water stress effects. *Ecology* 59, 431-433.

ITABARI J.K., GREGORY P.J., JONES R.K., 1993. Effects of Temperature Soil Water Status and Depth of Planting on Germination and Emergence of Maize (*Zea mays*) Adapted to Semi-Arid Eastern Kenya. *Experimental Agriculture* 29, 351-364.

ITIER B., BRISSON N., DOUSSAN C., TOURNEBIZE R., 1997. Bilan hydrique en agrométéorologie. *In du couvert végétal à la petite région agricole.* Tome 2, Ecole chercheurs INRA en bioclimatologie, 1996/04/03-07 Le Croisic, France. Cruiziat P. and Lagouarde J.P. (eds), 383-398.

JACQUEMOUD, S., BARET, F. 1992. Modeling spectral and directional soil reflectance. *Remote Sensing Environ.* 41, 123-132.

JAMAGNE M., BÉTRÉMIEUX R., BÉGON J.C., MORI A., 1977. Quelques données sur la variabilité dans le milieu naturel de la réserve en eau des sols. *Bull Tech Inf.,* 324-325, 119-157.

JAMIESON P.D., BROOKING I.R., PORTER J.R., WILSON D.R., 1995. Prediction of leaf appearance in wheat: a question of temperature. *Field Crop Res.* 41, 35-44.

JAYASUNDARA H.P.S., THOMSOM B.D., TANG C., 1998. Responses of cool season grain legumes to soil abiotic stresses. *Advances in Agron.* 63, 77-151.

JENSEN E.S., 1987. Seasonnal patterns of growth and nitrogen fixation in field-grown pea. *Plant and Soil* 101, 29-37.

JEUFFROY M.H., BARRE C., BOUCHARD C., DEMOTES-MAINARD S., DEVIENNE-BARRET F., GIRARD M.L., RECOUS S., 2000. Fonctionnement d'un peuplement de blé en conditions de nutrition azotée sub-optimale. *In Fonctionnement des peuplements végétaux sous contraintes environnementales,* 20-21/01/1998, Paris, France. INRA (ed.), 289-304.

JEUFFROY M.H., RECOUS S. 1999. AZODYN: a simple model simulating the date of nitrogen deficiency for decision support in wheat fertilisation. *Eur. J. Agron.* 10, 129-144.

JEUFFROY M.H., WAREMBOURG F., 1991. Carbon transfer and partitioning between vegetative and reproductive organs in *Pisum sativum* L. *Plant Physiol.* 97, 440-448.

JONES J.W., DAYAN E., ALLEN L.H., VAN KEULEN H., CHALLA H., 1991. A dynamic tomato growth and yield model (TOMGRO). *Trans. ASAE* 34, 663-672.

JONES, J.W., HOOGENBOOM, G., PORTER, C.H., BOOTE, K.J., BATCHELOR, W.D., HUNT, L.A., WILKENS, P.W., SINGH, U., GIJSMAN, A.J., RITCHIE, J.T., 2003. The DSSAT cropping system model. Eur. J. Agron. 18: 235-265.

JUSTES E., JEUFFROY M.H., MARY B., 1997. II. The nitrogen requirement of major agricultural crops. Wheat, barley and durum wheat. *In Diagnosis of the nitrogen status in crops.* G. Lemaire (ed.), Springer-Verlag, 73-91.

JUSTES E., MARY B., MEYNARD J.M., MACHET J.M., THELIER-HUCHÉ L., 1994. Determination of a critical nitrogen dilution curve for winter wheat crops. *Annals of Botany* 74, 397-407.

KAGE H., EHLERS W., 1996. Does transport of water to roots limit water uptake of field crops? *Zeitschrift Fur Pflanzenernahrung und Bodenkunde* 159, 583-590.

KANNEGANTI V.R., FICK G.W., 1991. A Warm-Season Annual Grass Growth Model Parameterized for Maize and Sudangrass. *Agricultural Systems* 36, 439-470.

KEATING B.A., CARBERRY P.S., HAMMER G.L., PROBERT M.E., ROBERTSON M.J., HOLZWORTH D., HUTH N.I., HARGREAVES J.N.G., MEINKE H., HOCHMAN Z., MCLEAN G., VERBURG K., SNOW V., DIMES J.P., SILBURN M., WANG E., BROWN S., BRISTOW K.L., ASSENG S., CHAPMAN S., MCCOWN R.L., FREEBAIRN D.M. & SMITH C.J., 2003. An overview of APSIM, a model designed for farming systems simulation. *European Journal of Agronomy* **18**, 267-288.

KERSEBAUM K.C., RICHTER J., 1991. Modelling nitrogen dynamics in a plant-soil system with a simple model for advisory purposes. *Fertilizer Research* 27: 273-281.

KHALIL K., MARY B., RENAULT P., 2004. Nitrous oxide production by nitrification and denitrification in soil aggregates as affected by O_2 concentration. *Soil Biology & Biochemistry* 36, 687-699.

KINIRY J. R., BLANCHET R., GASSMAN P. W., DEBAEKE P., 1992. A general process-oriented model for two competing plant species. *Trans. ASAE* 35, 801-810.

KIRSCHBAUM M.U.F., 2006. The temperature dependence of organic matter decomposition – still a topic of debate. *Soil Biology & Biochemistry* 38, 2510-2518.

KOUCHI H., NAKAJI K., 1985. Utilization and metabolism of photoassimilated ^{13}C in soybean roots and nodules. *Soil Sci. Plant Nutr.* 31 (3), 323-334.

LACAPE M.J., WERY J., ANNEROSE D.J.M., 1998. Relationships between plant and soil water status in five field-grown cotton (*Gossypium hirsutum* L.) cultivars. *Field Crop. Res.* 57: 29-43.

LAUNAY M., BRISSON N., 2004. STICS adaptability to a novel crop as an application of modularity in crop modelling: example of sugarbeet. *In Proceedings of 8th ESA Congress "European Agriculture in a Global Context"*, Copenhagen, Denmark, 11-15 July, 2004, 283-284.

LAUNAY M., BRISSON N., SATGER S., HAUGGAARD-NIELSEN H., DIBET A., CORRE-HELLOU G., KASYANOVA E., MONTI M., DAHLMANN C., 2008. "Investigating intercrop management options in European organic cropping systems thanks to crop modelisation". In preparation.

LAWRIE A.C., WHEELER C.T., 1974. The effect of flowering and fruit formation on the supply of photosynthetic assimilates to nodules of *Pisum sativum* L. in relation to the fixation of nitrogen. *New Phytol. 73,* 1119-1127.

LECOEUR J., SINCLAIR T.R., 2001. Analysis of nitrogen partitioning in field pea resulting in linear increase in nitrogen harvest index. *Field Crop Res.* 71:151-158.

LEENHARDT D., VOLTZ M., RAMBAL S., 1995. A survey of several agroclimatic soil water balance models with reference to their spatial application. *Eur. J. Agron.* 4, 1-14.

LEMAIRE G., GASTAL F., 1997. N uptake and distribution in plant canopies. *In Diagnosis of the nitrogen status in crops*, Lemaire G. (ed.), Springer-Verlag, 3-44.

LEMAIRE G., SALETTE J., 1984. Relation entre dynamique de croissance et dynamique de prélèvement d'azote pour un peuplement de graminées fourragères. I. Etude de l'effet du milieu. *Agronomie* 4, 423-430.

LIENNARD M.E., 2002. Contribution à l'étude de la prévision de la précocité de floraison et du déterminisme climatique des nécroses florales de l'Abricotier, *Prunus armeniaca* L., dans le contexte des changements climatiques. Mémoire INH, Angers, France, 44 p.

LIMAUX F., RECOUS S., MEYNARD J.M., GUCKERT A., 1999. Relationship between rate of crop growth at date of fertiliser N application and fate of fertiliser N applied to winter wheat. *Plant and Soil* 214: 49-59.

LUDWIG B., JOHN B., ELLERBROCK R., KAISER M., FLESSA H., 2003. Stabilization of carbon from maize in a sandy soil in a long-term experiment. *Eur. J. Soil Sci.* 54, 117-126.

LUDWIG B., SCHULZ E., RETHEMEYER J., MERBACH I., FLESSA H., 2007. Predictive modelling of C dynamics in the long-term fertilization experiment at Bad Lauchstädt with the Rothamsted carbon model. *Eur. J. Soil Sci.*, 58, 5, 1155-1163.

LYON D.J., HAMMER G.L., MCLEAN G.B. & BLUMENTHAL J.M., 2003. Simulation supplements field studies to determine no-till dryland corn population recommendations for semiarid western Nebraska. *Agronomy Journal* **95**, 884-891.

MAAS S.J., 1993. Parameterized Model of Gramineous Crop Growth: 1 Leaf Area and Dry Mass Simulation. *Agronomy Journal* 85, 348-353.

MACDUFF J.H., DAVIS S.C., DAVIDSON I.A., 1996. Inhibition of N_2 fixation by white clover (*Trifolium repens* L.) at low concentrations in NO_3^- in flowing solution culture. *Plant and Soil* 180, 287-295.

MAILHOL J.C., RUELLE P., REVOL P., DELAGE L., LESCOT J.M., 1996. Operative modelling for actual evapotranspiration assessment: calibration methodology. *In Evapotranspiration and irrigation scheduling*. Proceedings of the International Conference, Nov 3-6 1996, San Antonio, Texas, 474-479.

MAKOWSKI D., HILLIER J., WALLACH D., ANDRIEU B., JEUFFROY M.H., 2006. Parameter estimation for crop models. *In Working with crop models*, Wallach D., Makowsky D., J.J. Jones (eds). Elsevier, 101-149.

MARY B., BEAUDOIN N., JUSTES E., MACHET J.M., 1999. Calculation of nitrogen mineralization and leaching in fallow soil using a simple dynamic model. *Eur. J. Soil Sci.* 50, 549-566.

MARY B., GUERIF M., 2005. Effet du stress azoté sur la plante : définition d'un indice de nutrition instantané. *In Séminaire STICS*, Carry-le-Rouet, 17-18 mars 2005, 23-27.

MCCANN I.R., MCFARLAND M.J., WITZ J.A., 1991. Near-Surface Bare Soil Temperature Model for Biophysical Models. *Transactions of the ASAE* 34, 748-755.

MCMASTER G.S., MORGAN J.A., WILHELM W.W., 1992. Simulating Winter Wheat Spike Development and Growth. *Agricultural and Forest Meteorology* 60, 193-220.

MCMILLAN W.D., BURGY R.H., 1960. Interception loss from grass. *J. Geophys. Res.* 65, 2389-2394.

MEYER, W.S., GREEN, G.C., 1981. Plant indicators of wheat and soybean crop water stress. *Irrig. Sci.* 2, 167-176.

MEYNARD J.M., DORÉ T., HABIB R., 2001. L'évaluation et la conception de systèmes de culture pour une agriculture durable. *Compte rendu Acad. Agric. Fr.*, 223-236.

MILROY S.P., GOYNE P.J., 1995. Leaf area development in barley-model construction and response to soil moisture status. *Australian Journal of Agricultural Research* 46, 845-860.

MONOD H., NAUD C., MAKOWSKI D., 2006. Uncertainty and sensitivity analysis for crop models. *In Working with crop models*. Wallach D., Makowsky D., Jones J.J. (eds). Elsevier, 55-99.

MONTEITH J.L., 1965. Evaporation and Environment. *Symp. Soc. Exp. Biol.* 19, 205-234.

MONTEITH J.L., 1972. Solar radiation and productivity in tropical ecosystems. *J. Appl. Ecol.* 9, 747-766.

MORVAN T., 1999. Quantification et modélisation des flux d'azote résultant de l'épandage de lisier. Thèse, université Pierre et Marie Curie, Paris, France, 157 p.

MORVAN T., LETERME P., 2001. Vers une prévision opérationnelle des flux d'azote résultant de l'épandage de lisier : paramétrage d'un modèle dynamique (STAL). *Ingénieries* 26, 17-26.

MUCHOW R.C., CARBERRY P.S., 1990. Phenology and Leaf-Area Development in a Tropical Grain Sorghum. *Field Crops Research* 23, 221-237.

MUCHOW R.C., SINCLAIR T.R., BENNETT J.M., 1990. Temperature and Solar Radiation Effects on Potential Maize Yield Across Locations. *Agronomy Journal* 82, 338-343.

MUMEN M., 2006. Caractérisation du fonctionnement hydrique des sols à l'aide d'un modèle mécaniste de transferts d'eau et de chaleur mis en oeuvre en function d'informations disponibles sur le sol. Thèse, université d'Avignon, France, 181 p.

MUNIER-JOLAIN N.G., MUNIER-JOLAIN N.M., ROCHE R., NEY B., DUTHION C., 1998. Seed growth rate in grain legumes. I: seed growth is not affected by assimilates availability on seed growth rate. *J. Exp. Bot.* 49, 1963-1969.

NELDER J.A., 1961. The fitting and generalization of the logistic curve. *Biometrics* 17, 89-110.

NICOLARDOT B., PARNAUDEAU V., CHABALIER P. F., 2006. Effet de la température sur la minéralisation de différents produits organiques apportés dans les sols. *In Actes de clôture de l'ATP (CD-Rom)*, CIRAD.

NICOLARDOT B., RECOUS S., MARY B., 2001. Simulation of C and N mineralization during crop residue decomposition: a simple dynamic model based on the C:N ratio of the residues. *Plant and Soil* 228, 83-103.

NICOULLAUD B., COUTURIER A., BEAUDOIN N., MARY B., COUTADEUR C., KING D., 2004. Modélisation spatiale à l'échelle parcellaire des effets de la variabilité des sols et des pratiques culturales sur la pollution nitrique agricole. *In Organisation spatiale des activités agricoles et processus environnementaux*. Monestiez P., Lardon S. et Seguin B. (eds), Coll. Science Update, INRA Editions, 143-161.

NICOULLAUD B., DARTHOUT R., DUVAL O., 1995. Etude de l'enracinement du blé tendre d'hiver et du mais dans les sols argilo-limoneux de petite beauce. *Etude et gestion des sols* 2,183-200.

NICOULLAUD B., KING D., TARDIEU F., 1994. Vertical distribution of maize roots in relation to permanent soil characteristics. *Plant and Soil* 159, 245-254.

NOVAK, V. 1989. Determination of critical moisture contents of soils in the course of evapotranspiration. *Sov. Soil Sci.,* 21: 122-126.

ONG C.K., 1983. Response to temperature in a stand of pearl millet (*Pennisetum typhoides* S & H): I vegetative development. *Journal of Experimental Botany* 34, 332-336.

OORTS, K., LAURENT, F., MARY, B., 2007. Experimental and simulated soil mineral N dynamics for long-term tillage systems in northern France. *Soil & Tillage Research* 94, 441-456.

OTEGUI M.E., NICOLINI M.G., RUIZ R.A., DODDS P.A., 1995. Sowing date effects on grain yield components for different maize genotypes. *Agronomy Journal* 87, 29-33.

OZIER-LAFONTAINE H., LECOMPTE F., SILLON J.F., 1999. Fractal analysis of the root architecture of Gliricidia sepium for the spatial prediction of root branching, size and mass: model development and evaluation in agroforestry. *Plant and Soil* 209: 167-180.

PALACIOS E.V., QUEVEDO A.N., 1996. Irrigation scheduling using small evaporation pan and computer program for developing countries. *In Evapotranspiration and irrigation scheduling* Proceedings of the International Conference, nov. 3-6, 1996, San Antonio, Texas (USA), 1029-1034.

PARARAJASINGHAM S., HUNT L.A., 1991. Wheat Spike Temperature in Relation to Base Temperature for Grain Filling Duration. *Canadian Journal of Plant Science* 71, 63-69.

PARTON W.J., SCHIMEL D.S., COLE C.V., OJIMA D.S., 1987. Analysis of factors controlling soil organic matter levels in Great Plains grasslands. *Soil Sci. Soc. Am. J.* 51: 1173-1179.

PASSIOURA J.B. (1996) Simulation models: science, snake oil, education, or engineering? *Agronomy Journal* 88, 690-694.

PEART R. M., JONES J.W., CURRY R.B., BOOTE K., ALLEN L.H., 1989. Impact of climate change on crop yield in the south-eastern USA: a simulation study. *In The Potential effects of Global Change on the United States,* Smith J.B., Tirpak D.A. (eds), vol. 1.

PELLEGRINO A., WERY J., LEBON E., 2002. Crop management adaptation to water-limited environments. *In Proceedings of the VII ESA congress.* July15-18/2002, Cordoba (Spain), 313-314.

PENA-CABRIALES J.J., CASTELLANOS J.Z., 1993. Effect of water stress on N_2 fixation and grain yield of *Phaseolus vulgaris* L. *Plant and Soil,* 152, 151-155.

PENMAN H.L., 1948. National evaporation from open water, bare soil land grass. *Proceedings of the Royal Society* 193, Londres.

POWLSON D.S., HART P.B.S., POULTON P.R., JOHNSTON A.E., JENKINSON D.S., 1992. Influence of soil type, crop management and weather on the recovery of ^{15}N-labelled fertilizer applied to winter wheat. *J. Agric. Sci. Camb.* 118, 83-100.

POWLSON D.S., PRUDEN G., JOHNSTON A.E., JENKINSON D.S., 1986. The nitrogen cycle in the Broadbalk Wheat Experiment: recovery and losses of ^{15}N-labelled fertilizer applied in spring and inputs of nitrogen from the atmosphere. *J. Agric. Sci. Camb.* 107, 591-6609.

PRIESTLEY C.H.B., TAYLOR R.J., 1972. On the assessment of surface heat flux and evaporation using large-scale parameters. *Monthly Weat. Rev.* 100, 81-92.

REBIÈRE B., 1996. Effet d'un excès d'eau sur la croissance d'une culture de blé d'hiver. Thèse, université de Strasbourg, 214 p.

RECOUS S., FRESNEAU C., FAURIE G., MARY B., 1988. The fate of labelled ^{15}N urea and ammonium nitrate applied to a winter wheat crop. I. Nitrogen transformations in the soil. *Plant and Soil* 112, 205-214.

RECOUS S., LOISEAU P., MACHET J.M., MARY B., 1997. Transformation et devenir de l'azote de l'engrais en système de grande culture et sous prairie. *In Maîtrise de l'azote dans les écosystèmes cultivés*. Lemaire G., Nicolardot B. (eds.). Les Colloques de l'INRA, Paris, France, 83, 105-120.

RECOUS S., MACHET J.M., 1999. Short term immobilisation and crop uptake of fertiliser nitrogen applied to winter wheat: effect of date of application in spring. *Plant and Soil* 206, 137-149.

RECOUS S., MACHET J.M., MARY B., 1992. The partitioning of fertilizer-N between soil and crop: comparison of ammonium and nitrate applications. *Plant and Soil* 144, 101-111.

RECOUS S., ROBIN D., DARWIS S., MARY B., 1995. Soil inorganic N availability: effect on maize residue decomposition. *Soil Biology and Biochemistry* 27, 1529-1538.

RENNIE R.J., KEMP G.A., 1980. Dinitrogen fixation in pea beans (*Phaseolus vulgaris*) as affected by growth stage and temperature regime. *Can. J. Bot.* 59, 1181-1188.

RICHARD G., BRISSON N., LEBONVALLET S., RIPOCHE D., BOIZARD H., DÉFOSSEZ P., CHANZY A., ROGER-ESTRADE J., 2007. Some characteristics of soil structure that have been introduced in STICS. *In Séminaire STICS,* Session 6, 20-22 mars 2007, Reims, France.

RICHARD G., CELLIER P., 1998. Effect of soil tillage on bare soil energy balance and thermal regime: an experimental study. *Agronomie* 18, 163-181.

RICHARDSON E.A, SEELEY S.D., WALKER D.R., 1974. A model for estimating the completation of rest for Redhaven and Elberta peach trees. *HortScience* 9, 331-332.

RICKMAN R.W., WALDMAN S.E., KLEPPER B., 1996. MODWht3: A development-driven wheat growth simulation. *Agronomy Journal* 88, 176-185.

RIOU C., ITIER B., SEGUIN B., 1988. The influence of surface roughness on the simplified relationship between daily evaporation and surface temperature. *Int. J. Remote Sensing* 9, 1529-1533.

RIPOCHE, D., WEISS M., PREVOT L. 2001. Driving the STICS crop model by exogenous values of leaf area index. Application to remote sensing. *In Proceedings of the second international symposium on modelling cropping system*. July 16-18, 2001, Florence (Italy), 169-170.

RITCHIE J.T., 1972. Model for predicting evaporation from a row crop with incomplete cover. *Water Resource Research* 8, 1204-1213.

RITCHIE J.T., 1981. Soil water availability. *Plant and soil* 58, 327-338.

RITCHIE J.T., 1985. A user-oriented model for the soil water balance in wheat. *In Wheat Growth and Modeling*. Fry E., Atkin T.K. (eds). *Plenum publishing corporation NATO-ASI Series*, 293-30.

RITCHIE J.T., Otter S., 1984. Description and performance of CERES-Wheat a user-oriented wheat yield model. *In USDA-ARS-SR Grassland Soil and Water Research Laboratory Temple RX*, 159-175.

ROBERTSON M.J., FUKAI S., LUDLOW M.M., HAMMER G.L., 1993. Water Extraction by Grain Sorghum in a Sub-Humid Environment: 2 Extraction in Relation to Root Growth. *Field Crops Research* 33, 99-112.

RODRIGO A., RECOUS S., NÉEL C., MARY B., 1997. Modelling temperature and moisture effects on C-N transformations in soils: comparison of nine models. *Ecological modelling* 102, 325-339.

ROSENTHAL, W.D., ARKIN, G.F., SHOUSE, P.J., 1985. Water deficit effects on sorghum transpiration. *In Advances in evapotranspiration,* Proceedings of National Conference, Chicago, ASAE PublI, 159-169.

RUGET F., BETHENOD O., COMBE L., 1996. Repercussions of increased atmospheric CO_2 on maize morphogenesis and growth for various temperature and radiation levels. *Maydica* 41, 181-191.

RUGET F., BRISSON N., 2007. Taking into account the perennity of grass. *In Séminaire STICS,* Session 3, 20-22 mars 2007, Reims, France.

RUGET F., DELÉCOLLE R., LE BAS I., DURU M., BONNEVIALE N., RABAUD E., DONET I., PÉRARNAUD V., PANIAGUA C. 2002. L'estimation régionale des productions fourragères. *In Modélisation des agro-écosystèmes et aide à la décision,* 281-300, CIRAD, Collection Repères, Malézieux, Trébuil, Jaeger eds.

RUGET F., NOVAK S., GRANGER S., 2006. Use of the ISOP system, based on the STICS model, for the assessment of forage production. Adaptation to grassland and spatialized application. *Fourrages* 186, 241-256.

SAFFIH-HDADI K., MARY B., 2008. Modeling consequences of straw residues export on soil organic carbon. *Soil Biology & Biochemistry,* 40, 3, 594-607.

SATGER S., RUGET F., BRISSON N., VOLAIRE F., LELIÈVRE F., 2007. STICS crop model adaptation to modelise mediterranean meadow functioning during summer drought, STICS Workshop, Session 3, 20-22 mars 2007, Reims, France.

SAU F., BOOTE K.J., BOSTICK W.M., JONES J.W., MÍNGUEZ M.I., 2004. Testing and improving evapotranspiration and soil water balance of the DSSAT crop models. *Agron. J.* 96, 1243-1257.

SCOPEL E., MULLER B., ARREOLA TOSTADO J.M., CHAVEZ GUERRA E., MARAUX F., 1998. Quantifying and modelling the effects of a light crop residue mulch on the water balance: an application to rainfed maize in western Mexico. *In XVI World Congress of Soil Science.* 20-26 August 1998, Montpellier, France.

SEGHIERI J., FLORET C., PONTANIER R., 1995. Plant Phenology in Relation to Water Availability – Herbaceous and Woody Species in the Savannas of Northern Cameroon. *Journal of Tropical Ecology* 11, 237-254.

SEGUIN B., ITIER B. 1983. Using midday surface temperature to estimate daily evaporation from satellite thermal IR data. *Int. J. Remote Sensing* 4, 371-383.

SELLERS W.D., 1965. Physical climatology. *University of Chicago (ed), 272 p.*

SENE K.J., 1994. Parameterizations for energy transfers from a sparse vine crop. *Agricul. For. Meteorol.* 71: 1-18.

SHUTTLEWORTH W.J., WALLACE J.S., 1985. Evaporation from sparse canopy-An energy combination theory. *Quarter Journal of the Royal Meteorological Society* 111, 839-855.

SIERRA J., BRISSON N., RIPOCHE D., NOEL D., 2003. Application of the STICS crop model to predict N availability and nitrate transport in a tropical acid soil cropped with maize. *Plant and Soil* 256: 333-345.

SINCLAIR T.R., 1986. Water and nitrogen limitations in soybean grain production I – Model development. *Field Crops Research* 15, 125-141.

SINCLAIR T.R., DE WIT, C.T., 1976. Analysis of the carbon and nitrogen limitations to soybean yield. *Agron. J.* 68, 319-324.

SINCLAIR T.R., MOSCA G., BONA S., 1993. Simulation Analysis of Variation Among Seasons in Winter Wheat Yields in Northern Italy. *Journal of Agronomy and Crop Science – Zeitschrift Fur Acker und Pflanzenbau* 170, 202-207.

SINCLAIR T.R., SELIGMAN N.G., 1996. Crop modeling: from infancy to maturity. *Agronomy Journal* 88 698-704.

SINGELS A, DE JAGER J.M., 1991. Refinement and validation of the PUTU wheat crop growth model 3 Grain growth . *South African Journal of Plant and Soil* 8, 73-77.

SLABBERS P.J., 1980. Practical prediction of actual evapotranspiration. *Irrigation Science* 1, 185-196.

SMITH M., ALLEN R., PEREIRA L., 1996. Revised FAO methodology for crop water requirements *In Proceedings of the International Conference on Evapotranspiration and irrigation scheduling.* Camp C.R., Sadler E.J., Yoder R.E. (eds), 116-124.

SONOHAT G., SINOQUET H., VARLET-GRANCHER C., RAKOCEVIC M., JACQUET, A., SIMON J. C., ADAM, B., 2002. Leaf dispersion and light partitionning in three-dimensionally digitized tall fescue-white clover mixtures. *Plant Cell Envir.* 25, 529-538.

SPAETH S.C., SINCLAIR T.R., 1985. Linear increase in soybean harvest index during seed-filling. *Agronomy Journal* 77, 207-211.

SPITTERS C.J.T., TOUSSAINT H.A.J.M., GOUDRIAAN J., 1986. Separating the diffuse and direct component of global radiation and its implications for modelling canopy photosynthesis. Part I. Components of incoming radiation. *Agric. Forest Meteorol.* 38, 217-229.

SPRENT J.I., STEPHENS J.H., RUPELA O.P., 1988. Environmental effects on nitrogen fixation. *In* RJ Summerfield, ed, World Crops: Cool Season Food Legumes. Kluwer Academic Publishers, Dordrecht, The Netherlands, 801-810.

STOCKLE C.O., KJELGAARD J., 1996. Parameterizing Penman-Monteith surface resistance for estimating daily crop ET. *In Proceedings of the International Conference on Evapotranspiration and irrigation scheduling.* Camp C.R., Sadler E.J., Yoder R.E. (eds), 697-703.

STOCKLE C.O., WILLIAMS J.R., ROSENBERG C.A., JONES C.A., 1992. A method for estimating direct and climatic effects of rising atmospheric carbon dioxide on growth and yield crops. *Agric. Syst.* 38, 225-238.

STOCKLE O.S., DONATELLI M., NELSON R., 2003. CropSyst, a cropping systems simulation model. *Eur. J. Agron.* 18, 289-307.

STRUIK P.C., CASSMAN K.G., KOORNEEF M., 2007. *A dialogue of interdisciplinary collaboration to bridge the gap between plant genomics and crop sciences.* In *Scale and complexity in Plant Systems reserach, gene plant crop relations* p 319-328, JHJ Spiertz, PC Struik and HH Van Laar eds. Springer Dordrecht, Netherlands, 329 p.

TARDIEU F. (2003) Virtual plants: modelling as a tool for the genomics of tolerance to water deficit. *Trends in Plant Science* **8,** 9-14.

TARDIEU F. GRANIER C., MULLER B., 1999. Research review Modelling leaf expansion in a fluctuating environment: are changes in specific leaf area a consequence of changes in expansion rate? *New Phytol.,* 143, 33-43.

TEITTINEN M., KARVONEN T., PELTONEN J., 1994. A Dynamic Model for Water and Nitrogen Limited Growth in Spring Wheat to Predict Yield and Quality. *Journal of Agronomy and Crop Science – Zeitschrift Fur Acker und Pflanzenbau* 172, 90-103.

TEIXEIRA, J.L., LIU, Y., ZHANG, H.J., PEREIRA, L.S., 1996. Evaluation of the ISAREG irrigation scheduling model in the north China plain. *In Evapotranspiration and irrigation scheduling*. Proceedings of the International Conference, nov 3-6 1996, San Antonio, Texas: 632-638.

THORNLEY J.H.M., 1996. Modelling water in crops and plant ecosystems. *Ann. Bot.* 77, 261-275.

TOLK J.A., HOWEL T.A., STEINER J.L., KRIEG D.R., 1996. Corn canopy resistance determined from whole plant transpiration. *In Proceedings of the International Conference on Evapotranspiration and irrigation scheduling*. C.R. Camp, Sadler E.J., Yoder R.E. (eds), 664- 672.

TRAPANI N., HALL A.J., SADRAS V.O., VILELLA F., 1992. Ontogenetic Changes in Radiation Use Efficiency of Sunflower (*Helianthus annuus* L) Crops. *Field Crop. Res.* 29, 301-316.

VACHAUD G., VAUCLIN M., ADDISCOTT T.M., 1993. Solute transport in the vadose zone: a review of models. *In Technologies for Environmental Cleanup: soil and groundwater.* Avogadro A., Ragaini R.C. (eds), ECSC Netherlands, 157-185.

VALÉ M., 2006. Quantification et prédiction de la minéralisation nette de l'azote du sol in situ, sous divers pédoclimats et systèmes de culture français. Thèse INPL Toulouse, France, 182 p.

VAN BAVEL C.H.M., 1953. A drought criterion and its application in evaluating drought incidence and hazard. *Agron. J.* 45, 167-172.

VAN DER PLOEG R.R., RINGE H., MACHULLA G., 1995. Late fall site-specific soil nitrate upper limits for groundwater protection purposes. *Journal of Environmental Quality* 24, 725-733.

VAN ITTERSUM M.K., LEFFELAAR P.A., VAN KEULEN H., KROPFF M.J., BASTIAANS L., GOUDRIAAN J., 2003. On approaches and applications of the Wageningen crop models. *Eur. J. Agron.* 18, 201-234.

VAN KEULEN H., SELIGMAN N.G., 1987. Simulation of water use nitrogen nutrition and growth of a spring wheat crop. *In Simulation monograph* Pudoc Wageningen, The Netherlands. 310 p.

VARLET GRANCHER C., BONHOMME R., 1979. Application aux couverts végétaux des lois de rayonnement en milieu diffusant. II – Interception de l'énergie solaire par une culture. *Ann. agron.* 30, 1, 1-26.

VARLET GRANCHER C., BONHOMME R., CHARTIER M., ARTIS P., 1981. Evolution de la réponse photosynthétique des feuilles et efficience théorique de la photosynthèse brute d'une culture de canne à sucre (*Saccharum officinarum* L.). *Agronomie* 1, 6, 473-481.

VARLET-GRANCHER C., BONHOMME R., CHARTIER M., ARTIS P., 1982. Efficience de la conversion de l'énergie solaire par un couvert végétal. *Oecologia Plantarum* 3, 3-26.

VARLET-GRANCHER C., BONHOMME R., SINOQUET H. 1993. Crop structure and light microclimate. Varlet-Grancher C., Bonhomme R., Sinoquet H. (eds.). INRA Editions, Versailles, France, p. 359-372.

VARLET-GRANCHER C., GOSSE G., CHARTIER M., SINOQUET H., BONHOMME R., ALLIRAND J.M., 1989. Mise au point: rayonnement solaire absorbé ou intercepté par un couvert végétal. *Agronomie* 9, 419-439.

VILOINGT T., 2005. Identification et quantification des risques liés à l'implantation des cultures de printemps en semi direct derrière une culture intermédiaire : importance de l'humidité du sol. Mémoire de fin d'études ESA Angers, France, 122 p.

VINTEN A.J.A., REDMAN M.H., 1990. Calibration and validation of a model of non-interactive solute leaching in a clay-loam arable soil. *Journal of Soil Science* 41, 199-214.

VOISIN A.S., SALON C., JEUDY C., WAREMBOURG F.R., 2003. Symbiotic N2 fixation activity in relation to C economy of *Pisum sativum* L. as a function of plant phenology. *Journal of Experimental Botany* 54, 393, 2733-2744.

VOISIN A.S., SALON C., MUNIER-JOLAIN N., NEY B., 2002. Quantitative effects of soil nitrate, growth potential and phenology on symbiotic nitrogen fixation of pea (*Pisum sativum* L.). *Plant and soil* 243, 1, 31-42.

WALLACE J.S., 1995. Towards a coupled light partitioning and transpiration model for use in intercrops and agroforestry. *In Ecophysiology of Tropical Intercropping*. Sinoquet H.,Cruz P. (eds.). INRA Editions, Versailles, France, 153-162.

WAREMBOURG F.R., 1983. Estimating the true cost of dinitrogen fixation by nodulated plants in undisturbed conditions. *Can. J. Microbiol.* 29, 930-937.

WARREN-WILSON J., 1972. Control of crop processes. *In Crop processes in controlled environment*, Rees A.R., Cockshull K.E., Hand D.W., Hurd R.G. (eds), Academic Press, London, 7-30.

WATERER J.G., VESSEY J.K., 1993. Effect of low static nitrate concentrations on mineral nitrogen uptake nodulation, and nitrogen fixation in field pea. *J. Plant Nutr.* 16 (9), 1775-1789.

WEAICH K., CASS A., BRISTOW K.L., 1996. Pre-emergent shoot growth of maize (*Zea mays* L) as a function of soil strength. *Soil and Tillage Research* 40, 3-23.

WEIR A.H., BRAGG P.L., PORTER J.R., RAYNER J.H., 1984. A winter wheat crop simulation model without water or nutrient limitations *Journal of the Agriculture Science, Cambridge* 102, 371-382.

WHISLER J.R., ACOCK B., BAKER D.N., FYE R.E., HODGES H.F., LAMBERT J.R., LEMMON H.E., MCKINION J.M., REDDY, V.R., 1986. Crop simulation models in agronomy systems, *Adv. Agron.* 40, 141-208.

WILLIAMS J.R., JONES C.A., DYKE P.T., 1984. A modelling approach to determining the relationship between erosion and soil productivity. *Transactions of the ASAE* 27, 129-144.

WILLIAMS J.R., JONES C.A., KINIRY J.R., SPANEL D.A., 1989. The EPIC Crop Growth Model. *Transactions of the ASAE* 32, 497-511.

WILSON,J.B.,1988.A Review of Evidence on the Control of Shoot:Root Ratio, in Relation to Models *Ann Bot.* 61, 433-449.

WÖSTEN J.H.M., NEMES A., LILLY A., LE BAS C., 1999. Development and use of a database of hydraulic properties of European soils. *Geoderma* 90, 169-185.

YIN X., STRUIK P.C. & KROPFF M.J., 2004. Role of crop physiology in predicting gene-to-phenotype relationships. *Trends in Plant Science* **9**, 426-432.

YIN X., VAN LAAR H.H., 2005. *Crop Systems dynamics. An ecophysiological simulation model for genotype-by-environment interactions*. Wageningen Academic Publishers. Wageningen (Netherlands), 153 p.

ZIMMER D., 2001. Mémoire d'habilitation à diriger des recherches. Université Paris VI, France, 119 p.

Figure list

Table list

Definition of symbols

Parameter or variable	Definition	unit
A	soil evaporation parameter combining climatic and soil aspects	mm
$AANGST_G$	coefficient of the Angstrom's relationship for extraterrestrial radiation	–
$ABSCISSION_P$	sensescent leaf proportion falling on the soil	–
ABSO	nitrogen absorption rate by plant	kg N ha^{-1} day^{-1}
ABSODRP	N demand during grain filling as a proportion of the dilution curve demand	–
ABSZ	profile of N uptake	kg N ha^{-1} day^{-1} cm^{-1}
$ACLIM_C$	climatic component of A	mm
$ADENS_V$	interplant competition parameter	–
$ADFOL_P$	parameter determining the leaf density evolution within the chosen shape	m^{-1}
ADIA	function estimating temperature in altitude	°C
$ADIL_P$	parameter of the critical curve of nitrogen requirements	N %
$AFPF_P$	parameter of the logistic function defining sink strength of fruits (indeterminate growth): relative fruit age at which growth is maximal	–
$AFRUITSP_V$	potential number of fruits per inflorescence and per degree.day	nb inflo^{-1} degree.days^{-1}
AIRG	daily irrigation	mm
AKS_G	parameter of calculation of the energetic lost between the inside and the outside of a greenhouse	Wm^{-2}K^{-1}
A_{LAI}	parameter describing the shape of the LAI curve when it is considered as a driving variable	–
ALBEDOLAI	albedo of the crop combining soil with vegetation	–
$ALBEDOMULCH_G$	albedo of crop mulch	–

ALBEDOMULCH$_T$	albedo of plastic cover	–
ALBEDO$_S$	albedo of the bare dry soil	–
ALBSOL	albedo of the soil	–
ALBVEG$_G$	albedo of the vegetation	–
ALLOCFRMX$_P$	maximal daily allocation towards fruits as a proportion of daily growth	–
ALLOCFRUIT	allocation ratio of assimilats to the fruits	–
ALPHACO2$_P$	coefficient allowing the modification of radiation use efficiency in case of atmospheric CO2 increase	–
ALPHAPT$_C$	coefficient of the PriestleyTaylor evaporatiion formulae	–
ALTINVERSION$_G$	altitude of the thermal inversion when calculating altiutude temperature	m
ALTISIMUL$_C$	altitude of the simulation	m
ALTISTATION$_C$	altitude of the clilmatic station (and the climatic variables)	m
AMM	profile of ammoniacal nitrogen	kg N ha^{-1} cm^{-1}
AMMSURF	ammonium inputs	kg N ha^{-1}
AMPFROID$_P$	semi thermal amplitude thermique for vernalising effect	°C
AMPLSURF	daily thermal amplitude at the soil surface	°C
AMPLZ	profile of daily thermal amplitude	°C
ANIT	daily nitrogen provided	kgN.ha^{-1} j^{-1}
ANITCOUPE$_T$	amount of mineral fertilizer applications at each cut (forage crop)	kg N ha^{-1}
ANOX	profile of the index of anoxia	
AO	index for a variable defined in the shade (intercropping)	–
ARGI$_S$	percentage of clay in the surface layer	%
AS	index for a variable defined in the sun (intercropping)	
AZOMES	amount of mineral nitrogen in the soil between surface and PROFMES$_T$	kgN.ha^{-1}
AZOZRAC0$_P$	parameter of the influence of nitrates on legume nodules	kg N ha^{-1} cm^{-1} soil
AZOZRAC100$_P$	parameter of the influence of nitrates on legume nodules	kg N ha^{-1} cm^{-1} soil
BANGST$_G$	coefficient of the angstrom's relationship for extraterrestrial radiation	–
BDENSD$_P$	BDENSp for the dominant crop in case of intercrop	plants m^{-2}
BDENS$_P$	minimal density from which interplant competition starts	plants m^{-2}
BDENSU$_P$	BDENSp for the understorey crop in case of intercrop	plants m^{-2}
BDIL$_P$	parameter of the nitrogen critical dilution curve	–
BELONG$_P$	parameter of the curve of coleoptile elongation	degree.days^{-1}

Symbol	Definition	Unit
$BETA_G$	parameter of increase of maximal transpiration when occurs a water stress	–
$BFORMNAPPE_S$	shape parameter of the water table	–
$BFPF_P$	parameter of the logistic curve defining sink strength of fruits (indeterminate growth): rate of maximum growth proportionately to maximum weight of fruits	–
$BIOROGNEM_T$	minimal biomass to be removed when tipping (automatic calculation)	t ha^{-1}
BKS_G	parameter of calculation of the energetic lost between the inside and the outside of a greenhouse	Wm^{-2}K^{-1}
B_{LAI}	parameter describing the shape of the LAI curve when it is considered as a driving variable	–
BOUCHON	index showing if the shrinkage slots are opened (0) or closed (1)	0-1
$CADENCEREC_T$	number of days between two harvests	day
$CAILLOUX_S$	volumetric stone content	%
$CALC_S$	calcareous content	%
$CAPILJOUR_S$	capillary rises	mm day^{-1}
$CELONG_P$	parameter of the subsoil plantlet elongation curve	–
$CFES_S$	parameter defining the soil contribution to evaporation as a function of depth	–
$CFPF_P$	parameter of the first potential growth phase of fruit, corresponding to an exponential type function describing the cell division phase.	–
CHARGEFRUIT	amount of filling fruits on the plant	nb fruits.m^{-2}
$CIELCLAIR_G$	fraction of insolation defining the "clear weather" notion	–
CNBIO	C//N ratio of the zymogeneous biomass	–
$CNGRAINREC_T$	minimal grain nitrogen content for harvest	0-1
CNHUM	C/N ratio of the newly formed humified matter	–
CNPAILLRAC	nitrogen concentration of the stems	%
CNPLANTE	nitrogen concentration of entire plant	%
CNRESIDU	C/N ratio of falling leaves	–
$CO2_C$	atmospheric CO2 content above 330 ppm	ppm
$CODEADRET_C$	code defining the slope orientation	
$CODEDENIT_G$	code activating the calculation of denitrification	
$CODEFENTE_S$	code for swelling soils	
$CODEFRMUR_G$	code defining the maturity status of the fruits in the variable CHARGEFRUIT	
$CODENITRIF_G$	code activating the calculation of nitrification	
$CODERES_T$	code defining the type of organic residues	
$CODLOCFERTI_T$	code defining the location of fertilisation	
$CODLOCIRRIG_T$	code defining the location of irrigation	
$COEFAMFLAX_P$	multiplier coefficient of the development phase AMFLAX to use crop temperature	–

COEFB$_G$	parameter defining radiation effect on conversion efficiency	–
COEFDEVIL$_G$	multiplier coefficient of the exterior radiation to compute potential evapotranspiration inside of a greenhouse	–
COEFLEV	ratio between the emerged and the germinated density	–
COEFLEVAMF$_P$	multiplier coefficient of the development phase LEVAMF to use crop temperature	–
COEFLEVB	ratio between the emerged and the germinated density due to crusting	–
COEFMSHAUT$_P$	ratio biomass/ useful height cut of crops	t ha^{-1} m^{-1}
COEFRNET$_G$	coefficient of calculation of the net radiation under greenhouse	–
CONCIRR$_T$	nitrate concentration in irrigation water	kg N ha^{-1} mm^{-1} water
CONCN	soil nitrate concentration	kg N ha^{-1} mm^{-1} water
CONCNODSEUIL$_P$	threshold soil nitrate concentration for nodulation	kg N ha^{-1} mm^{-1} water
CONCRR$_G$	rainfall mean nitrogen concentration	kg N ha^{-1} mm^{-1} water
CONCSEUIL$_S$	threshold soil nitrate concentration for lixiviation	kg N ha^{-1} mm^{-1} water
CONTRDAMAX$_P$	maximal root growth reduction due to soil strenghtness (high bulk density)	–
CONV	convection flow of mineral N	kg N ha^{-1} day^{-1}
CORECTROSEE$_G$	temperature to substract to Tmin to estimate dew point teñperature (in case of missing air humidity data)	°C
COUVERMULCH	proportion of soil covered by the vegetal cover	–
COUVERMULCH$_T$	proportion of soil covered by a plastic mulch	–
CRES	amount of c in the soil organic residues	kg C .ha^{-1}
CROIFRUIT	fruit growth	g m^{-2}
CROIRAC$_V$	growth rate of the root front	cm degree day^{-1}
CRUST	indicator of crust conditions at the soil surface	0/1
CSURNRESSUITE	C/N ratio of calculated crop residue for the next crop	–
CSURNRES$_T$	C/N ratio of the preceeding crop residues	–
CU	chill units accumulation	–
CUH	hourly chill unit	–
CUMLARCZ	sum of the effective root lengths over the profile	cm root.cm^{-2} soil
CUMOFFRN	sum of nitrogen soil supply over the profile	kg N ha^{-1}
CVENT$_G$	constant for greenhouse thermal calculation	–
D	dominant crop in case of intercrop	m
DA	bulk density	g cm^{-3}
DACHISEL$_T$	bulk density after soil tillage by a chisel	g cm^{-3}
DACOHES$_G$	bulk density under which root growth is reduced due to a lack of cohesion	g cm^{-3}
DAF$_S$	bulk density of fine earth	g cm^{-3}
DALABOUR$_T$	bulk density after soil ploughing	g cm^{-3}
DAREC$_T$	bulk density after harvest compaction	g cm^{-3}

DASEM$_T$	bulk density after sowing compaction	g cm^{-3}
DASEUILBAS$_G$	threshold of bulk density of soil below that the root growth is not limited	g cm^{-3}
DASEUILHAUT$_G$	threshold of bulk density of soil below that the root growth no more possible	g cm^{-3}
DCBIO	change in microbial soil biomass	kg C ha^{-1} day^{-1}
DCHUM	change in humus soil biomass	kg C ha^{-1} day^{-1}
DCRES	change in residue soil biomass	kg C ha^{-1} day^{-1}
DE	equivalent depth of the aquifer below the level of drains	m
DEBSENRAC$_P$	life span of a root	degree.days
DECOMPOSMULCH$_G$	decomposition rate of crop cover	day^{-1}
DELTA	extinction coefficient used for evaporation calculation	–
DELTABSO	nitrogen dependant biomass growth	t ha^{-1}.day^{-1}
DELTAI$_1$	daily increase of the green leaf index for determinate rops	m^2 leaf.m^{-2} soil
DELTAI$_2$	daily increase of the green leaf index for indeterminate rops	m^2 leaf.m^{-2} soil
DELTAI$_{dens}$	density component of DELTAI	plant m^{-2}
DELTAI$_{dev}$	phasic development component of DELTAI	m^2 plant^{-1} degree-day^{-1}
DELTAI$_{dev}$MAX	maximal value of DELTAI$_{dev}$	m^2 plant^{-1} degree-day^{-1}
DELTAIMAXI	maximum increase in leaf expansion	m^{2}m^{-2}day^{-1}
DELTAI$_{stress}$	stress component of DELTAI	–
DELTAI$_T$	thermal component of DELTAI	degree-days
DELTAMSRESEN	daily sensecence of residual dry matter of a forage crop	t ha^{-1} day^{-1}
DELTAT	gradient of the relationship between saturation vapour pressure and temperature	mbars °C^{-1}
DELTATEMP	difference in mean daily temperature inside and outside a greenhouse	°C
DELTAZ	deepening of the root front	cm day^{-1}
DELTAZ$_{stress}$	stress component of DELTAZ	–
DELTAZ$_T$	thermal componenet of DELTAZ	cm day^{-1}
DEMANDE	daily nitrogen need of the plant	kgN.ha^{-1} day^{-1}
DENENG$_G$	maximal proportion of N losses for each fertilizer type by denitrification	–
DENSITE	plant density	plants.m^{-2}
DENSITE$_D$	plant density of the dominant crop in case of intercropping	plants.m^{-2}
DENSITE$_T$	sowing planty density	plants.m^{-2}
DENSITE$_U$	plant density of the understorey crop in case of intercropping	plants.m^{-2}
DENSITEUeq	equivalent plant density of the understorey crop in case of intercropping accounting for the dominant plant	plants.m^{-2}
DESHYDBASE$_P$	phenological rate of evolution of fruit water content (>0 or <0)	g water.g fresh matter^{-1}. °C^{-1}

DFOL	"Within the shape" leaf area density	m^2 leaf m^{-3}
DFOLBAS$_p$	minimal value for DFOL	m^2 leaf m^{-3}
DFOLHAUT$_p$	maximal value for DFOL	m^2 leaf m^{-3}
DFPF$_p$	parameter of the first potential growth phase of fruit, corresponding to an exponential type function describing the cell division phase.	–
DFR	fruit development stage	–
DH	displacement height	m
DIFF	diffusion flow of mineral nitrogen	kg N ha^{-1} day^{-1}
DIFN$_G$	diffusion coefficient bof nitrogen at field capacity	cm^2 day^{-1}
DIFTHERM$_G$	soil thermal diffusivity	cm^2 s^{-1}
DLAIMAXBRUT$_p$	maximum rate of DELTAI	m^2 leaf plant^{-1} degree day^{-1}
DLTAGN	daily increase of grain nitrogen content	kg N ha^{-1} day^{-1}
DLTAGS	growth rate of the grains	t ha^{-1}.day^{-1}
DLTAISEN	daily increase of the senescent leaf index	m^2leaf.m^{-2} day^{-1}
DLTAMS	growth rate of the plant	t ha^{-1}.day^{-1}
DLTAMSEN	senescence rate of the plant	t ha^{-1}.day^{-1}
DLTAMSTOMBE	daily sensescent biomass falling on the soil	t ha^{-1}.day^{-1}
DLTAREMOBIL	amount of reserves remobilised	g.m^{-2}.day^{-1}
DN	net mineralization rate	kg N ha^{-1} day^{-1}
DOS	saturation deficit within the canopy	mbars
DOSIMXN$_T$	maximum nitrogen amount authorised at each time step (mode automatic fertilization)	kg N ha^{-1} day^{-1}
DOSIMX$_T$	maximum water amount of irrigation authorised at each time step (mode automatic irrigation)	mm day^{-1}
DPHVOL	increase in surface soil ph	–
DPHVOLMAX$_G$	maximum increase in surface soil ph	–
DRACLONG$_p$	maximum rate of root length production	cm root plant^{-1} degree.days^{-1}
DRAIN	water flux drained at the base of the soil profile	mm day^{-1}
DSAT	saturation deficit at the reference level	mbars
DURAGE	natural life span of leaves	Q10
DUREEFRUIT$_V$	total growth period of a fruit at the setting stage to the physiological maturity	degree.days
DURVIE	actual life span of the leaf surface	°C
DURVIEF$_V$	maximal lifespan of an adult leaf	Q10
DURVIEI$_p$	lifespan of a young leaf (at the AMF stage) expressed in proportion of DURVIEF	–
DURVIESUPMAX$_p$	proportion of additional lifespan due to an overfertilization	–
EAI	equivalent leaf area index for ear	m^2m^{-2}
EAURES$_T$	water content of residue	%
EBMAX	maximum value of radiation use efficieny	g MJ^{-1}
EDIRECT	water amount evaporated by the soil + intercepted by leafs + intercepted by the mulch	mm day^{-1}
EDIRECTM	maximum value of EDIRECT	mm day^{-1}
EFCROIJUV$_p$	maximum radiation use efficiency during the juvenile phase(LEV-AMF)	g MJ^{-1}

EFCROIREPRO$_P$	maximum radiation use efficiency during the grain filling phase (DRP-MAT)	g MJ^{-1}
EFCROIVEG$_P$	maximum radiation use efficiency during the vegetative stage (AMF-DRP)	g MJ^{-1}
EFDA	bulk density effect on root distribution in the profile	–
EFFEUIL$_T$	proportion of daily leaf removed at thinning	–
EFFIRR$_T$	irrigation efficiency: proportion of water effectively entering the crop water balance compared to the water coming out the irrigation system	–
EFFN	nitrogen use efficiency	–
EFNRAC	mineral nitrogen effect on the root distribution in the layers	–
ELMAX$_P$	maximum elongation of the coleoptile in darkness condition	cm
ELONG	elongation of the coleoptile	cm
EMD	direct evaporation of water intercepted by leafs	mm day^{-1}
EMISSA	emissivity of the atmosphere	–
EMPD	direct water evaporated from leaf interception	mm day^{-1}
EMPD$_D$	direct water evaporated from leaf interception of the dominant crop in case of intercrop	mm day^{-1}
EMPD$_U$	direct water evaporated from leaf interception of the understorey crop in case of intercrop	mm day^{-1}
EMULCH	direct evaporation of water intercepted by the mulch	mm day^{-1}
ENGAMM$_T$	ammonium proportion in the fertilizer	–
ENVFRUIT$_P$	proportion of the envelop weight relative to the maximum grain weight	–
EO	crop evaporation value if none of the soil or plant surfaces had limited water	mm day^{-1}
EOP	maximum transpiration flux	mm day^{-1}
EOP$_D$	maximum transpiration fluxof the dominant crop in case of intercrop	mm day^{-1}
EOP$_U$	maximum transpiration fluxof the understorey crop in case of intercrop	mm day^{-1}
EOS	maximum soil evaporation flux	mm day^{-1}
EP	actual plant transpiration flux	mm day^{-1}
EPD$_S$	thickness of the mixing cell in the soil	cm
EPT	potential evapotranspiration according to priestley-taylor formula	mm day^{-1}
EPZ	plant uptake soil profile	mm day^{-1} cm^{-1}
ESOL	actual soil evaporation flux	mm day^{-1}
ESTIMET	evapotranspiration estimated from the water balance for the previous day and the climatic demand of the day in case of greenhouse	mm day^{-1}
ESZ	evaporation soil profile	mm day^{-1} cm^{-1}
ET	daily evapotranspiration (ES+EP)	mm day^{-1}

ETMAX	maximum value of daily evaporation flux	Wm^{-2}
ETMIN	minimum value of daily evaporation flux	Wm^{-2}
EXOBIOM	index of water logging active on radiation use efficiency and transpiration	–
EXOFAC	variable for water logging	–
EXOLAI	index of water logging active on surface growth	–
EXTIN	extinction coefficient of photosynthetically active radiation calculated in case of radiative transfer	–
$EXTIN_P$	extinction coefficient of photosynthetic active radiation prescribed in case of Beer 's law analog	–
FAPAR	proportion of radiation intercepted by the canopy	–
$FAPAR_D$	proportion of radiation intercepted by the dominant crop in case of intercrop	–
$FAPAR_U$	proportion of radiation intercepted by the understorey crop in case of intercrop	–
FBIO	factor accounting for nitrogen microbial requirement	–
FCO2	specie-dependant CO2 effect on radiation use efficiency	–
FCO2S	specie-dependant CO2 effect on stomate closure	–
FDENNO3	Nitrate content effect on denitrification	–
FDENT	thermal effect on denitrification	–
FDENW	soil water content effect on denitrification	–
FGELFLO	frost index acting on thefruit (or grain) number	–
FGELJUV	frost index acting on LAI during the juvenile phase	–
FGELLEV	frost index acting on plant density during the plantlet phase	–
FGELVEG	frost index acting on LAI during the vegetative phase	–
FH	soil water content effect on mineralization	–
$FINERT_G$	proportion of inactive organic nitrogen	–
FIXMAX	maximal nitrogen fixation capacity by legume nodules	kg N ha^{-1} day^{-1}
$FIXMAXGR_P$	FIXMAX component due to grain growth	kg N ha^{-1} day^{-1}
$FIXMAXVEG_P$	FIXMAX component due to vegetative growth	kg N ha^{-1} day^{-1}
FIXPOT	potential n_2 fixation	kg N ha^{-1} day^{-1}
FIXREEL	actual rate of symbiotic uptake	kg N ha^{-1} day^{-1}
FLUXRAC	profile of nitrogen uptake associated with the limiting absorption capacity of the plant	kg N ha^{-1} day^{-1}
FLUXSOL	profile of nitrogen uptake associated with the limiting transfer from soil to roots	kg N ha^{-1} day^{-1}
FM	fresh matter	t ha^{-1}
$FMIN1_G$	parameter of the potential mineralization rate as a function of clay and the CaCO3 contents	day^{-1}
$FMIN2_G$	parameter of the potential mineralization rate as a function of clay and the CaCO3 contents	–

FMIN3$_G$	parameter of the potential mineralization rate as a function of clay and the CaCO3 contents	–
FNX$_G$	daily maximum fraction of ammonium transformed in nitrates	–
FP	cumulative foliage produced	m²m⁻²
FPFT	sink strength of fruits	g.m⁻² day⁻¹
FPV	sink strength of growing leaves	g.m⁻² day⁻¹
FRACINSOL	fraction of insolation	–
FRUIS	proportion of run-off water above the activation threshold (PMINRUIS$_G$)	–
FSNH3	volatilisation of NH3	µg.m⁻².day⁻¹
FSTRESSGEL	frost index	–
FTEMHA$_G$	parameter of the thermal effect on mineralization (FTH)	K⁻¹
FTEMHB	parameter of the thermal effect on mineralization (FTH)	–
FTEMH$_G$	parameter of the thermal effect on mineralization (FTH)	–
FTEMP	temperature-related radiation use efficiency reduction factor	–
FTEMPREMP	temperature-related grain filling reduction factor	–
FTEMRA$_G$	parameter of the thermal effect on residue mineralization (FTR)	–
FTEMR$_G$	parameter of the thermal effect on residue mineralization (FTR)	K⁻¹
FTH	thermal effect on basal mineralization	–
FTR	thermal effect on residue mineralization	–
FXA	anoxic effect on symbiotic uptake	–
FXN	nitrogen effect on symbiotic uptake	–
FXT	temperature effect on symbiotic uptake	–
FXW	water effect on symbiotic uptake	–
GAMMA	psychrometric constant	mbars °C⁻¹
GMAX	maximum value of daily soil heat flux	Wm⁻²
GMIN	miniimum value of daily soil heat flux	Wm⁻²
GRADTN$_G$	thermal gradient with altitude for minimal temperature	°C 100m⁻¹
GRADTNINV$_G$	thermal gradient with altitude for minimal temperature below the latitude of gradient inversion	°C 100m⁻¹
GRADTX$_G$	thermal gradient with altitude for maximal temperature	°C 100m⁻¹
H2OFEUILJAUNE$_P$	water content of yellow leaves	g water gFM⁻¹
H2OFEUILVERTE$_P$	water content of green leaves	g water gFM⁻¹
H2OFRVERT$_P$	water content of fruits before the beginning of hydrous evolution (IDEBDES)	g water gFM⁻¹
H2OGRAINMAX$_T$	maximal water content allowed at harvest	g water gFM⁻¹
H2OGRAINMIN$_T$	minimal water content allowed at harvest	g water gFM⁻¹
H2ORESERVE$_P$	water content of reserves	g water gFM⁻¹

H2OTIGESTRUC$_p$	structural stem part water content	g water gFM^{-1}
HA	residual soil water content	mm water cm soil^{-1}
HAUTBASE$_p$	base height of crop foliaige	m
HAUTCOUPEDEFAUT$_T$	cut height for forage crops (calendar calculated)	m
HAUTCOUPE$_T$	cut height for forage crops (calendar prescribed)	m
HAUTEUR	height of canopy	m
HAUTEUR$_D$	height of the dominant crop in case of intercrop	m
HAUTEUR$_U$	height of the understorey crop in case of intercrop	m
HAUTMAX$_p$	maximum height of the crop genetically determined	m
HAUTMAXTEC$_T$	maximum height of the crop technically determined	m
HAUTROGNE$_T$	cutting height	m
HB	soil layer between the seedbed and the root front	cm
HCCX$_G$	field capacity of pebbles	g water g soil^{-1} x 100
HCUM	available water over the rooting zone	mm water cm soil^{-1}
HMAX	maximum height of water table between drains	cm
HMINCX	wilting point of pebbles	g water g soil^{-1} x 100
HMINM$_G$	minimal soil water content for mineralization expressed as a proportion of field capacity	–
HMINN$_G$	minimal soil water content for nitrification expressed as a proportion of field capacity	–
HNAPPE	height of water table with active effects on the plant	cm
HN$_s$	wilting point water content in the seed bed	mm water cm soil^{-1}
HOPTM$_G$	optimal soil water content for mineralization expressed as a proportion of field capacity	–
HOPTN$_G$	optimal soil water content for nitrification expressed as a proportion of field capacity	–
HUCC	field capacity water content	mm water cm soil^{-1}
HUILREC$_T$	minimal oil content allowed for harvest	g oil gFM^{-1}
HUM	hourly air humidity	–
HUMCAPIL$_s$	threshold soil water content under which caoillary rises occur	g eau g^{-1} sol x 100
HUMIDITE	daily moisture in the canopy	–
HUMIN	wilting point water content	mm water cm soil^{-1}
HUMIRAC	influence of soil water content on germination and root growth	–
HUMSEUILTASSREC$_T$	soil water threshold above which harvest machines damage soil by compaction expressed in ratio of field capacity	–
HUMSEUILTASSSEM$_T$	soil water threshold above which sowing machines damage soil by compaction expressed in ratio of field capacity	–
HUR	soil water content	mm water cm soil^{-1}
HX$_s$	water content at field capacity in the seed bed	mm water cm soil^{-1}
I	current day	–

IAMF	day of the stage AMF (maximal of leaf growth, end of juvenile phase)	julian day
IDEBDES	date of onset of water dynamics in harvested organs	julian day
IDEBDORM	day of the dormancy entrance calculated	julian day
IDEBDORM$_P$	day of the dormancy entrance prescribed	julian day
IDNO	beginning of fixation	julian day
IDRP	day of the stage DRP: beginning of grain/fruit filling	julian day
IFINDORM	dormancy break day	julian day
IFLO	date of anthesis	Julian day
IFNO	stop of fixation	julian day
IFVINO	death of nodules	julian day
IGER	date of germination	julian day
ILAT	date of the beginning of the latence phase for grain calculation	julian day
ILAX	day of the stage LAX: maximal leaf area index	julian day
ILET	date of the plantlet stage	julian day
ILEV	day of the stage LEV: emergence	julian day
IMAT	day of the stage MAT: physiological maturity	julian day
IMB	date of the beginning of seed moistening	julian day
INFIL$_S$	infiltrability parameter at the base of the horizon	mm day^{-1}
INFLOMAX$_P$	maximal number of inflorescences per plant	nb plant^{-1}
INFRECOUV$_P$	ulai at the stage AMF (inflexion point of the soil cover rate increase)	–
INN	Nitrogen nutrition index (cumulative INN)	–
INNGRAIN1$_P$	threshold of INNS defining plant demand during grain filliing	–
INNGRAIN2$_P$	threshold of INNS defining plant demand during grain filliing	–
INNI	instantaneous nitrogen nutrition index	–
INNIMIN$_P$	INNI (instantaneous INN) corresponding to INNmin	–
INNLAI	index of nitrogen stress active on leaf growth	–
INNMIN$_P$	minimum value of INN allowed for the crop	–
INNS	index of nitrogen stress active on growth in biomass	–
INNSENES	index of nitrogen stress active on leaf death	–
INNSEN$_P$	parameter of the nitrogen stress function active on senescence (INNSENES)	–
INNTURGMIN$_P$	parameter of the nitrogen stress function active on leaf expansion (INNLAI)	–
INOU	date of end of setting of harvested organs	julian day
INTERRANG$_T$	width of the interrang	m
INV	function estimating temperature in altitude	°C
IPLT$_T$	sowing or planting prescribed date	julian day
IPLT	calculated sowing date	julian day

IRAZO	nitrogen harvest index	N grain N plant^{-1}
IRCARB	carbon harvest index	g grain g plant^{-1}
IREC	date of harvest (first if several)	julian day
IRECBUTOIR$_\text{T}$	latest allowed harvest date	julian day
IRMAX$_\text{P}$	maximum carbon harvest index	–
IRRIGN	N inputs by irrigation	kg N ha^{-1}
IRRLEV$_\text{G}$	amount of irrigation applied automatically on the sowing day when the model calculates irrigation, to allow germination	mm
IZRAC	index of water logging stress on roots	–
JULAPPLMULCH$_\text{T}$	date of the mulch application	julian day
JULECLAIR$_\text{T}$	date of fruits removal	julian day
JULEFFEUIL$_\text{T}$	date of leaf removal	julian day
JULROGNE$_\text{T}$	date of plant shapening	julian day
JULTAILLE$_\text{T}$	pruning day	julian day
JULVERNAL$_\text{P}$	date of vernalisation entering for perennial grasses	julian day
JVC$_\text{V}$	number of vernalizing days required	days
JVCMINI$_\text{P}$	minimum vernalizing days required	nb days
JVI	vernalizing contribution of a given day	–
K		
K2	actual mineralisation rate	kg N day^{-1}
K2HUM	potential mineralization rate	kg N day^{-1}
KBIO$_\text{G}$	decomposition rate constant for soil biomass	day^{-1}
KCOUVMLCH$_\text{G}$	parameter of the relationship between vegetal mulch and soil cover rate	–
KGDIFFUS	proportion of diffusive radiation reaching the soil	–
KGDIRECT	proportion of direct radiation reaching the soil	–
KH	coefficient of heat transfer in the cold shelter	Wm^{-2}K^{-1}
KHAUT$_\text{G}$	Parameter of the relationship between LAI and crop height	–
K$_\text{LAI}$	parameter describing the shape of the LAI curve when it is considered as a driving variable	m^2 leaf m^{-2} soil
KM1$_\text{P}$	constant of nitrate affinity by the root uptake system 1 (high affinity)	μmole. cm root^{-1}
KM2$_\text{P}$	constant of nitrate affinity by the root uptake system 2 (low affinity)	μmole. cm root^{-1}
KMAX$_\text{P}$	maximum crop coefficient for water requirements	–
KREPRAC$_\text{P}$	parameter of biomass root partitioning: evolution of the root/total biomass ratio	–
KRES	decomposition rate for soil residues	day^{-1}
KS	coefficient of energy losses between the outside and inside of the shelter	W m^{-2} K^{-1}
KSOL$_\text{S}$	hydraulic conductivity in the soil above and below the drains	cm day^{-1}
KSTEMFLOW$_\text{P}$	parameter of ther relationship between LAI and stemflow	–

KTROU$_P$	extinction coefficient of PAR through the crop, used in case of radiation transfers.	–
L	latent heat of vaporisation	MJ kg^{-1}
LAI	Leaf Area index	m^2 leaf m^{-2} soil
LAI0$_I$	initial leaf area index	m^2 leaf m^{-2} soil
LAICOMP$_P$	LAI from which starts competition inbetween plants	m^2 leaf m^{-2} soil
LAIDEBEFF$_T$	LAI of the beginning of leaf removal	m^2 leaf m^{-2} soil
LAIEFFCUM	LAI removed	m^2 leaf m^{-2} soil
LAIEFFEUIL$_T$	LAI of the end of leaf removal	m^2 leaf m^{-2} soil
LAIPLANTULE$_P$	plantlet leaf area index at the plantation	m^2 leaf m^{-2} soil
LAIRESIDUEL$_T$	residual leaf index after each cut	m^2 leaf m^{-2} soil
LAIROGNECUM	LAI removed by shapening	m^2 leaf m^{-2} soil
LAISEN	Leaf Area Index of senescent leaves	m^2 leaf m^{-2} soil
LARGEUR	width of the plant shape	m
LARGROGNE$_T$	width of shapening	m
LARTEC$_T$	width of canopy due to techniques	m
LAT$_C$	latitude	radians
LDRAIN$_S$	between drain ½ spacing	cm
LOCFERTI$_T$	depth of fertilizer apply (when applied in depth of soil)	cm
LOCIRRIG$_T$	depth of water apply (when applied in depth of soil)	cm
LONGSPERAC$_P$	root length/root mass ratio	cm g^{-1}
LRACSENTOT	total length of senescent roots	cm root.cm^{-2} soil
LRACZ	efficient root density profile	cm root.cm^{-3} soil
LVFRONT$_P$	root density at the root front	cm root.cm^{-3} soil
LVOPT$_G$	optimum root density for water and nitrogen uptake	cm root.cm^{-3} soil
MABOIS	prunning dry weight	t.ha^{-1}
MACROPOR	poral space corresponding to macroporosity	mm
MAENFRUIT	dry matter of harvested organ envelopes	t.ha^{-1}
MAFEUILJAUNE	dry matter of yellow leaves	t.ha^{-1}
MAFEUILTOMBE	dry matter of fallen leaves	t.ha^{-1}
MAFEUILVERTE	dry matter of green leaves	t.ha^{-1}
MAFRAISFEUILLE	leaf fresh matter	t.ha^{-1}
MAFRAISRES	reserve fresh matter	t.ha^{-1}
MAFRAISTIGE	structural stem fresh matter	t.ha^{-1}
MAFRUIT	dry matter of harvested organs	t.ha^{-1}
MARGEROGNE$_T$	allowed quantity of biomass inbetween two shapenings when asking automatic shapening	t.ha^{-1}
MASEC	aboveground dry matter	t.ha^{-1}
MASEC0$_I$	initial biomass	t.ha^{-1}
MASECABSO	biomass accounting for nitrogen uptake	t.ha^{-1}
MASECMETA$_G$	biomass of the plantlet when all nitrogen is assumed as metabolic	t.ha^{-1}

MASECPLANTULE$_P$	initial shoot biomass of plantlet	t ha^{-1}
MASVOLCX$_G$	pebbles bulk density	g cm^{-3}
MATIGESTRUC	dry matter of stems (only structural parts)	t.ha^{-1}
MAXAZORAC$_P$	parameter of the effect of soil nitrogen on root soil partitioning	kg N ha^{-1} mm^{-1}
MINAZORAC$_P$	parameter of the effect of soil nitrogen on root soil partitioning	kg N ha^{-1} mm^{-1}
MINEFNRA$_P$	parameter of the effect of soil nitrogen on root soil partitioning	–
MOUILL	water retained on the foliage	mm
MOUILLABILMULCH$_G$	maximum wettability of vegetal mulch	mm t^{-1} ha
MOUILLABIL$_P$	maximum wettability of leaves	mm LAI^{-1}
MOUILLMULCH	water retained by the vegetal mulch	mm
MSRAC	root biomass	t ha^{-1}
MSRESIDUEL$_T$	residual dry matter after a forage cut	t ha^{-1}
MULCHBAT$_G$	mulch depth from which a crust occurs	cm
N2ODENIT	N$_2$O produced by denitrification	kgN ha^{-1} day^{-1}
N2ONIT	N$_2$O produced by nitrification	kgN ha^{-1} day^{-1}
NBFEUILLE	number of leaves on main stem	nb
NBFEUILPLANT$_P$	leaf number per plant when planting	nb plant^{-1}
NBFGELLEV$_P$	leaf number at the end of the juvenile phase (frost sensitivity)	nb plant^{-1}
NBGRAINS	grain number	grains m^{-2}
NBGRMAX$_V$	maximum number of grains	grains m^{-2}
NBGRMIN$_P$	minimum number of grains	grains m^{-2}
NBINFLO	number of inflorescences	–
NBINFLOECL$_T$	number of fruits or inflorescences removed per plant	nb
NBINFLO$_P$	prescribed potential number of inflorescence	nb
NBJGERLIM$_P$	threshold number of day after grain imbibition without germination lack	nb days
NBJGRAIN$_P$	period before IDRP to compute grain number	nb days
NBJGRAUTO	days of autothrophy for a moistened seed	nb days
NBJHUMEC	maximal period that seed can be in a moist status without seed death occurs	nb days
NBJMAXAPRESRECOLTE$_T$	number of days until harvest is launched when it's postponed by the «harvest decision» option activation	nb days
NBJMAXAPRESSEMIS$_T$	number of days from IPLT defining the period for sowing when "sowing decision" option is activated	nb days
NBJSEUILTEMPREF$_T$	number of days allowing significant growth to decide sowing	nb days
NBOITE$_P$	number of age classes of fruits to discretise fruit growth for the indeterminate crops	nb
NDENENG	daily denitrification of nitrogen from fertiliser or soil (if option «denitrification» is activated)	kg.ha^{-1}.day^{-1}
NFRUITNOU	number of set fruits	fruits.m^{-2}

NH3REF$_C$	atmospheric ammonia concentration	µg N m^{-3}
NH3SURF	ammonia concentration at the soil surface	µg N m^{-3}
NHUM	amount of active nitrogen of the humus pool in the soil	kg N.ha^{-1}
NHUMT	total quantity of N humus (active + inert fractions) in the soil	kg N.ha^{-1}
NIT	profile of soil nitrates	kg N.ha^{-1} cm^{-1}
NITRIF	nitrate productyion from nitrification	kg.N ha^{-1}day^{-1}
NLEVLIM1$_P$	number of days after germination decreasing the emerged plants if emergence has not occur	nb days
NLEVLIM2$_P$	number of days after germination after which the emerged plants are null	nb days
NMAX	maximal crop nitrogen content	%
NMETA$_P$	metabolic nitrogen content	%
NMINRES$_T$	mineral nitrogen quantity in residue	kg N ha^{-1}
NODN	soil nitrate concentration affecting nodule functionning	kg N ha^{-1} mm^{-1}
NORG$_S$	soil organic nitrogen content in the upper layer (PROFHUM$_S$)	%
NORGENG	amount of N immobilized	kg N ha^{-1}
NRES$_P$	nitogen reserve content when the canopy weight is MASECDIL$_P$	%
NVOLATORG	part of the mineral nitrogen in the residue that can be volatilized	kg.N ha^{-1}
NVOLENG	daily volatilisation of nitrogen from fertiliser	kgN.ha^{-1}.day^{-1}
NVOLORG	volitilized nitrogen	kgN.ha^{-1}.day^{-1}
OBSTARAC$_S$	soil depth which will block the root growth	cm
OFFRN	profile of mineral N available for root uptake	kgN.ha^{-1}.cm^{-1}
OMBRAGETX$_G$	difference in maximum temperatures between south and north facing slopes	°C
ORGENG$_G$	maximal amount of microbial immobilized N from the fertilizer	kgN.ha^{-1}
ORIENTRANG$_T$	direction of ranks	rd (0=NS)
PARAZOMORTE$_P$	parameter qualifing the N content of dead leaves	–
PARSURRG$_C$	PAR/total radiation ratio	–
PENTINFLORES$_P$	parameter of the calculation of the inflorescences number	–
PENTLAIMAX$_P$	parameter of the logistic curve of LAI growth	–
PENTRECOUV$_P$	parameter of the logistic curve of the soil cover rate increase	–
PFEUILVERTE	proportion of green leaves in total non-senescent biomass	–
PGRAIN	grain weight	g
PGRAINGEL	frozen grain weight	gm^{-2}
PGRAINMAXI$_V$	maximum weight of one grain (0% water content)	g
PHIV0$_G$	parameter allowing the calculation of the under shelter climate	–

PHMAXNIT$_G$	soil pH threshold above which nitrification is maximal	–
PHMAXVOL$_G$	soil pH threshold above which volatilization is maximal	–
PHMINNIT$_G$	soil pH threshold under which nitrification is nill	–
PHMINVOL$_G$	soil pH threshold above which volatilization is minimal	–
PHOBASE$_P$	base photoperiod for development	hours
PHOI	current photoperiod	hours
PHOSAT$_P$	saturating photoperiod for development	hours
PH$_S$	soil pH	–
PHVOL	soil pH at soil surface varying with mineral N level	–
PHVOLS$_G$		
PHYLLOTHERME$_P$	thermal duration between the apparition of two successive leaves on the main stem	degree.days
PLNMIN$_G$	minimal rainfall to apply N fertilizer in case of calculation	mm
PLUIEBAT$_S$	minimal rain quantity for the crust occurrence	mm day^{-1}
PMINRUIS$_G$	minimal amount of precipitation to start a drip	mm day^{-1}
POTCROIFRUIT	potential growth of a fruit	g fruit^{-1} day^{-1}
POTGERMI$_P$	soil water potential inducing grain moistening	MPa
PRECIP	daily amount of water (precipitation + irrigation)	mm day^{-1}
PRECIPN	N inputs by rainfall	kg N ha^{-1}
PROFDENIT$_S$	thickness of the denitrifying layer	cm
PROFDRAIN$_S$	drain depth	cm
PROFHUMREC$_T$	thickness of the layer potentially affected by compaction at harvest	cm
PROFHUM$_S$	thickness of the active layer for mineralization	cm
PROFHUMSEM$_T$	thickness of the layer potentially affected by compaction at sowing	cm
PROFIMPER$_S$	depth of the impermeable floor	cm
PROFMES$_T$	thickness of the soil layer for water and mineral nitrogen integrated meausrements	cm
PROFRES$_T$	minimal value of the depth where residue are incorporated	cm
PROFSEM$_T$	sowing depth	cm
PROFSOL$_S$	soil thickness	cm
PROFTRAV$_T$	maximal value of the depth where residue are incorporated	cm
PROP	demand/offer n ratio	–
PROPFIXPOT	phenology-dependent coefficient affecting n$_2$ fixation	–
PROPNBJGERLIM$_G$	coefficient reducing the period of moistened seed autotrophy due to high temperature	–
PROPRAC$_G$	proportion of root /shoot biomass	–
PROPVOLAT	proportion of volaitizable n in residue	–
PSIBASE	predawn leaf water potential foliaire de base	MPa

PSISOL	soil water potential profile	MPa
$PSISTO_p$	absolute value of the potential of stomatal closing	bars
$PSITURG_p$	absolute value of the potential of the beginning of decrease of the cellular extension	bars
Q		
$Q0_s$	parameter of the end of the maximum evaporation stage	mm
Q10	exponential thermal unit for development calculation: used for senescence ($2.0^{UDEVCULT/10}$)	
$Q10_p$	q10 used for the dormancy break calculation	–
QDRAIN	daily water outflow from the drain	mm day^{-1}
QLES	cumulative n-no3 leached at the base of the soil profile	kgN.ha^{-1}
QLESD	cumulative N-NO3 leached into drains	kgN.ha^{-1}
QMINH	cumulative mineral nitrogen arising from humus	kgN.ha^{-1}
QMINR	cumulative mineral nitrogen arising from organic residues	kgN.ha^{-1}
QMULCH	quantity of plant mulch	t.ha^{-1}
$QMULCH0_T$	amount of initial mulch	t ha^{-1}
$QMULCHRUIS0_G$	amount of mulch to annul the drip	t ha^{-1}
QNDENENG	cumulative denitrification of nitrogen from fertiliser or soil	kgN.ha^{-1}
QNORGENG	cumulative organisation of nitrogen from fertiliser	kgN.ha^{-1}
QNPLANTE	amount of nitrogen taken up by the plant	kgN.ha^{-1}
$QNPLANTE0_I$	initial nitrogen amount in the plant	kgN.ha^{-1}
QNPLANTULE	amount of nitrogen in the plantlet	kgN.ha^{-1}
QNPLMAX	maximal amount of N possible in the crop	kgN.ha^{-1}
QNVOLENG	cumulative volatilisation of nitrogen from fertiliser	kgN.ha^{-1}
QRESSUITE	crop residues returning to the soilfor the following crop	t ha^{-1}
RAA	aerodynamic resistance between the cover and the reference level zr	s.m^{-1}
RAAMAX	daily maximum value of RAA	s.m^{-1}
RAAMIN	daily minimum value of RAA	s.m^{-1}
RAC	resistance of the canopy boundary layer	s.m^{-1}
RAC_D	RAC for dominant crop in case of intercropping	s.m^{-1}
RAC_U	RAC for understorey crop in case of intercropping	s.m^{-1}
RA_G	default value of RAA	s.m^{-1}
RAINT	photosynthetic active radiation intercepted by the canopy	MJ.m^{-2}
$RAPSENTURG_p$	threshold soil water content active to simulate water sensecence stress as a proportion of the turgor stress	–
RAS	aerodynamic resistance between the soil and the canopy	s.m^{-1}
$RATIODENIT_s$	ratio between N_2O emission and total denitrification	

$RATIOLN_T$	nitogen nutrtion stress index below which can be triggered a fertilization in automatic mode	–
$RATIOL_T$	Water stress index below which we start an irrigation in automatic mode	–
$RATIONIT_S$	ratio between N_2O emission and total nitrification	–
$RATIOSEN_P$	fraction of senescent biomass as the ratio at the total biomass	–
$RAYON_S$	mean root radius	cm
RC	resistance of canopy	$s.m^{-1}$
RC_D	resistance of the dominant crop in case of intercrop	$s.m^{-1}$
RC_U	resistance of the dominant crop in case of intercrop	$s.m^{-1}$
RDIF	ratio between diffuse radiation and global radiation	–
RDIFFUS	diffusive radiation	$MJm^{-2}day^{-1}$
RDIRECT	direct radiation	$MJm^{-2}day^{-1}$
$RDRAIN_G$	drain radius	cm
RDROIT	radiation not intercepted by the crop	$MJm^{-2}day^{-1}$
REMOBILJ	amount of biomass remobilized on a daily basis from the reserves	$g.m^{-2}\ day^{-1}$
$REMOBRES_P$	maximal proportion of carbon reserve remobilizable daily	–
REMONTEE	capillary rise at the base of the soil profile	$mm\ day^{-1}$
REPRAC	aboveground / underground partitioning coefficient of biomass	–
$REPRACMAX_P$	maximum value for REPRAC	–
$REPRACMIN_P$	minimum value for REPRAC	–
RESMES	amount of soil water over $PROFMES_T$ depth	mm
RESPERENNE	carbon reserve during the cropping season, or during the intercrop period (for perenial crops)	$t\ ha^{-1}$
$RESPERENNE0_I$	initial reserve biomass	$t\ ha^{-1}$
$RESPLMAX_P$	maximal size of the reserve compartment	$kg\ plant^{-1}$
RESRAC	soil water reserve in the root zone	mm
$RESSUITE_T$	parts of the plant recycled for the next crop	name
RFPI	slowing effect of the photoperiod on plant development	–
RFVI	slowing effect of the vernalization on plant development	–
RGEX	extraterrestrial radiation	$MJm^{-2}day^{-1}$
RGLO	long wave radiation	$MJm^{-2}day^{-1}$
RLJ	root length growth	$m\ root\ m^{-2}\ day^{-1}$
RLJ_{dens}	plant density component of RLJ	$plant\ m^{-2}$
RLJ_{dev}	plant density component of RLJ	$m\ root\ plant^{-1}\ degree\text{-}day^{-1}$
RLJFRONT	growth at the root front	$m\ root\ m^{-2}\ day^{-1}$
RLJ_{stress}	stress component of RLJ	–
RLJ_T	thermal component of RLJ	degree.days
RLTOT	total length of roots	$cm\ root.cm^{-2}\ soil$

RMAXI	maximum soil water reserve utilised	mm
RNET	net radiation	MJ m^{-2}day^{-1}
RNETMAX	maximal daily value of RNET	Wm^{-2}
RNETMIN	minimal daily value of RNET	Wm^{-2}
RNETP$_D$	net radiation affecting the dominant crop in case of intercropping	MJ m^{-2}day^{-1}
RNET$_{PE}$	net radiation estimation in the penman formula	MJ m^{-2}day^{-1}
RNET$_{PT}$	net radiation estimation in the priestley-taylor formula	MJ m^{-2}day^{-1}
RNETP$_U$	net radiation affecting the understorey crop in case of intercropping	MJ m^{-2}day^{-1}
RNETS	net radiation affecting the soil	MJ m^{-2}day^{-1}
ROMBRE	radiation fraction in the shade	–
RSMIN$_p$	minimal stomatal resistance of leaves	s m^{-1}
RSOLEIL	radiation fraction in the full sun	–
RSRSO	total to extraterrestrial radiation ratio	–
RSURRU	soil water status as a proportion of readily available water	–
RTRANSMIS	radiation transmitted through the crop	MJm^{-2}day^{-1}
RUGOCHISEL$_T$	soil surface rugosity after soil tillage without soil inverting	m
RUGOLABOUR$_T$	soil surface rugosity after soil tillage with soil inverting	m
RUISOLNU$_S$	fraction of drip rainfall (by ratio at the total rainfall) on a bare soil	–
RUISSEL	daily run-off (surface + overflow)	mm day^{-1}
RUISSELSURF	daily surface run-off	mm day^{-1}
SAT	amount of water remaining in the soil macroporosity	mm
SB	seed bed	cm
SBV	specific surface area of biomass	cm^2 g^{-1}
SBVMAX	leaf expansion allowed per unit of biomass accumulated	cm^2 g^{-1}
SENFAC	water stress index on senescence	–
SENSANOX$_p$	anoxia sensitivity (0=insensitive)	–
SENSIPHOT	photoperiod sensitivity (1=insensitive)	–
SENSRSEC$_p$	root sensitivity to drought (1=insensitive)	–
SLA	specific leaf surface area	cm^2 g^{-1}
SLAMAX$_p$	maximal sla of green leaves	cm^2 g^{-1}
SLAMIN$_p$	minimal sla of green leaves	cm^2 g^{-1}
SOMGER	current growing degree.days in the seed bed	degree.days
SOMSEN	current thermal time for sensecence	Q10
SOURCEPUITS	sources/sink ratio	–
SOURCEPUITS$_1$	first calculation of SOURCEPUITS	–
SPFRMAX$_p$	maximal sources/sinks value allowing the trophic stress calculation for fruit onset	–
SPFRMIN$_p$	minimal sources/sinks value allowing the trophic stress calculation for fruit onset	–

SPFRUIT	index of trophic stress applied to the number of fruits	–
SPLAI	source/sink ratio applied to leaf growth	–
SPLAIMAX$_P$	maximal source/sink value allowing the trophic stress calculation for the leaf growth	–
SPLAIMIN$_P$	minimal value of ratio source/sink for the leaf growth	–
STADECOUPEDF$_T$	stage of automatic cut	–
STAMFLAX$_V$	duration between IAMF and ILAX	degree.days
STDEBSENRAC$_P$	life span of roots	degree.days
STDNOFNO$_P$	duration between IDNO and IFNO	degree.days
STDORDEBOUR$_P$	duration between the dormancy break and the bud break	degree.days
STDRPDES$_P$	duration between IDRP and IDEBDES	degree.days
STDRPMAT$_V$	duration between IDRP and IMAT	degree.days
STDRPNOU$_P$	duration between IDRP and INOU	degree.days
STEMFLOW	amount of water running along the stem	mm
STEMFLOWMAX$_P$	maximal fraction of rainfall which flows out along the stems	–
STFLODRP$_V$	duration between IFLO and INOUIDRP	degree.days
STFNOFVINO$_P$	duration between IFNO and IFVINO	degree.days
ST$_{LAI}$	variable describing the shape of the LAI curve when it is considered as a driving variable	degree.days
STLEVAMF$_V$	duration between ILEV and IAMF	degree.days
STLEVDNO$_P$	duration between ILEV and IDNO	degree.days
STLEVDRP$_V$	duration between ILEV and IDRP	degree.days
STOPRAC$_P$	stage when root growth stops	name
STPLTGER$_P$	duration between IPLT and IGER	degree.days
STRESSDEV$_P$	maximum phasic delay allowed due to stresses	–
SUCREREC$_T$	minimal sugar rate at harvest	g sugar g FM^{-1}
SURFAO	fraction of surface in the shade	–
SURFAS	fraction of surface in the sun	–
SURFOUVRE$_T$	proportion of vents related to the total surface area of the greenhouse	–
SWFAC	index of stomatal water stress	–
SWRMIN	soil water factor for denitrification	–
TAIR	mean air temperature	°C
TAUXCOUV	ground cover	–
TCKMAX$_P$	value of TAUXCOUV corresponding to KMAX$_P$	–
TCMAX$_P$	maximum temperature of leaf expansion	°C
TCMIN$_P$	minimum temperature of leaf expansion	°C
TCULT	crop surface temperature (daily average)	°C
TCULTMAX	crop surface temperature (daily maximum)	°C
TCULTMIN	crop surface temperature (daily minimum)	°C
TCXSTOP$_P$	high temperature stopping phasic development, leaf expansion and senesecence	°C
TDENREF1$_G$	cardinal temperature for denitrification	°C

Symbol	Definition	Unit
TDENREF2_G	cardinal temperature for denitrification	°C
TDEW	dewpoint temperature	°C
TDMAX_P	maximum threshold temperature for development	°C
TDMIN_P	minimum threshold temperature for development	°C
TEAUGRAIN	fruit(grain) water content	g water gFM^{-1}
TEMAX_P	maximal threshold temperature for net photosynthesis	°C
TEMIN_P	minimum threshold temperature for net photosynthesis	°C
TEMPDESHYD_P	increase in the fruit dehydration due to the increase of crop temperature (TCULT-TAIR)	% water °C^{-1}
TEMPNOD1_P	cardinal temperature driving N_2 fixation	°C
TEMPNOD2_P	cardinal temperature driving N_2 fixation	°C
TEMPNOD3_P	cardinal temperature driving N_2 fixation	°C
TEMPNOD4_P	cardinal temperature driving N_2 fixation	°C
TEOPTBIS_P	end of the thermal optimal plateau for net photosynthesis	°C
TEOPT_P	beginning of the thermal optimal plateau for net photosynthesis	°C
TETA	available water content in the root zone	cm^3 water cm^{-3} soil
TETP	reference potential evapotranspiration (entered or calculated)	mm day^{-1}
TETSEN	threshold soil water content accelerating sensecence	cm^3 water cm^{-3} soil
TETSTOMATE	threshold soil water content limiting transpiration and photosynthesis	cm^3 water cm^{-3} soil
TETURG	threshold soil water content limiting leaf expansion	cm^3 water cm^{-3} soil
TF_{LAI}	parameter describing the shape of the lai curve when it is considered as a driving variable	degree.days
TFROID_P	optimal temperature for vernalisation	°C
TGELFLO10_P	temperature corresponding to 10 % of frost damages on the flowers or the fruits	°C
TGELFLO90_P	temperature corresponding to 90 % of frost damages on the flowers or the fruits	°C
TGELJUV10_P	temperature corresponding to 10 % of frost damage on the LAI (juvenile stage)	°C
TGELJUV90_P	temperature corresponding to 90 % of frost damage on the LAI (juvenile stage)	°C
TGELLEV10_P	temperature corresponding to 10% of frost damage on the plantlet	°C
TGELLEV90_P	temperature corresponding to 90% of frost damage on the plantlet	°C
TGELVEG10_P	temperature corresponding to 10 % of frost damage on the LAI (adult stage)	°C
TGELVEG90_P	temperature corresponding to 90 % of frost damage on the LAI (adult stage)	°C

TGMIN_P	minimum threshold temperature for germination and emergence	°C
TIGEFEUILLE_P	stem (structural part)/leaf proportion	–
TI_{LAI}	parameter describing the shape of the LAI curve when it is considered as a driving variable	degree.days
TLETALE_P	lethal temperature for the plant	°C
TMAX	maximum daily air temperature	°C
TMAXREMP_P	maximal temperature for grain filling	°C
TMIN	minimum daily air temperature	°C
TMINREMP_P	minimal temperature for grain filling	°C
TNITMAX_G	cardinal temperature for nitrification	°C
TNITMIN_G	cardinal temperature for nitrification	°C
TNITOPT_G	cardinal temperature for nitrification	°C
TNITRIF	profile of nitrified ammonia	kg N ha^{-1} cm^{-1}
TOTAPN	total fertilizer application	kg N ha^{-1} cm^{-1}
TPM	vapour pressure in air	mbars
TRANSPLASTIC_G	translission coefficient of the shelter plastic	–
TRECOUVMAX_P	proportion of the soil covered by an isolated plant,	–
TREF_G	reference temperature for soil mineralisation processes	°C
TRG	global radiation affecting the crop (entered or calculated)	MJ.m^{-2} day^{-1}
TRGEXT	exterior radiation in case of a shelter	MJ.m^{-2} day^{-1}
TRR	daily rainfall	mm. day^{-1}
TRSOLVOLAT	code for manure volatilization	$^{-1}/+1$
TSOL	profile of soil temperature	°C
TURFAC	index of turgescence water stress	–
TURSLA	mean water stress TURFAC experienced since emergence	–
TVAR	saturating vapor pressure as a function of temperature	mbars
TVENT	daily mean speed of wind	m.s^{-1}
U	understorey crop in case of intercrop	m
UDEVCULT	effective temperature for the development, computed with TCULT	degree.days
UDLAIMAX_P	ULAIfrom which the rate of leaf growth decreases	–
ULAI	daily relative development unit for LAI	–
UPVT	daily development unit	degree.days
URAC	daily relative development unit for root growth	–
USM	unit of simulation	
VABS2_G	crop N uptake rate at which denitrification losses reach 50% of their maximum.	kgN ha^{-1} day^{-1}
VABSN	nitrogen accumulation rate in the plant (uptake and fixation)	kg N ha^{-1} day^{-1}
VABSMOY	nitrogen uptake rate	kgN ha^{-1} day^{-1}

Symbol	Definition	Units
$VIGUEURBAT_p$	proportion of plants succeeding to emerge through the crust	–
$VITIRAZO_p$	rate of increase of the nitrogen harvest index	g grain g plant $^{-1}$ day^{-1}
$VITIRCARB_p$	rate of increase of the carbon harvest index	g grain g plant $^{-1}$ day^{-1}
VITMOY	average growth rate during the latence phase (ILAT-IDRP)	g m^{-2}day^{-1}
$VITNO_p$	potential rate of nodule set up	Nb degree.days $^{-1}$
$VITPROPHUILE_p$	increase rate of oil harvest index	g oil g MS^{-1} day^{-1}
$VITPROPSUCRE_p$	increase rate of sugar harvest index	g sugar g MS^{-1} day^{-1}
$VLAIMAX_p$	ULAI at the inflexion point of the function DELTAI=f(ULAI)	–
$VMAX1_p$	maximal nitrate uptake rate by the uptake system 1 (high affinity) of roots	µmole cm^{-1} h^{-1}
$VMAX2_p$	maximal nitrate uptake rate by the uptake system 2 (low affinity) of roots	µmole cm^{-1} h^{-1}
VMINH	basal mineralization rate	kg N ha^{-1} day^{-1}
$VOLENG_G$	maximal proportion of N losses by volatilization of the fertilizer	–
$VPOTDENIT_s$	total denitrification potential rate	kg N ha^{-1} day^{-1}
W	biomass variable of dilution curves	t ha^{-1}
WFPS	profile of saturation soil status	–
WH_G	N/C ratio of humified organic matter	–
XMULCH	thickness of mulch created by evaporation from the soil	cm
$XORGMAX_G$	N rate at which this maximum microbial immobilization is reached for a given fertilizer	kg N ha^{-1}
XSH	inter-row points of radiative transfer calculation	nb
$YRES_G$	partition coefficient for residue mineralisation	–
Z0	crop roughness	m
Z0S	soil or understorey crop roughness	m
$Z0SOLNU_s$	bare soil roughness	m
ZDEMI	root depth that ensures at least an extraction near the soil surface of 20% of the water available	cm
$ZESX_s$	maximal depth of soil affected by soil evaporation	cm
$ZLABOUR_p$	depth of ploughing	cm
ZNONLI	root depth if no obstacle	cm
$ZPENTE_p$	depth where the root density is ½ of the surface root density for the reference profile	cm
$ZPRLIM_p$	maximum depth of the root profile for the reference profile	cm
ZRAC	depth reached by root system	cm
ZRAC0	initial depth of root front	cm
$ZRACPLANTULE_p$	depth of the initial root front of the plantlet	cm
ZR_c	reference height of meteorological data measurement	m

Index of parameters and variables